Homosexual Behaviour in Animals

Behavioural observations from both the field and captivity indicate that same-sex sexual interactions are widespread throughout the animal kingdom, and occur quite frequently in certain non-human species. Proximate studies of these phenomena have yielded important insights into genetic, hormonal and neural correlates. In contrast, there has been a relative paucity of research on the evolutionary aspects. *Homosexual Behaviour in Animals* seeks to readdress this imbalance by exploring animal same-sex sexual behaviour from an evolutionary perspective. Contributions focus on animals that routinely engage in homosexual behaviour and include birds, dolphin, deer, bison and cats, as well as monkeys and apes, such as macaques, gorillas and bonobos. A final chapter looks at human primates. This book will appeal to graduate students and researchers in evolutionary biology, biological anthropology, zoology, evolutionary psychology, animal behaviour and anyone interested in the current state of knowledge in this area of of sexology.

VOLKER SOMMER obtained his Ph.D. from the University of Göttingen, Germany and is now Professor of Evolutionary Anthropology at University College London. His empirical research focuses on social and sexual behaviour in primates. He is involved in long-term eco-ethological studies of langur monkeys in India (since 1981), of gibbons in Thailand (since 1990) and of chimpanzees in Nigeria (since 1999).

PAUL L. VASEY obtained his Ph.D. from the Université de Montréal, Canada and is now an Associate Professor in the Department of Psychology & Neuroscience at the University of Lethbridge, Lethbridge, Alberta, Canada. His behavioural and neuro-anatomical research focuses on understanding the development and evolution of female homosexual behaviour in Japanese macaques. He currently conducts research on free-ranging macaques at Arashiyama, Japan.

Homosexual Behaviour in Animals

An Evolutionary Perspective

Edited by

VOLKER SOMMER
Department of Anthropology,
University College London, UK

PAUL L. VASEY
Department of Psychology & Neuroscience,
Lethbridge, Canada

CAMBRIDGE UNIVERSITY PRESS
Cambridge, New York, Melbourne, Madrid, Cape Town, Singapore,
São Paulo, Delhi, Dubai, Tokyo, Mexico City

Cambridge University Press
The Edinburgh Building, Cambridge CB2 8RU, UK

Published in the United States of America by Cambridge University Press, New York

www.cambridge.org
Information on this title: www.cambridge.org/9780521182300

First published 2006
First paperback edition 2010

A catalogue record for this publication is available from the British Library

ISBN 978-0-521-86446-6 Hardback
ISBN 978-0-521-18230-0 Paperback

Contents

Contributors

Ludek Bartoš
Ethology Group, Research Institute of Animal Production (VUZV),
10-Uhrineves CZ 104 00 Praha, Czech Republic

Barbara Fruth
Max-Planck-Institut für Verhaltensphysiologie, 82319 Seewiesen, Germany

Josef Hemetsberger
Konrad-Lorenz-Forschungsstelle für Ethologie, A-4645 Grünau 11, Austria

Gottfried Hohmann
Max-Planck-Institut für Evolutionäre Anthropologie, Deutscher Platz 6,
04103 Leipzig, Germany

Jana Holečková
Department of Zoology, Faculty of Science, Charles University, Prague &
Global Relation Department, Ministry of Environment of The Czech
Republic, Prague, Czech Republic

Rodney L. Johnson
Mannheimer Foundation, 20155 SW 360th St., Homestead, FL 33034, USA

Ellen Kapsalis
University of Miami, Division of Veterinary Resources, South Campus
Research Facility, 12500 SW 152nd St., Miami, FL 33177, USA

Catherine E. King
Diergaarde Blijdorp, Royal Rotterdam Zoological and Botanical Gardens,
Postbus 532 3000 AM Rotterdam, The Netherlands

Kurt Kotrschal
Konrad-Lorenz-Forschungsstelle für Ethologie, A-4645 Grünau 11, Austria

Diana Kyriazis
Department of Anthropology and Department of Experimental Psychology, University of Oxford, South Parks Road, Oxford OX1 3UD, United Kingdom

Janet Mann
Department of Psychology & Department of Biology, Georgetown University, 37th and O Street NW, Washington, DC 20057, USA

Catherine Roden
Royal Zoological Society of Antwerp, Center of Research and Conservation, K. Astridplein 26, B-2018 Antwerp, Belgium

Peter Schauer
Institute of Archaeology, University College London, Gower Street, London WC1E 6BT, United Kingdom

Volker Sommer
Department of Anthropology, University College London, Gower Street, London WC1E 6BT, United Kingdom

Paul L. Vasey
Department of Psychology and Neuroscience, University of Lethbridge, 4401 University Drive, Lethbridge, Alberta, T1K 3M4, Canada

Hilde Vervaecke
University of Antwerp, Department of Biology, Universiteitsplein 1, B-2610 Wilrijk and Royal Zoological Society of Antwerp, Center of Research and Conservation, K. Astridplein 26, B-2018 Antwerp and Bison Project, Domaine de Bois St. Jean 7, 6982 Samré, Belgium

Brigitte Weiss
Konrad-Lorenz-Forschungsstelle für Ethologie, A-4645 Grünau 11, Austria

Dennis Werner
Depto. de Antropologia/CFH, Universidade Federal de Santa Catarina, Campus Universitário, Florianópolis, Santa Catarina, 88.000, Brazil

Juichi Yamagiwa
Laboratory of Human Evolution Studies, Division of Biological Sciences, Graduate School of Sciences, Kyoto University, Sakyo, Kyoto 606-8502, Japan

Akihiro Yamane
Kitakyushu Museum of Natural History and Human History, 2-4-1 Higashida, Kitakyushu 805-0071, Japan

Acknowledgements

We wish to thank those colleagues who, as experts on certain taxonomic groups, reviewed individual chapters: Ruth Thomsen, Gerulf Reiger, Martha M. Robbins, Craig Stanford, Anne Storey and Lynne A. Fairbanks. Thanks are due to Alejandra Pascual Garrido for preparing the index.

PART I   INTRODUCTION

1

Homosexual behaviour in animals: topics, hypotheses and research trajectories

PAUL L. VASEY AND VOLKER SOMMER

A paradox for Darwinian theory?

At first glance, homosexual behaviour seems to violate a basic 'law' of nature: that of procreation. The notion that organisms exist to reproduce themselves is a staple of pre-scientific worldviews, and evolutionary theory, from its very beginnings, has also elaborated upon this idea. In a strict Darwinian view, individuals should seek to maximize reproductive success; organisms are predicted to strive to maximize the number of viable offspring and with this the representation of their own genetic information passed down to future generations.

Sexual selection is the key theoretical framework for interpreting sexual behaviour from an evolutionary perspective. Darwin described sexual selection as a process of differential reproduction that occurs because males vary in their ability to acquire female mates (that is, reproductive partners). He identified two basic mechanisms that influence mate acquisition. Mate competition occurs intra-sexually among males for females and encompasses physical fights and threats as well as ritualized displays of courtship aimed at attracting discriminating females. Mate choice occurs inter-sexually, and typically involves females selecting the most attractive male competitor. More recently, sexual coercion has been proposed as an additional mate acquisition mechanism that males can employ if they are unsuccessful at competing for, or attracting, female reproductive partners (Smuts and Smuts, 1991).

Sex differences in patterns of mate acquisition were later explained by Trivers (1972) in terms of parental investment theory. In 'typical' species, females invest more into offspring than males, since they not only gestate and lactate, but also provide most of the post-parturition care to offspring. This difference limits their potential reproductive rate, so that, at any one time, there will be fewer fertile females than reproductively active males in a population. Thus, males are, in theory, expected to compete intra-sexually for reproductive access to females, sexually coerce females if necessary, and copulate relatively indiscriminately. Females should discriminate among potential male mates in favour of those that contribute the most to offspring quality and survival. In 'atypical' species – such as sea-horses and various birds – the roles are reversed: males provide the bulk of parental investment and are choosy about the females with which they copulate, while females compete among themselves for male mates (Gywnne, 1991).

From the perspective of sexual selection and parental investment theories, one might be left with the impression that sex is synonymous with reproduction. Parental investment theory clearly predicts that individuals should choose and compete for sexual partners that confer the greatest reproductive advantage. As such, when given a choice, reproductive sex should be preferred over non-reproductive sex. It follows that opposite-sex mates should be preferred over same-sex sexual partners, and mate competition should occur intra-sexually.

In reality, many species engage in homosexual behaviours as well. However, animals which engage in sexual interactions with members of their own sex are obviously not in immediate pursuit of reproductive goals (that is, conception). Because homosexual behaviour appears to undermine reproduction, it seems appropriate to ask why animals engage in these behaviours at all.

Earlier studies of animal behaviour tended to dismiss occurrences of same-sex sexual behaviour as mere quirks or such instances were classified as pathological manifestations. The use of caged subjects was prevalent and meant that these interactions were invariably characterized as abnormal products of captivity, unlikely to be found in 'nature'. As early as the 1700s biologists such as George Edwards (1758–64) were speculating on the causes of such behavioural 'abnormalities'. He stated that 'three or four young [bantam] cocks remaining where they could have no communication with hens . . . each endeavoured to tread his fellow, though none of them seemed willing to be trodden. Reflections on this odd circumstance hinted to me, why the natural appetites, in some of our own species, are diverted into wrong channels' (p. xxi).

Research conducted throughout the 1890s purported that an absence of opposite-sex partners and artificial confinement could 'force' individuals to choose same-sex mates. It is interesting to note that the same researchers who reported such findings showed that pigeons (Columbia livia) will participate in same-sex sexual interactions even if they are housed in mixed-sex groups. Moreover, some researchers even demonstrated that certain same-sex pairs stopped their homosexual activity upon being isolated from their opposite-sex group mates. For example, Whitman (1919) stated that 'a number of pairs of mature males were isolated; some of these were observed for several months, but no real matings resulted from any of these cases'. He then went on to conclude, however, that 'confinement will thus force matings which would not otherwise occur. Pairings between like sexes are secured in this manner'. Clearly, such an interpretation is more reflective of the opinions of the observer than an objective observation of the behavioural phenomenon under investigation.

It was not long before fieldworkers, confronted with evidence of homosexual behaviour in 'nature', were integrating this perspective into their lexicon and summarily dismissing such interactions as idiosyncratic pathological manifestations or worse (for example, Mute swans: Ritchie, 1926). Studies of domestic livestock that followed asserted that the 'artificial' effects of domestication produced homosexual behaviour. The economic benefits associated with livestock reproduction may have further promoted the view that homosexual behaviour is an undesirable 'problem' (for example, domestic chickens; Guhl, 1948).

Later research in the domain of behavioural endocrinology sought a causal link between the perinatal hormonal environment and adult homosexual behaviour in a variety of species, including rodents and primates. This work, too, may have inadvertently contributed to the perspective of animal homosexual behaviour as a disordered condition (for example, Phoenix et al., 1959; Pomerantz et al., 1986).

Nevertheless, more and more detailed studies of animals in their natural environments made it increasingly difficult to discount all sexual interactions in animals among members of the same sex as exceptions, as idiosyncrasies, or as pathologies. Slowly, but steadily, a quite different picture emerged. A recent encyclopaedic volume by Bruce Bagemihl (1999) on animal homosexual behaviour provides evidence that hundreds of mammals, birds, reptiles, amphibians, fishes, insects, spiders and other invertebrates engage in same-sex sexual activity. Clearly, what was once thought to be an aberration appears to be a behavioural pattern that is broadly, albeit unevenly, distributed across the animal kingdom (see also Dagg, 1984; Sommer, 1990; Vasey, 1995). Indeed, within a select number of species, homosexual activity is widespread and occurs at levels that approach or sometimes even surpass heterosexual activity.

A number of excellent reviews currently exist, which explore the hormonal and neural mechanisms underlying same-sex sexual behaviour (Adkins-Regan, 1988, 1998; Adkins-Regan et al., 1997; Paredes and Baum, 1997). Important insights have been gained from such research. Nevertheless, it is difficult not to conclude that same-sex sexual behaviour is aberrant when viewed through the lens of this type of invasive experimental work. Many of the animal models used in such studies do not appear to spontaneously exhibit homosexual activity as part of their species-typical behavioural repertoires. Instead, such behaviour has to be elicited experimentally, either by destroying areas of the brain or by exposing the subjects to abnormal levels of steroid hormones perinatally. For those researchers interested in spontaneously expressed same-sex sexual behaviour, the information gleaned from such studies may be limiting. Reflecting on this research emphasis, Adkins-Regan (1988) stated that more studies of same-sex sexual behaviour in intact, untreated animals are needed.

Nevertheless, to date, there has been a relative paucity of research on the evolutionary aspects of homosexual behaviour in intact, untreated animals. Moreover, locating the information that does exist has not always been an easy task, for it is often scattered throughout obscure journals, technical reports and unpublished dissertations, or as Bagemihl has noted 'buried even further under out-dated value judgments and cryptic terminology' (1999, p. 87). The current volume seeks to address this gap in our knowledge by exploring the extent to which homosexual behaviour in animals can be understood from an evolutionary perspective. Why does such behaviour persist when it appears to conflict with an individual's reproductive imperative and, as such, flout the expectations of a Darwinian world view?

An evolutionary perspective on the topic of animal homosexual behaviour was first articulated in a landmark paper by the ethologist Wolfgang Wickler (1967), who suggested that homosexual behaviour in animals served some adaptive social function. He proposed that same-sex mounting was a ritualized gesture that individuals used to communicate about their dominance relationships. Wickler saw mounting as an expression of dominance, while allowing oneself to be mounted expressed subordinate status *vis-à-vis* the mounter. He reasoned that same-sex mounting commonly occurred in response to incipient aggression because it reduced the chances of escalated fighting by reiterating the dominance status quo. Wickler termed such behaviours which were sexual in form, but which served some social function, 'socio-sexual'. The concept of socio-sexual behaviour is significant because – although Wickler never explicitly stated it as such – it suggested a possible adaptive basis for homosexual behaviour.

The adaptionist perspective was greatly stimulated by the emergence of sociobiology in the years preceding and following the publication of Edward O. Wilson's (1975) book *Sociobiology: The New Synthesis*. Sociobiology aims to understand the evolutionary forces that shape social behaviour in animals, including humans. This paradigm shift also resulted in an alternative view of homosexual behaviour. Namely, this type of behaviour was no longer seen as an abnormality, but rather, a product of evolutionary processes and explicable in adaptive terms. For example, Wilson (1975) suggested that homosexual behaviour might be maintained in the population because the actors, while foregoing direct reproduction, would help their kin reproduce and in doing so reproduce indirectly any genes they shared with those kin (also see Weinrich, 1980; Ruse, 1981). The sociobiological perspective generated a number of adaptive hypotheses for homosexual behaviour, but was much less successful in establishing supporting evidence in either humans or animals (Wilson, 1975; Kirsch and Rodman, 1982; Ruse, 1988; Dickemann, 1993; Kirkpatrick, 2000). Despite this shortcoming, the wave of adaptationist thinking brought on with the emergence of sociobiology provided an important stimulus for ethologists to explore additional socio-sexual functions for homosexual behaviour in animals, and we describe some of these in greater detail below.

Early on in this project, we decided we would not attempt to cover all taxa in the animal kingdom. Bruce Bagemihl aimed for, and achieved, a much broader degree of comprehensiveness in his book, *Biological Exuberance: Animal Homosexuality and Natural Diversity* (1999). His compilation made it clear, however, that the evidence for homosexual behaviour in animals is overwhelmingly sketchy and anecdotal.

A number of reasons might account for the paucity of research on this topic. The most often-cited rationale as to why so little research is conducted in this area is that researchers are apprehensive about homophobic reactions (Wolfe, 1991; Bagemihl, 1999). Some researchers might fear being correctly, or mistakenly, labelled as gay or lesbian. Others imagine that their careers will be negatively impacted if their names become associated with this sort of subject matter. Linda Wolfe (1991, p. 130) reports that 'several (anonymous at their request) primatologists . . . have told me that they have observed both male and female homosexual behaviour during field studies. They seemed reluctant to publish their data, however . . . because they feared homophobic reactions ('my colleagues might think that I am gay').' She concludes: 'If anthropologists and primatologists are to gain a complete understanding of primate sexuality, they must cease to allow the folk model (with its accompanying homophobia) to guide what they see and report' (Wolfe, 1984, p. 130).

It is possible, however, that more theoretically motivated considerations dissuade the majority of researchers from ever undertaking research on homosexual behaviour in animals. If, for example, reproduction is the engine that drives evolution, then some investigators working on issues pertaining to the evolution of behaviour, particularly sexual behaviour, might reason that non-reproductive modes of sexuality, such as homosexual activity, are biologically irrelevant. Moreover, it is possible that scientists simply lack a theoretical framework for interpreting homosexual behaviour in animals and, as such, avoid studying the phenomenon because they 'don't know what it means' (Wolfe, 1984, p. 130).

Finally, the lack of research on this topic may be because, despite media hype and the claims of some researchers, relatively few species habitually engage in homosexual behaviour. One can hardly be expected to undertake a research project (let alone an entire research program that spans one's career) aimed at studying a particular behaviour if doing so necessitates observing the study species for hundreds of hours before a single instance of the behaviour is observed. No funding agency would support such work and no investigator would be able to sustain such a research agenda. Thus, most studies on homosexual behaviour are generated in conjunction with, or as a sideline to, research on other topics.

Here, we examine the extent to which prevailing evolutionary approaches to this subject are sufficient by concentrating on species that engage in homosexual activity on a routine basis. All of the current contributors were struck by the frequency of same-sex sexual interactions exhibited by their study species and were thus able to accumulate relatively large sets of data that lend themselves to quantitative analyses. The chapters contained in this volume are therefore unique in that they were written by a small group scientists who have enough data at hand so that they could test some of the current theories about the functions and evolutionary history of homosexual behaviour in animals.

The first half of the volume includes contributions about birds (Chapter 2, Kotrschal, Hemetsberger and Weiss: *geese*; Chapter 3, King: *flamingos*), cetaceans (Chapter 4, Mann: *bottle-nosed dolphins*), ungulates (Chapter 5, Vervaecke and Roden: *bison*; Chapter 6, Bartoš and Holečková: *deer*) and carnivores (Chapter 7, Yamane: *feral cats*). The second half focuses on primates such as Old World monkeys (Chapter 8, Vasey: *Japanese macaques*; Chapter 9, Kapsalis and Johnson: *rhesus macaques*; Chapter 10, Sommer, Schauer and Kyriazis: *langur monkeys*) and apes (Chapter 11, Yamagiwa: *mountain gorillas*; Chapter 12, Fruth and Hohmann: *bonobos*). The bias towards primates is due to the fact that homosexual behaviour has been reported more often in this order of mammals. At present, it is unclear whether this reflects a true phylogenetic difference in the frequency with which homosexual behaviour is expressed.

Because we are committed to a broad comparative perspective on the topic of homosexual activity, this volume also includes a chapter on the human 'animal' (Chapter 13, Werner). It is our conviction that evolutionary treatises should not be 'homocentric' in that they either focus on humans, while excluding a comparison with other animals, or that they focus on animals, while excluding our species, *Homo sapiens*. Such boundaries, when maintained for reasons of orthodoxy and dogmatism, are meaningless and counterproductive to scientific understanding – a point which we will reiterate below. Of course, humans are unique and the behaviour of humans does, therefore, require unique explanations – but so does the behaviour of bottlenose dolphins and bisons. Darwinian theory, unlike anthropocentric philosophies, recognizes the need to identify the unique characteristics of a taxon, while not ignoring the many similarities with other organisms that are likely to exist.

Finally, we include a brief discussion of recommendations for future research trajectories in a field that is clearly still in its infancy (Chapter 14, Vasey).

Even a strictly scientific treatment of the topic is likely to become an issue of moral and political debate. It would be disingenuous to suggest that this research has no sociopolitical implications for humans because animals are often used as a gauge for measuring what constitutes 'natural' versus 'unnatural' behaviour worthy of legal protection or persecution (Boswell, 1980; Weinrich, 1980; Haraway, 1989; Sommer, 1990; Travis and Yeager, 1991). We therefore dose the volume with a brief essay on how comparisons of human homosexuality and animal homosexual behaviour have been used and abused throughout the centuries (Chapter 15, Sommer).

Defining 'homosexual behaviour' in animals

Any useful discussion of homosexual behaviour in animals requires an explicit statement concerning what counts as 'sexual behaviour'. For the purposes of this volume, an ethological definition of sexual behaviour is needed that can be easily operationalized. As such, sexual behaviour is defined as including courtship displays (or sexual solicitations), mounting, and any interaction involving genital contact between one individual and another. Although stimulation of the genitals or other erogenous zones can result in orgasm, orgasmic response is not a necessary criterion for labeling a behaviour as sexual, nor is penetration.

Other definitions of sexual behaviour are much broader than the one we employ in this paper. For example, Bagemihl (1999) characterizes sexual behaviour in terms of five sweeping categories that include courtship, affection, interactions involving mounting and genital contact, pair bonding and parenting activities. Our concern with such an approach is that by casting the net too widely, there is a risk of including social interactions under the rubric of sexual behaviour. This, in turn, could result in interactions with particular social partners being labeled as sexual behaviour. This is not to say that relationships such as pair living, or affectionate behaviours such as kissing, cannot be sexual. However, labeling them as such typically occurs because they exist in close temporal association with courtship displays, mounting or genital contact. In the absence of such sexual markers, the sexual or social character of a particular behaviour is often open for debate. Clearly, this conservative definition of sexual behaviour risks excluding potential examples of sexual activity and some researchers will find it decidedly narrow. Nevertheless, a much stronger case can be made that those behaviours that are consistent with the definition we present here are, indeed, sexual.

The subjectivity involved in defining *homo*sexual behaviour according to context, function and motivation has been repeatedly called into question and criticized as ignoring the multifaceted nature of these interactions (Hanby, 1974; Reinhardt *et al.*, 1986; Srivastava *et al.*, 1991). For the purposes of this volume then, homosexual behaviour refers to courtship displays, mounting and/or genital contact and stimulation between same-sex individuals. Thus, homosexual behaviour or activity refers to discrete acts or interactions. As such, this term does not imply some sort of life-long pattern of homosexual activity or exclusivity, nor does it denote any particular form of enduring sexual relationship, monogamous or otherwise.

It deserves to be stressed that homosexual behaviour is not and should not, be taken as synonymous with sexual orientation, sexual orientation identity, sexual partner preference or categories of sexual beings.

- *Sexual orientation* (for example, heterosexuality, homosexuality, bisexuality, autosexuality, asexuality) refers to an individual's overall pattern of sexual attraction and arousal (or lack thereof) during some defined time period (for example, adolescence, adulthood, the previous year). Typically, this pattern is characterized by multiple parameters, including sexual solicitations, actual sexual contact and genital blood flow. In humans, sexual fantasy is often used to characterize sexual orientation as well, but researchers studying animals have no means by which to assess sexual fantasy in their subjects, assuming that such a phenomenon exists.
- *Sexual orientation identity* refers to the sexual orientation that an individual considers themselves to have (Cass, 1983/84). This may or may not dovetail with the various parameters that characterize an individual's sexual orientation. With the exception of the great apes, evidence for self-recognition in animals is weak (Tomasello and Call, 1997). As such, there is no reason to expect that animals would develop personal identities based on some introspective sense of sexual orientation.
- *Sexual partner preference* refers to an individual's predilection for sexual partners of one sex, or the other, or both, when given a choice (Vasey, 2002).

- Finally, some cultures, but not others, have created *categories of sexual beings* by classifying people based on their sexual behaviours and/or sexual orientations. The deployment of such classifications schemes typically involves the implicit presupposition that particular types of sexual behaviour and sexual orientations encapsulate the very essence of an individual. Classifying individuals in this manner results in culturally constructed categories of sexual beings (for example, the homosexual, the heterosexual, queer, lesbian, gay, etc.; cf. Foucault, 1990; Katz, 1995; for non-Western examples see Nanda, 2000). Homosexual behaviour in animals should not be taken as evidence for membership in one of these culturally constructed categories. There is no evidence that non-human animals classify their conspecifics in this manner, nor is there any particular reason to expect that they would do so.

The term 'homosexual behaviour in animals': critiques and responses

Some specialists have criticized the use of 'homosexual behaviour' when used in reference to non-human animals such as primates (for example, Dixson, 1998; Wallen and Parsons, 1997). They argue that the term is misleading and a mischaracterization of the behavioural phenomenon in question. Several reasons might account for the critics reticence to use this descriptor when labelling certain same-sex interactions in animals.

To begin with, some researchers may treat the term as synonymous with related terms described above. But, as we have pointed out, care must be taken not to make conceptual leaps from terms that denote behaviour to those that denote sexual orientation, sexual orientation identity, sexual partner preference, categories of sexual beings. Nor should the term 'homosexual behaviour' be taken to imply sexual exclusivity. Failure to make such distinctions can lead to unnecessary disagreements over interpretation of the relevant behavioural data.

Some critics might be unhappy with the term 'homosexual behaviour' because they consider this to be a uniquely human phenomenon (see Weinrich, 1980). In other words, in the minds of some critics, same-sex sexual interactions in animals are not identical in expression to human homosexual behaviour and, therefore, cannot be properly labelled 'homosexual behaviour'. This criticism is flawed on several counts. To begin with, what human behaviours have an exact counterpart in the animal kingdom? Paternal caregiving? Alliance formation? Social structures? Culture? Tool use? These behavioural patterns, as manifested in humans, have no identical counterpart in nature, yet they have all been productively discussed using a cross-species perspective (Taub and Mehlman, 1991; Chapais, 1991; McGrew, 1992; Rodseth et al., 1991). Such cross-species comparisons function, minimally, to highlight what makes humans truly unique. Most animals do not engage in heterosexual behaviour within the context of pair living, nor do they employ the ventro-ventral ('missionary') position, although many humans do. Does this mean that such animals do not engage in heterosexual behaviour? Clearly, attempting to identify some sort of *exact* animal counterpart to human behaviour is an unrealistic goal (Weinrich, 1980).

In addition, we would do well to ask: which human homosexual behaviour is it that the critics are referring to? Human homosexual behaviour is manifested in a myriad of ways such as mutual masturbation, oral–genital and oral–anal contact, anal penetration, body-to-body rubbing (LeVay and Valente, 2002). These behaviours do not necessarily co-occur and their prevalence varies between the sexes and cross culturally. In addition, there is a vast cross-cultural literature demonstrating that homosexual interactions in humans can be characterized as age differentiated, gender differentiated, role structured or egalitarian (for example, Greenberg, 1988; Herdt, 1997; Murray, 2000; Nanda, 2000). These patterns also appear to vary inter-sexually and cross culturally in terms of their prevalence. In light of this diversity, which of these same-sex activities and patterns count as 'homosexual behaviour'? Moreover, who gets to decide? And based on what criterion?

At a very foundational level, the perspective that human behaviour provides the benchmark by which animal behaviour is defined and measured represents a profound misuse of the comparative method. The comparative approach to the study of behaviour intends to draw comparisons between: (a) closely related species living in different habitats, which are therefore subject to different selection pressures, or (b) distantly related species living in similar habitats, which therefore experience similar selection pressures (Harvey and Pagel, 1991; Stamp-Dawkins, 1995). The comparative approach recognizes the need to identify the unique characteristics of a particular species, while not ignoring the many similarities with other species that are likely to exist. Such overlaps in the patterning of morphology and behaviour exist either as a result of shared common ancestry (homology) or as the result of similar selection pressures, which produced comparable traits to cope with similar environments (analogy).

The validity of the comparative method rests on the ability of researchers to identify the correct level of behavioural analysis. Discrete behaviours or cognitive mechanisms represent an appropriate level of cross-species analysis, whereas culturally constructed behavioural assemblages do not. Thus, for individuals interested in the evolution of the sexual division of labour, discrete behavioural components of this behavioural conglomerate are the appropriate level of analysis (for example, sex differences in foraging, food transport, inter-individual food transfer, etc.) but not the culturally constructed concept of sexual division of labour itself. Thus, we might imagine a critic arguing that homosexual behaviour cannot be studied in animals because unlike homosexual men, animals do not form egalitarian sexual relationships, engage in anal intercourse and engage in exclusive homosexual behaviour over the lifespan. But this kind of reasoning (apart from being sexist) is profoundly flawed because it is framed by a particular historically and culturally specific notion of what it means to be a male homosexual. As such, this reasoning takes place at the wrong level of analysis; that is, at the level of culturally constructed behavioural categories ('the homosexual') and not at the level of discrete behaviours and cognitive mechanisms.

Some critics may be unhappy with the term 'homosexual behaviour' because it implies that the behaviours in question are sexually motivated, when, in fact, they might be 'socio-sexual' in character. That is, they might be enacted to facilitate adaptive social goals and not because they are sexually pleasurable, per se. Indeed, sexual motivation is rarely ascribed to socio-sexual interactions because their adaptive social functions are often seen as their primary *raison d'être*, thus downplaying or annihilating any sexual component that such interactions might have. As such, it is typical that these behaviours are conceptualized as existing only as desexualized vehicles by which fitness-enhancing goals can be achieved. However, there is no reason why sexual motivation and social function should be mutually exclusive. Indeed, Wickler (1967) himself recognized this point and argued that same-sex mounting could be sexually motivated or occur in a sexual context and still serve some social function. Thus, the label 'homosexual behaviour' does not negate the possibility that a behaviour might have a socio-sexual function. The term 'same-sex socio-sexual behaviour' would be an inappropriate alternative to 'homosexual behaviour' because it explicitly assumes that the same-sex interactions in question have some socio-sexual function. It is common practice among animal behaviourists to attribute socio-sexual functions to animal homosexual activity, in a *post-hoc* fashion. However, socio-sexual functions have to be demonstrated empirically and not simply tacked on to behaviours as afterthoughts simply because they fit with our idea of how a Darwinian world is organized. Although the literature on animal homosexual behaviour is replete with socio-sexual hypotheses, it is surprising how few investigators actually follow through with rigorous, empirical tests of these adaptive propositions.

Other terms have been proposed, including 'same-sex sexual behaviour', 'iso-sexual behaviour' and 'ambisexual behaviour'. Just like 'homosexual behaviour', all of these labels imply that the behaviour under investigation is sexual and, as such, none of them circumvents this supposed concern. The term 'same-sex behaviour' has also been proposed as an alternative to homosexual behaviour, but that would be even more problematic. For example, in using 'same-sex behaviour', how would we distinguish the behaviours that interest us in this volume from, say, female–female aggression or male–male grooming?

'Same-sex pseudocopulation' seems an inappropriate descriptor as well, since its core assumption is that male–female copulatory patterns are the template against which all other sexual interactions should be measured. In essence, the implication is that same-sex partners are trying to mimic heterosexual patterns of copulation. In many instances, however, the form homosexual interactions take are distinctly different from heterosexual copulation (review in Bagemihl, 1999). For example, although female Japanese macaques routinely mount other females, they employ a much wider variety of mount postures than do hetero-sexual couples (Vasey et al., 2006). Moreover, in comparison to male mounters, females execute per pelvic thrust: (a) more movement, (b) a greater variety of movements and (c) more complex movements. Apart from these concerns, a term such as 'pseudocopulation' serves as a complete misnomer for same-sex interactions involving courtship.

The final alternative with which we are left is 'same-sex pseudosexual behaviour'. The problem with this term, however, is that it implies that same-sex interactions are not sexual. This is something that needs to be demonstrated, not assumed, particularly in light of the fact that various aspects of these inter-actions appear to be sexual. Conversely, the assumption that such interactions *are* sexual needs also to be demonstrated, and not assumed, given the possibil-ity that same-sex individuals might engage in such activity for reasons that are exclusively socio-sexual.

There are at least three means by which the sexual nature of same-sex courtship, mounting and genital contact might be established using non-invasive observational techniques. First, it can be determined if the same-sex behaviours under investigation are patterned in similar ways to male–female courtship and copulation. Although, as we have mentioned, male–female courtship and copu-lation should not be the yardsticks by which all other types of sexual behaviour are measured, many researchers consider these behaviours to be sexual, *a priori*. Consequently, demonstrating that same-sex courtship, mounting and genital contact parallel their male–female counterparts in terms of their expression represents strong evidence (at least in the minds of some researchers) that the former are, indeed, sexual behaviours. Secondly, when examining the issue of sexual motivation, it is also important to determine whether patterns of same-sex courtship, mounting and genital contact differ in their expression from more generalized patterns of social affiliation. If we subscribe to the idea that sexual behaviours are related to, but ultimately distinct from, ordinary social behaviours, then, presumably, patterns of sexual affiliation should not completely mimic patterns of affiliation that are widely recognized as social. Thirdly, prolonged and directed genital stimulation, or orgasm, or both, during same-sex interaction represent strong evidence that these behaviours are sexually motivated.

Ideally, we could refer to the behaviours that interest us here as 'same-sex interactions in animals involving courtship and genital contact and stimulation'. However, this would clearly be very linguistically cumbersome. We recognize that homosexual behaviour is not a perfect descriptor, insofar as the term implies some degree of sexual motivation. We hope, however, that we have demonstrated that all of the terms that have been proposed as alternatives are far more problematic and also carry with them implicit, untested assumptions. Moreover, the assumption of sexual motivation embedded in the label 'homosexual behaviour', is, as we have shown, amenable to empirical analysis. At the very least, observers can quickly determine if an interaction is 'sexual' in terms of the superfical form it takes, but demonstrating a socio-sexual function for such behaviour is not quite so easy.

If none of the terms discussed is perfect, why do we favour 'homosexual' behaviour above all the other alternatives? The value of adopting this term is that it leaves the door open for evolutionary comparisons between humans and other species, provided that researchers focus on the appropriate level of analysis (that is, discrete behaviours and cognitive mechanisms). The alternative would be to endorse the notion that only humans engage in homosexual behaviour. This, of course, promotes a conceptualization of human homosexual behaviour as a phenomenon that exists outside the explanatory framework of evolutionary theory. In other words, the behaviour is perceived as so idiosyncratic and unique – so aberrant and bizarre – that it can only be understood on its own terms, as something that exists separate from the rest of the natural world and the processes that govern that world.

It is important to note that we are not necessarily suggesting homology when we use the term homosexual behaviour in reference to various species, including humans. For example, if we talk about tool use in sea otters or infanticide in dung beetles, we do not presuppose that these behaviours are homologous with human tool use or infant killing. Likewise, we can talk about flight in insects without presupposing that it is homologous to flight in birds or bats. As such, our aim here is, in part, to investigate if similar behaviours have been shaped by similar evolutionary forces or processes, and if they are analogous or homologous.

The analytical framework: 'How?' and 'Why?'

Evolutionary thinking was inspired, in part, by eighteenth- and nineteenth-century naturalists who attempted to bring order to the rather mind-boggling diversity of organisms by classifying them according to similarity. Scientists intent on fleshing out the history of life constructed phylogenetic trees by describing the differences and similarities in morphology among species. This approach was later extended to behaviour, particularly as more and more animals were studied in their natural environments. A distinct school of comparative ethology began to blossom in the first half of the twentieth century when Karl Ritter von Frisch, Konrad Lorenz and Nikolas Tinbergen focused on cross-taxa comparison of behaviour.

Nikolas Tinbergen stressed that biological traits, whether morphological or behavioural, can be subjected to various levels of inquiry: the individual's development, the physiology which enables the execution of the behaviour, the purpose of the behaviour for survival and reproduction and the historic pathways that led to the behaviour (Tinbergen, 1963). Each of these dimensions contributes to different types of 'understanding'.

Our book attempts to apply elements of this method of analysis to a particular pattern of activity, namely same-sex sexual behaviour. It might therefore be helpful if we illustrate the Tinbergen approach with a fictional inquiry into the natural history of white rabbits – some of which, as it happens, also exhibit a preference for sexual interactions with members of their own sex (cf. also Alcock, 1989).

Let us first explore the cause for the rabbits' whiteness. A physiologist, whom we approach, explains that these rabbits have dark fur when born, but that their fur quickly loses its pigment. Next we ask an ecologist. She points out that these creatures live in snow-bound regions of the Northern hemisphere and that their white fur might have to do with camouflage, that is, protection from predators such as eagles and foxes. Rabbits further South have a darker pelt.

Both explanations are correct and both complement each other. Each explanation addresses a specific question: the physiologist's explanation addresses the question 'how' a particular trait came to be, while the ecologist's explanation addresses the question 'why' a particular trait exists.

The line of inquiry stimulated by the question 'how' or 'how come' explores the conditions within an organism which bring the particular trait about, in our example the lack of pigment. The first dimension of the answer – that rabbits are born with dark fur – is related to the 'ontogeny' or development of the trait. The second dimension of the answer – that the hair lacks pigment – is related to the 'mechanism' of the trait. A more detailed study would perhaps reveal the effects of heredity on the trait, and the exact metabolic pathways that turn dark hair into white. Both dimensions of the 'how' question aim reasons at immediate for the expression of a trait, revealing so-called 'proximate causes'.

The other line of inquiry refers to 'why' questions (or 'what for') and explores the value a particular trait had for the survival and reproduction of an organism. The first dimension of the answer – that whiter rabbits benefit from camouflage – is related to the 'function' of the trait or its evolutionary 'purpose'. A more detailed study would perhaps demonstrate the differential reproduction of rabbits with different degrees of whiteness. The second dimension of the answer – that rabbits further South have darker hair – is related to the 'phylogeny' of the trait or its evolutionary history. By comparing different morphs of rabbits across the globe, we might be able to reconstruct the fur colour of the common ancestor of all extant rabbits, and the adaptive radiation of the trait across the globe. Both dimensions of the 'why' question aim to explain the evolutionary basis of the trait under investigation, revealing so-called 'ultimate causes'.

Let us now apply this scheme to the same-sex sexual behaviour exhibited by our rabbits. A physiologist will again focus on *proximate causes*. She might first point out that such behaviour is correlated with prenatal exposure to certain levels of hormones that influence developmental pathways in the brain (*ontogeny*). A more detailed study of the sensory-motor processes would reveal the exact interplay between the nervous system and the muscles that move the skeleton. This would help us understand how some male rabbits assume a posture otherwise displayed by females during heterosexual copulations, thus enabling another male to mount them (*mechanism*).

A behavioural ecologist who focuses on *ultimate* causes might discover that these rabbits have a system of communal rearing of young which includes some individuals who refrain from breeding and instead act as 'helpers-at-the-nest' of younger siblings. They keep the brood warm, and warn against approaching predators. The same-sex sexual behaviour ensures that there are enough non-breeders around (*function*). Rabbits further South live under less harsh conditions, which does not require that older individuals warm the brood, and same-sex sexual behaviour did not therefore evolve in these regions (*phylogeny*).

Of course, the biological world is extremely complex, and in most instances we will have only a partial understanding of the proximate and ultimate causes that bring about traits in morphology and behaviour. This is certainly also true with respect to many of the homosexual interactions of animal species portrayed in this volume. Still, what we can do is to derive predictions from functional and phylogenetic hypotheses, test them against the data and reject at least some of them. Such a process subscribes to the Popperian mode of achieving scientific progress: through falsification of proposed explanations. We should therefore emphasize that the Tinbergian framework is largely 'heuristic', that is a technique designed to provide direction when attempting to solve a problem. The heuristic method does not claim that one could ever come up with definitive answers. It is foremost a way of structuring our thinking, of bringing order to phenomena, in very much the same way as the early naturalists ordered their collections of beetles and exotic plants. The heuristic method is thus more modest than its etymology from the Greek word 'heuriskein' for 'to discover' might suggest, and definitely more modest than the famous exclamation 'Heureka!'.

Proximate causes: ontogeny and mechanisms

To date, no complete Tinbergian analysis of homosexual behaviour has been conducted on a single non-human species. Various researchers have, however, focused on particular proximate levels of analysis for their specific study species. For example, genetic research on the fruit-fly *Drosophila* demonstrates that ectopic expression of a gene known as the white gene induces male–male courtship (Zhang and Odenwald, 1995). Hormonal research shows that the female zebra finch (*Taeniopygia guttata*) treated with estradiol or fadrozole (an oestrogen synthesis inhibitor) as nestlings or embryos prefer to pair with other females in adulthood, even when potential male partners are available (Adkins-Regan, 2002). Neurobiological research indicates that male ferrets (*Mustela putorius*) whose brains are subjected to excitotoxic lesions of the pre-optic area/anterior hypothalamus prefer male sexual partners (Paredes and Baum, 1995).

Although genetic, hormonal and neurobiological levels of analysis are outside the scope of this volume, a number of contributors touch on issues pertaining to developmental mechanisms by showing that homosexual behaviour is more frequent during particular life history stages. In some species, juveniles and sub-adults appear to engage in homosexual behaviour most frequently (bison, Chapter 5; bottlenose dolphins, Chapter 4; gorillas, Chapter 11), whereas, in others, the behaviour is manifested most commonly in late adulthood (geese, Chapter 2). Still others show that the role individuals adopt during homosexual mounting interactions is influenced by age, with younger participants taking the role of the mountee and older participants taking the role of the mounter (gorillas, Chapter 11; feral cats, Chapter 7).

Some contributions to this volume also bring evidence to bear on the proximate issue of how social/demographic factors play an immediate role in the expression of homosexual behaviour. For example, same-sex sexual interactions among male bottlenose dolphin calves are influenced, in part, by the sociality of the males' mothers and the local cohort of male calves that is available

(Chapter 4). The loss of individuals in a population can trigger social instability and this, in turn, can result in increased levels of homosexual behaviour. Just such a case is described for a free-ranging population of female rhesus macaques (Chapter 9).

Under demographic conditions characterized by skewed sex ratios, members of the more abundant sex may have reduced access to members of the scarcer sex and this may result in an increased expression of homosexual activity. Homosexual pairings between male geese are clearly promoted by demographic conditions in which female sexual partners are relatively scarce (Chapter 2). Moreover, sex segregation in all-male groups seems to create a social context in which homosexual behaviour is more readily expressed. For example, sexual behaviour between males occurs relatively frequently in all-male groups of bison (Chapter 6), gorillas (Chapter 11) and langur monkeys (Chapter 10).

One might ask why individuals are compelled to seek out sexual partners at all in conditions involving skewed sex ratios. After all, abstinence is always an option. Moreover, the assumption that animals engage in homosexual behaviour in the context of skewed sex ratios because they are deprived of opposite-sex mates is an assumption that needs to be tested and not passively accepted in the absence of any data. For example, one study showed that female Japanese macaques engaged in more homosexual activity in the context of female–skewed adult sex ratios (Vasey and Gauthier, 2000). They did not do so, however, because they were unable to access a male. Even when a sexually motivated male with whom they had previously engaged in sexual activity was readily available to mate, females frequently ignored him in favour of same-sex sexual partners. As such, social/demographic causes are at best a partial explanation for the expression of homosexual behaviour, even in situations where opposite-sex individuals are rare or non-existent.

Finally, several authors stress that sexual reward by means of genital stimulation appears to be an important stimulus for the expression of homosexual behaviour in certain mammalian species (bottlenose dolphins, Chapter 4; feral cats, Chapter 7; Japanese macaques, Chapter 8; Hanuman langurs, Chapter 10; mountain gorillas, Chapter 11; bonobos, Chapter 12). In many instances, these researchers also attribute various socio-sexual functions to such behaviour, underscoring the point that sexual motivation and socio-sexual function need not be mutually exclusive.

Ultimate causes: functions

A whole array of hypotheses has been forwarded to understand the potential functions of homosexual behaviour. Our brief introduction will address some common merits and pitfalls of the various explanatory hypotheses by providing supportive as well as contradictory examples. We will illustrate our critique, where appropriate, by referring to contributions in this volume. Other reviews and critiques of these hypotheses, some of them more extensive than our own, can be found elsewhere (Dagg, 1984; Sommer, 1990; Kirsch and Weinrich, 1991; Vasey, 1995; Bagemih!, 1999).

Some explanations clearly seem far-fetched, whereas others appear more plausible. Not all evolutionary hypotheses for homosexual behaviour are applicable across all species. Moreover, there is hardly a taxon for which mono-causal explanations for homosexual behaviour seem to be appropriate. Still, it should be kept in mind that current functional explanations for heterosexual sex are also not mono-causal, as they do not necessarily focus on fertilization. For example, it has been proposed that females employ sex strategically to make males 'believe' that they are the fathers of their subsequent offspring. Females may, in this way, prevent male-committed infanticide or they may be able to extract investment from males with whom they have engaged in sex (review in Hrdy, 1981).

Most of the functional explanations for animal homosexual behaviour assume that these types of interactions are socio-sexual in character and that same-sex behaviour provides certain benefits to those engaging in it, in terms of facilitating some sort of fitness-enhancing social goal or breeding strategy. All but the first of these explanations (that is the 'controlling population size' hypothesis) assume that such benefits are accrued by the individual participants engaged in the homosexual interaction.

Controlling population size. *Homosexual behaviour becomes more frequent if population density increases, thus reducing the strain on resource availability.*
Aristotle, in his treatise, *Politics*, states that 'intercourse with males' was encouraged among Cretan men to reduce population pressure on the food supply (1272a, pp. 23–6, as cited in Kirsch and Weinrich, 1991). This line of reasoning is a classic case of a group selectionist argument, which assumes that behaviour serves the 'good of the group' – or, by extension, the 'good of the species'. Modern evolutionary biology considers this argument to be flawed because the system could easily be invaded by cheaters who will not care for the greater good of the group and continue to produce offspring for their personal benefit. These individuals that act for their own benefit would leave more copies of their genes in future generations than those that engage in same-sex behaviour. Same-sex behaviour would therefore disappear (review, for example, in Trivers, 1985).

Apart from this crucial theoretical pitfall, there is also little empirical support for the hypothesis. It rests on the assumption that individuals who engage in homosexual sex will not reproduce. This, however, is clearly not in line with the data, since bisexual interactions are widespread amongst animals. This means, that those individuals who engage in same-sex sexual interactions will normally also engage in opposite-sex sex. Homosexual behaviour does, therefore, not necessarily interfere with breeding. Exclusive homosexual behaviour – as reported in humans (Chapter 13) – does occur in animals, but is rather rare. Admittedly, the hypothesis could likewise imply that individuals switch from hetero- to homosexual behaviour when resources deplete. However, supporting data for this effect are likewise lacking. The hypothesis also predicts that members of declining populations should refrain from same-sex activities. Yet, homosexual behaviour is prevalent amongst various endangered species, for example gorillas (Chapter 11) and bonobos (Chapter 12).

Dominance expression. *Reaffirmation of the dominance hierarchy and expression of social status to prevent aggressive interactions. Mounting is a display of dominance, while being mounted is a display of submission.*

The dominance hypothesis is arguably the most cited 'explanation' for why animals engage in homosexual behaviour (reviews in, for example, Dagg, 1984; Srivastava *et al.*, 1991; Vasey, 1995; Bagemihl, 1999). We summarize the broad findings here. The hypothesis conceptualizes same-sex interactions as ritualized gestures to communicate social status. The participants thereby reduce the threat of incipient aggression associated with social tension (Wickler, 1967). According to this 'rank demonstration hypothesis', mounting is an expression of dominance, whereas allowing oneself to be mounted expresses subordinate status relative to the mounter. Consequently, a test of this hypothesis should demonstrate that the roles individuals adopt during same-sex mounting interactions are differentiated on the basis of their rank. In addition, mounting interactions among mounters should be particularly common at the outset of aggressive interactions, at which time reiterating an acceptance of the dominance status quo should function to reduce the threat of escalated fighting.

Some studies report a rather strict relationship between dominant individuals mounting subordinate individuals of the same sex (for example, female squirrel monkeys, Talmage-Riggs and Anschel, 1973; female stumptail macaques; Chevalier-Skolnikoff, 1976). However, most studies document that, while dominant individuals mount more often, the mounting of a dominant by a subordinate individual is not uncommon, depending on the social context (for example, langur monkeys, Chapter 10). In bonobos, subordinate females initiate homosexual interactions more often than their dominant counterparts, although the reverse pattern is sometimes observed (Chapter 12). In some species, such as deer, subordinates mount their dominant counterparts more often than the reverse (Chapter 6). Vice versa, in certain human societies, the roles men adopt during homosexual anal intercourse are strictly differentiated, with dominant, more aggressive males penetrating more passive subordinates (Chapter 13).

Other studies document that subordinates mount dominant individuals following agonistic interactions, which is exactly the opposite of what one would expect if these homosexual mounts were dominance displays. Furthermore, mounts by dominant individuals are sometimes associated with affiliative behaviour and solicitations by the mountee (langur monkeys, Chapter 10), which, in a heterosexual context, would be taken as a sign that indicates sexual interest.

The ambiguity of homosexual mounts in relation to rank has led many researchers to abandon a strict dominance interpretation of these interactions. Smuts and Watanabe (1990) suggested that a more productive way to interpret them might be as dominance negotiations during which the participants flesh out their relationships. Indeed, in bottle-nosed dolphins, homosexual interactions among male calves seems to be best described as exchanges involving dominance negotiation, not dominance assertion (Chapter 4). Reciprocal mounting between young male bison promotes inter-individual bonding, which may reduce the risk of predation (Chapter 5). In humans, male homosexual behaviour may function to reiterate dominance relationships, particularly in social situations that require male–male cooperation (Chapter 13).

An important limitation of the dominance hypothesis is that it is only applicable to mounting interaction in which one animal is positioned on top of another. Activities such as body rubbing, rump–rump rubbing, oral–genital or oral–anal contact and mutual masturbation do not lend themselves to categorization of 'mounter' and 'mountee' roles. The hypothesis is clearly of little value in interpreting such non-mounting interactions. For example, when bottle-nosed dolphins constantly roll and shift as they manipulate each other's genital area with their beaks, fins and genitalia (Chapter 4), no individual can be unambiguously labeled mounter or mountee. Similarly, the hypothesis is of little value in understanding same-sex courtship within the context of same-sex pairing in species such as greylag geese (Chapter 2) and flamingos (Chapter 3).

Finally, the stereotype of ascribing a submissive role to the mountee and a dominant role to the mounter is probably linked to the traditional heterosexist understanding that females (the quintessential 'mountee') should be viewed as 'submissive', whereas males (the quintessential 'mounter') are viewed as 'dominant'.

Social tension regulation. *Preventive conflict management: Getting another individual to have sex functions to counteract or buffer stressful modes of interactions.*
This hypothesis assumes that homosexual behaviour represents a type of preventive conflict management. The currency which renders this arrangement functional is the pleasure provided by same-sex interactions. Sex thus functions as a 'compensation payment' for agreeing to de-escalate conflicts, for example during conflicts over resources.

A typical example is the close temporal relationship between feeding and homosexual behaviour as observed amongst female bonobos (Kano, 1980; Kuroda, 1980; de Waal, 1987). Rates of the so-called gg-rubbing (genito–genital rubbing) are particularly high when food can be monopolized or when individuals are forced into close proximity because they occupy the same food patch. Indeed, bonobos entering occupied patches are more likely to acquire food after engaging in homosexual interactions (Kuroda, 1984; Furuichi, 1989; de Waal, 1987). Even in the midst of agonistic interactions, homosexual behaviour between female bonobos appears to terminate the conflict (Furuichi, 1989). Data on non-provisioned wild bonobos provide us with an even more complex picture of the relationship between social tension and same-sex sexual interactions, underscoring how food quality, accessibility and types of competition all interact to influence the expression of homosexual activity in this species (Chapter 12).

Homosexual behaviour can also reduce tension outside feeding contexts. Same-sex mounting in acorn woodpeckers, for example, is associated with the birds' dusk roosting activity (MacRoberts and MacRoberts, 1976) when inter-individual distances are reduced and competition for access to sleeping partners and sleeping sites occurs. In domestic cattle, mounting between steers increases in stressful situations associated with a lack of feed or changes in the group's social composition (Brower and Kiracofe, 1978; Lincoln, 1983). Contributors to this volume describe how same-sex interactions among male deer are most common during stressful situations such as herding (Chapter 5), and that mounting among female bison increases when group members are reintroduced after separation (Chapter 6).

Reconciliation. *Therapeutic conflict management: Pleasurable same-sex interactions can reestablish social bonds after mechanisms for regulating aggression failed.*
The hypothesis assumes that homosexual behaviour can be used to reestablish social bonds when tension reduction mechanisms for preventing aggression failed. The frequency with which homosexual behaviours are manifested should

therefore be very context specific. Namely, although homosexual behaviour may occur outside of a post-conflict period, it should be exhibited more frequently following an aggressive incident, as compared with a period preceding the conflict (Veenema, 2000; Sommer *et al.*, 2002).

Supporting evidence comes, for example, from studies of female–female mounting in various macaques (for example, Dixson, 1977; Oi, 1991). For other species such as bonobos, the evidence is more ambiguous. Genito-genital contact between captive females increased significantly following conflicts (de Waal, 1987), and homosexual interactions among males in a provisioned wild population also increased following aggressive interactions (Kano, 1980). However, although rates of genito-genital contacts among females of a wild unprovisioned population increased after conflicts, most genito-genital rubbing was unrelated to agonistic contexts (Chapter 12). The same is true for species such as dolphins (Chapter 4), deer (Chapter 6) or langur monkeys (Chapter 10). In female Japanese macaques, aggression even appears to inhibit, rather than facilitate, the expression of homosexual behaviour (Chapter 8).

Social bonding. *Same-sex behaviour can express mutual affection and affiliation, provide psycho-physical rewards and can thus reinforce long-term relationships.*

In an interesting merger of concepts from biological and social anthropology, Watanabe and Smuts (1999) pointed out that same-sex sexual behaviours often resemble ritualized signals because these interactions include stereotypical elements of vocalization, solicitation, mounting or mutual grooming. Although the context is non-linguistic, it is nevertheless socially so complex that it can be interpreted as a 'ceremony'. For example, in male olive baboon greetings, the males exchange a series of gestures that include occasional embracing, presenting and grasping the hindquarters with one or both hands, mounting, touching the scrotum and pulling the penis (Watanabe and Smuts, 1999). The significance of these greetings rest with the fact that virtually all other interactions between male baboons involve antagonistic exchanges and that they almost never engage in friendly behaviours outside these greetings. Mutual fondling of penis and testes – when one male places his vulnerable sex organs into the hands of another – is therefore an expression of mutual trust, analogous to oath-swearing in humans. If both parties subscribe to such formalism, which involves the exchange of 'costly' and thus 'honest' signals (Zahavi and Zahavi, 1997), then they convey an intention to cooperate.

There are numerous examples of avian and mammalian species in which individuals engage in homosexual behaviour within the context of pair living (review in Bagemihl, 1999). Same-sex individuals that are pair bonded frequently exhibit highly synchronized, bi-directional patterns of affiliation activities, which would seem to support the hypothesis that homosexual behaviour promotes social bonding. Such is the case with flamingos (Chapter 2) and greylag geese (Chapter 3).

Nevertheless, same-sex sexual interactions do not always appear to have a 'friendly' overtone. Instead, they often come across as extensions of rather antagonistic interactions (bison, Chapter 6; langur monkeys, Chapter 10). One also wonders why mounting should be a particularly effective way of re-enforcing social bonding since allo-grooming, with its added benefit of providing a hygienic service, is a rather straightforward possibility. Grooming might not be as 'costly' as allowing a potential competitor to touch one's genitals, but grooming is also an 'honest' signal. Indeed, in contrast to the ritual pattern of mounting between male baboons, grooming is much more common amongst male langur monkeys than is mounting (Chapter 10).

Alliance formation. A specific form of social bonding. Same-sex contacts promote bonding and allow partners to intimately explore each others psychological and physical attributes. Knowing the strengths and weaknesses of an ally will be advantageous during competitive interactions with others.

A number of behavioural scientists have taken the social bonding hypothesis one step further by proposing that homosexual behaviour promotes alliance formation among the participants. When one individual (the supporter) intrudes into an ongoing dyadic conflict and supports one of the opponents (the recipient of support) against the other (the target), the supporter and the recipient of support are said to be allies and to have formed an alliance. The rationale for this functional hypothesis is that same-sex contacts promote bonding and allow a potential ally to explore the psychological and physical strengths and weaknesses of a prospective partner, which will be helpful during conflicts with third parties. Sexual contacts may thus promote tit-for-tat type interactions (reciprocal altruism; Trivers, 1971).

Narratives about famous pairs of human males who jointly fought battles easily fit this pattern – be it Achilles and Patroklus or David and Jonathan. Both Greek Mythology and the Old Testament hint that the relationships of these men included sex (see also Kirkpatrick, 2000). It is conceivable that intimate knowledge of one's partner's psychological and physical characteristics was of advantage in combat situations.

Fairbanks *et al.* (1977) first explicitly articulated the alliance formation hypothesis for non-human primates. They observed high levels of homosexual behaviour between female rhesus macaques (*Macaca mulatta*) in newly formed groups

containing many unfamiliar individuals. They suggested that under such conditions, homosexual behaviour may restore social bonds in the context of group
social instability:

> In a natural troop, consort bonds between a male and a female are
> rapidly formed and broken, in contrast to typical female–female
> relationships, which are based on long-term familiarity . . . In the
> absence of the normal mechanisms for assuring female–female bonds,
> a few members of each group . . . turned to the behaviour pattern of
> the sexual consort relationship for rapid bond formation. The females
> who could form the first bonds joined in coalitions against their
> undefended peers and attempted to drive them from the group. This
> division of the social group into 'bonded females' and 'strangers' was
> apparently the first stage in the formation of a new group. (p. 248)

At first glance, it seems unlikely that the experimental situation created by
Fairbanks *et al.* (1977) has any parallel under free-ranging conditions. Nonetheless, a number of free-ranging species demonstrate homosexual behaviour in
situations involving less extreme social instability such as intergroup contact
and transfer (reviewed in Vasey, 1995). For example, same-sex mounting in blue
bellied rollers can function as a type of intergroup threat, which members of one
group direct towards members of another group (Moynihan, 1990). Male–male
mounting interactions in langurs monkeys can also be preceded by certain vocalizations and displays that are normally addressed to other groups with whom
the mounting males compete for food and female mates (Chapter 10).

The study by Smuts and Watanabe (1990) on male olive baboons deserves
special mention because it is the most detailed examination to date demonstrating a relationship between male–male alliance formation and homosexual
behaviour. Males that mounted and manipulated each other's genitals more frequently, formed the most cohesive and successful alliances against other males.
Often, as if to reaffirm their alliance bond, two males would engage in homosexual behaviour just before challenging a rival. The most intensively bonded
males encouraged behavioural symmetry in their relationship by actively soliciting each other for mountings and genital fondling. Smuts and Watanabe (1990)
suggest – in line with the 'costly signaling' rationale described above – that
males permit potential rivals intimate contact with their genitalia in order to
demonstrate a willingness to accept risk, and, thus, genuine interest in forming
a reciprocal alliance in spite of the short-term cost they entail.

Same-sex pairs amongst both greylag geese (Chapter 2) and flamingos (Chapter
3) seem to be better able to defend resources and support each other in conflicts
compared with unpaired individuals. Bottlenose dolphin males use high levels

of synchronous sexual behaviour to negotiate relationships and bond with each other, forming pairs or trios. In adulthood, it is believed that these bonded males cooperate with each other and compete against other bonded males for access to female mates (Chapter 4).

Likewise, female–female homosexual behaviour occurs amongst rhesus macaques in situations involving social instability and when they change groups, perhaps as a strategy to obtain allies (Chapter 9). Young bonobo females who immigrate into a new community appear to target dominant resident females (that is, potentially powerful allies) for homosexual sex (Idani, 1991). Female bonobos also develop sexual relationships with each other and employ these sex-founded coalitions to dominate males over access to resources (Parish, 1994). Human males may likewise use homosexual behaviour to faciliate alliance formation in situations requiring some degree of male–male cooperation (Chapter 13).

However, not all researchers subscribe to the idea that homosexual activity facilitates alliance formation in bonobos (Chapter 12; contra, for example, Furuichi, 1989). Evidence against the alliance formation hypothesis also comes from studies of Japanese macaques. These studies indicate that although same-sex sexual behaviour and alliance formation are causally related, the 'design features' of these interactions reveal that homosexual activity in this species is not a socio-sexual adaptation for alliance formation (Chapter 8). Moreover, same-sex activity in other taxa such as deer (Chapter 6) and female langur monkeys (Chapter 10) seems to also have little to do with alliance formation.

Acquisition of alloparental care. *Allows partners to gain help in rearing offspring. Sex may reward the helper with sexual pleasure or may simply increase the opportunity that a potential caregiver is exposed to a partner's offspring.*

As outlined above, homosexual behaviour can promote social bonding and may increase the likelihood that the actors cooperate. In line with this thinking, research on a variety of species indicates that same-sex pairs may cooperate in raising each others' offspring when the chances of raising offspring on their own is reduced or nil (see Diamond, 1989). Same-sex interactions might simply decrease spatial proximity between the participants so that a potential caregiver is exposed to a partner's offspring. Alternatively, individuals might reward helpers with sexual pleasure so as to encourage the so-called 'alloparents' to continue caring for their offspring.

Supporting evidence for this hypothesis has been found in several monogamous birds (reviewed in Diamond, 1989; Bagemihl, 1999; for example, Western gull, Hunt and Hunt, 1977; black-winged stilt, Kitagawa, 1988; lesser snow goose, Quinn *et al.*, 1989). In a clear refutation of this hypothesis, during homosexual consortships, female Japanese macaques are more aggressive to their same-sex sexual partners' offspring, compared to outside these relationships (Chapter 8).

Mate attraction. Mimicking heterosexual copulations may prompt sexual interest in hitherto reluctant heterosexual partners.

Researchers have suggested that by copying the copulatory pattern (position, pelvic thrusts) of rival males, a mounting female can potentially attract dominant males as sexual partners and increase her chances of insemination (Parker and Pearson, 1976). According to this hypothesis, female mountees do not gain access to the male sexual partner and, as such, they behave altruistically for the benefit of the female mounter. Consequently, if female homosexual behaviour functions in this manner, it may have evolved via kin selection or reciprocal altruism. Variations of this hypothesis have emphasized that either of the female partners (the mounter or the mountee) can attract males by engaging in homosexual behaviour. The mountee, for example, could demonstrate her receptivity, thus signalling her willingness to seek out a male mate.

The rationale embedded in this hypothesis makes most sense if both the mounter and the mountee are in a fertile reproductive stage. Indeed, female mammals will often mount each other during the ovulatory period (langur monkeys, Chapter 10). However, many cases of same-sex mounting take place outside the breeding season or when both mounter and mountee are infertile or out of sight of males (Japanese macaques, Chapter 8; langur monkeys, Chapter 10). Moreover, in both Japanese macaques and langur monkeys males are frequently disinterested in females that engage in homosexual interactions. When male Japanese macaques do exhibit an interest in homosexual consortships, female partners will sometimes threaten or attack them and attempt to drive the males away.

This hypothesis is more difficult to apply to male–male mounting, except if one speculates that males arouse each other in preparation for heterosexual copulation, instead of stimulating female partners.

Inhibition of competitor's reproduction. Mounters reduce the mountee's receptivity by providing alternative sexual stimulation, thus decreasing a competitor's reproductive success.

This hypothesis suggests that female homosexual mounting represents a form of intra-sexual competition, which minimizes the probability that rivals are inseminated (reviewed in Tyler, 1984). Mounting females are assumed to reduce the mountee's receptivity and access to male partners by providing alternative sexual stimulation. This decreases the mountee's probability of insemination, reduces the number of future competitors for the mounter, while at the same time increasing her relative reproductive success. Furthermore, limiting the mountee's copulations guards against male sperm depletion, which increases the mounting female's own chances of fertilization.

Researchers studying domestic turkeys (Hale, 1955; Hale and Schein, 1962) were among the first to suggest that females might mount other females as part of a breeding strategy aimed at reducing the mountee's sexual receptivity. The hypothesis draws also some support from studies of female primates, where oestrous (that is fertile) individuals are disproportionately more often mounted by others (for example, stumptail macaques, Chevalier-Skolnikoff, 1976; langur monkeys, Chapter 10).

A somewhat simpler version of the hypothesis assumes that homosexual mounting disrupts heterosexual activity. For example, male buff-breasted sandpipers are known to mount rival males who are in the midst of courting females (for example, Myers, 1989). There is also some evidence that homosexual mounts between female bison interfere with the bulls attempt to court oestrous cows (Chapter 5). Similarly, some male–male mounting interactions in feral cats may function to reduce the mountee's access to female mates (Chapter 7).

Still, one wonders why animals would want to disrupt copulations using mounting, when other methods such as pecking, chasing or slapping, seem to be much more effective. Indeed, direct aggression against copulating individuals is the preferred tactic of sexual harassment in species such as Indian langur monkeys (Sommer, 1989). Moreover, there are no empirical data which suggest that homosexual mounting does indeed reduce heterosexual activity or that fertilization rates are decreased – although such effects are difficult to measure.

Practice for heterosexual activities. *Homosexual activity represents learning opportunities for immature individuals, often during play. The behaviour functions to facilitate social and motor development for adult heterosexual roles.*

Numerous authors have suggested that homosexual interactions between immature animals during play interactions could function as practice for heterosexual activities such as courtship and copulation (reviewed in Dagg, 1984; Vasey, 1995; Bagemihl, 1999). According to this hypothesis, homosexual behaviour might facilitate social and motor skills that could then be employed in a heterosexual context in adulthood.

Experimental research shows that among macaques adequate opportunity to engage in mounting when young is necessary for competent performance of heterosexual copulation in adulthood, and the actual sex of immature mounting partners is irrelevant (Goy and Wallen, 1979). Observations such as these lend support to the hypothesis that homosexual behaviour can function as practice for future heterosexual interactions. In this volume, 'play-like' same-sex interactions involving immature individuals are reported from species as varied as dolphins (Chapter 4), bison (Chapter 6), langur monkeys (Chapter 10) and mountain gorillas (Chapter 11). These play interactions may function as practice for heterosexual activity in adulthood. Same-sex pairing in flamingos might also provide the participants with the opportunity to practices skills associated with defending and holding a breeding territory (Chapter 2).

There are serious problems, however, with the general application of this hypothesis. Its crucial pitfall is, of course, that adults will frequently engage in homosexual behaviour although they have no obvious need to gain sexual experience. Moreover, in some populations only certain individuals are observed to interact sexually with same-sex partners, which begs the question as to why some individuals should have no need for practice at all.

Kin selection. *Same-sex behaviour frees individuals from directing time and energy toward their own offspring and, instead, these individuals pass on their genes indirectly by helping kin reproduce.*

The concept of 'inclusive fitness' – as elaborated upon by W. D. Hamilton (1964) – holds that reproduction of one's genetic information can be achieved directly by producing offspring, or indirectly by enhancing the reproductive success of genetic relatives. This concept is based on the logic that one's own offspring carry, on average, half of one's own genetic information, whereas a niece or nephew, for example, will on average carry a quarter of one's own genes into the next generation. Thus, an individual who engages in homosexual behaviour might forego having an offspring (which would replicate 50% of its genes), but this individual would not reduce its inclusive fitness if it would help a brother or sister to produce two additional nieces or nephews (who would replicate 2 × 25% = 50% of its genes).

Various authors have amassed anecdotal evidence that supports the kin selection theory for homosexual behaviour in humans (Weinrich, 1980; Ruse, 1981; Kirkpatrick, 2000). For example, in the marriage resistance movement of southern China, two or three individuals would swear friendship, and this often led to homosexual behaviour. Women in the movement were relatively wealthy from work in the silk industry and regularly sent wages back to their natal families (Kirkpatrick, 2000). The 'little brother phenomenon' has also been taken as supporting the hypothesis, since the likelihood that a male is homosexual increases with the number of older brothers he has (Blanchard and Bogaert, 1996, as cited in Kirkpatrick, 2000).

The theory has, however, been criticized on theoretical grounds (Dickemann, 1995; Kirkpatrick, 2000; Bobrow and Bailey, 2001). Moreover, empirical support for the kin selection theory for homosexual behaviour in humans is entirely lacking. Indeed, in the only empirical test of this hypothesis to date, Bobrow and Bailey (2001) found that homosexual men were no more likely than heterosexual men to channel resources towards family members. In fact, heterosexual men tended to give more resources to siblings than homosexual men. Cross-cultural research shows that male homosexuals are no more common in societies with extended families where homosexuals live closest to their relatives and could presumably help them the most (Werner, 1979).

   Indirect reproduction linked to the paradigm of inclusive fitness is also considered to be the selective force behind numerous social systems of animals in which non-reproductive helpers raise younger siblings – for example, in social insects such as bees or ants and in many cooperatively breeding vertebrates such birds, wild dogs or marmoset monkeys (review in Wilson, 1975). However, apart from humans, there is very little evidence that 'helpers-at-the-nest' engage in homosexual behaviour. Often, they will not be sexually active at all, or they engage in clandestine heterosexual reproduction despite attempts by higher ranking individuals to suppress their direct reproduction (for example, Digby, 2000).

Ultimate causes: evolutionary history

   It should be clear from the preceding discussion that we consider it good science if hypothetical functional explanations for same-sex sexual behaviour can be tested and, on the basis of empirical findings, accepted or rejected. Readers will find that 'negative results' will be more commonly found throughout the contributions to this volume than claims for definitive explanations. This tendency does not, however, imply that the pattern of observed same-sex sexual behaviour cannot be explained within an evolutionary framework. It is possible that the behaviour under analysis is functional, in the evolutionary sense, but that the correct adaptive hypothesis has not yet been formulated or explored.

   However, in some species, homosexual behaviour may not be explicable in functional terms. If homosexual behaviour is not functional in some instances, then how might we account for the phenomenon? As mentioned previously, there are two principle types of evolutionary inquiries: questions related to *function* determine the current adaptive value of a behaviour; questions related to *evolutionary history*, in contrast, focus on reconstructing the historical (phylogenetic) steps that lead to the origin of a behaviour and how it changed over time in ancestral environments (Alcock, 1989).

   Importantly, all behaviours can be explained in terms of their evolutionary history, but not all behaviours can be explained in terms of adaptive functions. This is because adaptations are not the only products of the evolutionary process. Evolution also produces functionless by-products of adaptations. Moreover, there can be a mismatch between the current environment and the ancestral environment in which a species evolved.

Analyses of the evolutionary history of a behavioural pattern may therefore lead to three conclusions: (1) A behaviour represents an *adaptation* since it is *beneficial* for an organism's reproduction. Such behaviours will be selected for. (2) A behaviour represents a *functionless by-product* of an adaptation, but it prevails because it imposes *no costs* on an organism's reproductive performance. (3) A behaviour represents a *maladaptation* because it is *costly* for the organism since it impedes genetic reproduction. Such behaviours will be selected against.

For example, it is likely that it was beneficial for the reproductive fitness of an ancient primate male to develop an erection if a female presented herself in a sexually inviting position. Such erections would represent an adaptation. However, blood flow to a male genital will not increase reproductive success if it is stimulated by viewing a centerfold pictorial in a pornographic magazine. Erections, as manifested in the later context, are not the product of direct selection, but, rather, are by-products of selection reponses that were beneficial in an ancient environment. The evolved capacity of males to have a low threshold for sexual arousal in response to sexual imagery is a pattern that is now successfully exploited by the pornography industry. The behaviour may persist as a neutral by-product, as long as it does not interfere with reproductive performance. However, it is conceivable that a human male might become fixated by the supra-normal stimuli embodied by pornographic centerfolds and, as such, lose his appetite for real sex partners. At that point, the behaviour would be a maladaptation and would ultimately be selected against.

An additional goal of this volume is to raise the readers' consciousness with regard to the value of evolutionary history as a potential explanatory perspective for understanding homosexual behaviour in animals. In the following sections, we review phylogenetic explanations for same-sex sexual interactions, which assume that the behaviour has no current function. We include here a description of the somewhat eclectic 'biological exuberance' hypothesis proposed by Bruce Bagemihl (1999), which states that the single-minded search for functional explanations is completely misguided.

Functionless by-product of adaptations. Same-sex sexual behaviour is not the product of direct selection but a neutral by-product of selection for other traits.
Sub-adult male northern elephant seals mount juvenile males (1–2 years of age) during the breeding season (Rose et al., 1991). During these mounts, the sub-adult males display many of the behaviours that are characteristic of male

sexual behaviour towards adult females, although most of the juveniles respond by struggling and attempting to flee. From a functional perspective, there may not be any adaptive explanation for such behaviour. However, from the perspective of evolutionary history, the mounting of juvenile males by sub-adult males can be characterized as a low-cost by-product of an adaptation: high male sexual motivation. Sexual selection probably favoured ancestral male elephant seals with high sexual motivation because such individuals would, theoretically, have experienced enhanced lifetime reproductive success (Rose *et al.*, 1991).

Many instances of homosexual mounting in feral cats (Chapter 7) and deer (Chapter 6), respectively, could represent examples of functionless by-products of males' low threshold for sexual excitement – a tendency that is, on balance, adaptive. Female homosexual behaviour in Japanese macaques is likewise interpreted as a neutral sexual behaviour (Chapter 8). The evolutionary history of female–female mounting in this species might have included the following steps: (a) female–male mounting evolved as an adaptation that females use to solicit copulations from sluggish or disinterested males; (b) females derive immediate sexual reward from mounting via clitoral stimulation; and (c) female homosexual behaviour evolved as a neutral by-product of selection for female–male mounting and as a result of the proximate ability of females to derive immediate sexual reward from mounting. Similarly, male homosexual behaviour in mountain gorillas (Chapter 11) seems to have no socio-sexual function. Genital stimulation during play is a common behaviour of all the African apes during immaturity, and homosexual interactions among immature male gorillas seems to be merely one expression of this shared phylogenetic tendency.

Thus, animal homosexual behaviour is not always enacted to mediate adaptive social or reproductive functions, but it may exist simply for sexual gratification. Selection against the behaviour does not occur because it does not interfere with reproduction. Indeed, the vast majority of animals who engage in same-sex sexual behaviour will also engage in heterosexual sex and reproduce successfully (see this volume, Chapters 2–12).

It is worth mentioning that the earliest hypothesis that placed homosexual behaviour into a broader framework of evolutionary theory was likewise based on the idea that same-sex sexual activity is a by-product of an adaptation. The so-called '*balanced polymorphism*' *hypothesis* – originally proposed by George E. Hutchinson (1959) – states that homosexual behaviour is retained because it co-occurs with a second trait under positive selection. Therefore, several genetic forms (polymorphs) will co-exist in the population, and the balance of their frequency depends on the advantages and disadvantages conveyed to heterozygote and homozygote bearers of the trait. The explanation does not assume that same-sex sexual behaviour provides a benefit per se, or that it is a neutral trait, but that it is, in fact, reproductively costly to those engaging in same-sex behaviour.

The potential mechanisms behind this rationale can be illustrated by two scenarios:

(1) *'Heterozygote advantage'* – An allele that promotes homosexual behaviour in the homozygous state is maintained due to a postulated advantage it confers when in combination with the alternate allele, for example a superior immune system. The mechanism is similar to the often-cited interplay between sickle-cell anemia and malaria, a scenario particularly prevalent in West Africa. Some humans possess a gene that predisposes their red-blood cells to be sickle shaped, a phenotype which is maladaptive. When this trait is encoded in only one of the bearer's corresponding chromosomes, however, the phenotype is normal and the sickle-cell gene confers some immunity against malaria. Thus, in the heterozygous state, sickle-cell genes are therefore not selected against. As a consequence, however, there will always be individuals who are homozygous for the trait. Such individuals manufacture sickled red blood cells, suffer full-blown anemia and die at an early age. Thus, in the homozygous state, sickle-cell genes are selected against.

(2) The second potential mechanism is that of a *'sex-linked advantage'*. For example, a recessive genetic trait could be linked to the X-chromosome. In the XY combination, it would be disadvantageous for males because it leads to same-sex behaviour and thus reduces reproductive fitness. However, in the XX state, where one recessive allele would be combined with a dominant one, it would be beneficial because it enhances attributes such as attractivity or fertility in those females that possess the trait. The trait will thus be passed on maternally.

To date, one empirical test of Hutchinson's 'balanced polymorphism' hypothesis for homosexual behaviour has been conducted. Maternal relatives of homosexual men were found to have higher fecundity than the maternal relatives of heterosexual men (Camperio-Ciani *et al.,* 2004). This provides evidence for maternally inherited factors favouring male homosexuality *and* promoting female fecundity in humans. In other words, a gene for androphilia (that is, sexual attraction to males) might cause homosexuality in men and 'hyperheterosexuality' in women. Natural selection would favour the maintenance of this gene because the positive effect on the reproductive success of females would offset the negative effect on their homosexual male relatives. Other research suggests that male homosexual behaviour in humans may result from the heterozygous advantage conferred by genes for dominant and submissive behaviour (Chapter 13).

Maladaptation. Same-sex behaviour decreases the fitness of those who engage in it and will thus be subject to counter-selection. Often, it represents a pathology, for example as a consequence of unnatural circumstances of confinement.

A behaviour is considered to be maladaptive in an evolutionary sense, if it is detrimental for survival or reproduction, and will thus be subject to counter-selection. Anyone who thinks that the tendency to equate animal homosexual behaviour with all that is 'abnormal' is nothing more than a dusty historical footnote would do well to read Bagemihl's (1999, pp. 156–67) excellent treatment of homophobia and heterosexism in zoology over the last 200 years. The author marshals mountains of evidence to demonstrate that investigations into animal homosexual behaviour have been, and continue to be 'a nearly unending stream of preconceived ideas, negative "interpretations" or rationalizations, inadequate representations and omissions, and even overt distaste or revulsion towards homosexuality – in short, homophobia'. Bagemihl documents a litany of derogatory descriptions and heterocentric views widely used in more contemporary studies such as 'ludicrous', 'nonsense', 'perversion', 'inappropriate', 'disturbing', 'unnatural', 'bizarre' or 'malfunction'. These descriptors seem to be clearly rooted in the earlier conceptualizations of animal homosexual behaviour as pathology, but go one step further by suggesting such behaviour is, in the eyes of the observer, patently objectionable.

Captivity was often identified as the causal factor underlying the expression of this pathology. Implicit in this 'explanation' is the assumption of a mismatch between the current environment and the ancestral environment, and the logic underlying this perspective therefore presupposes some knowledge about the species' evolutionary history.

Undoubtedly, there are situations where captivity promotes the expression of homosexual behaviour. Hand-reared male goslings, for example, are more likely to form life-long homosexual pair bonds with their brothers, compared to goose-reared goslings (Chapter 2). Similarly, homosexual mounting in various species of deer increases dramatically under captive conditions involving crowding (Chapter 6). A study of two male rhesus monkeys reared exclusively together from eight to 27 months of age found that they consequently showed a high degree of sexual behaviour with each other as adults (Erwin and Maple, 1976, as cited in Wallen and Parsons, 1997). This would suggest that same-sex sexual behaviour can be produced by laboratory rearing conditions. On the other hand, studies of male and female rhesus monkeys reared in uni-sexual groups during the first year of life indicate that these monkeys displayed higher levels of homosexual behaviour, but they went on to engage in heterosexual behaviour as adults (Bercovitch *et al.*, 1988). Although the expression of homosexual behaviour can be causally related to captivity, such cases are relatively rare. Indeed, the frequency with which homosexual behaviour is expressed should be much higher if captive conditions predispose animals to engage in it (Vasey, 1995; Wallen and Parsons, 1997).

The strongest argument that same-sex sexual behaviour is not the subject of negative selection comes in the form of the numerous reports of such interactions among animals living in the wild. For example, most information concerning homosexual behaviour in primates comes from from field observations of naturally occurring groups (Vasey, 1995). Many of the contributors to this volume also provide evidence of homosexual behaviour in animals living under non-captive conditions. Indeed, there are few wilder places on earth than Shark Bay in Australia, the Virunga volcanoes in East Africa or the rainforests of the central Congo region. And, yet, the wild inhabitants of these areas, be they dolphins (Chapter 4), mountain gorillas (Chapter 11) or bonobos (Chapter 12) all engage in high levels of homosexual behaviour. In all these cases, homosexual activity does not reflect an abnormal response to captivity or a pathology, but, rather, it is a normal facet of the sexual repertoire of such animals in their natural environment.

Biological exuberance. *Solar energy exists in excess of what is needed for organisms to grow and stay alive. The surplus energy is used for 'extravagant activities' such as sexual reproduction and also for non-procreative behaviour including homosexual interactions.* Orthodox Darwinian theory assume that resources – including mates – are limited, and that the competition for access to these resources produces the diversity of anatomical and behavioural traits we observe in organisms. The 'biological exuberance' hypothesis assumes just the opposite.

It was proposed by Bruce Bagemihl who, after carefully reviewing numerous functional hypotheses, came to the conclusion that current evolutionary biology is prejudiced because of 'its single-minded attempt to find reproductive (or other) "explanations" for homosexuality' (1999, p. 213). He likened this enterprise to attempts to find functional explanations for kisses, which have been viewed to be vestiges of ritual food exchange, olfactory sampling or reconciliation. However, he argues that, even if the origins of the kiss could be traced to ancient functions, 'something ineffable' would still remain: '"The kiss" is a perfect symbol of the limitations of biological reductionism' (1999, p. 212).

The hypothesis of 'biological exuberance' rests on George Bataille's idea that 'solar energy is the source of life's exuberant development' (cited in Bagemihl, 1999, p. 255) and that homosexuality and other forms of sexuality that do not lead to fertilization are 'simply one of the many expressions of the natural intensity' of biological systems. Thus, the 'extravagance of life' is not the result of greater forces such as the laws of physics or evolution, but 'this relation is reversed: exuberance is the *source* and *essence* of life' (p. 255).

It is difficult to imagine how such an idea could be proven wrong, and it therefore seems as if the hypothesis exists outside the heuristic framework of scientific inquiry. This may well have been Bagemihl's intention in light of his attempt to merge concepts from diverse theoretical perspectives including the belief systems of indigenous peoples, chaos science, post-Darwinian evolution, Gaia theory, biodiversity studies, philosophy and economics. Interestingly, Bagemihl's reasoning has similarities with the often-criticized logic of group-selectionist thinking. Namely, a trait that is purely 'extravagant' and does not convey any reproductive advantage should, according to modern evolutionary theory, be selected against, since competitors who do not exhibit this extravagance would leave more offspring.

Key to Bagemihl's theory is the notion of diversity. There is certainly ample support for the idea that homosexual behaviour is a diverse phenomenon that requires multiple explanatory angles. In humans, for example, it has long been recognized that homosexual behaviour does not represent a uniform behavioural suite – a fact that is reflected in the aptly pluralized title *Homosexualities: A Study of Diversity among Men and Women* (Bell and Weinberg, 1978). Our volume adds substance to the considerable evidence that exists for both intra-specific and inter-specific variation of this behaviour. In this way, the contributions to this volume supplement the new picture of animal homosexual behaviour that has been emerging, slowly but steadily, and, by extension, a new picture of animal sexuality in general.

References

Adkins-Regan, E. (1988) Sex hormones and sexual orientation in animals. *Psychobiol.*, **16**, 335–47.

Adkins-Regan, E. (1998) Hormonal mechanisms of mate choice. *Am. Zool.*, **38**, 166–78.

Adkins-Regan, E. (2002) Development of sexual partner preference in the zebra finch: a socially monogamous, pair-bonding animal. *Arch. Sex. Behav.*, **31**, 27–33.

Adkins-Regan, E., Mansukhani, V., Thompson, R. and Yang, S. (1997) Organizational actions of sex hormones on sexual partner preference. *Brain Res. Bull.*, **44**, 497–502.

Alcock, J. (1989) *Animal Behavior*, 4th edn. Sunderland, MA: Sinauer Associates.

Bell, A. P. and Weinberg, M. (1979) *Homosexualities: A Study of Diversity in Men and Women*. New York: Simon & Shuster.

Bagemihl, B. (1999) *Biological Exuberance: Animal Homosexuality and Natural Diversity*. New York: St Martin's.

Bercovitch, F. B., Roy, M. M, Sladky, K. K. and Goy, R. W. (1988) The effects of iosexual rearing on adult sexual behaviour in captive male rhesus macaques. *Arch. Sex. Behav.*, **17**, 381–8.

Blanchard, R. and Bogaert, A. F. (1996) Homosexuality in men and number of older brothers. *Am. J. Psychiatry*, **153**, 27–31.

Bobrow, D. and Bailey, J. M. (2001) Is male homosexual behavior maintained via kin selection? *Evol. & Hum. Behav.*, **22**, 361–8.

Boswell, J. (1980) *Christianity, Social Tolerance and Homosexuality*. Chicago: University of Chicago Press.

Brower, G. R. and Kiracofe, G. H. (1978) Factors associated with the buller-steer syndrome. *J. Anim. Sci.*, **46**, 26–31.

Camperio-Ciani, A., Corna, F. and Capiluppi, C. (2004) Evidence for maternally inherited factors favouring male homosexuality and promoting female fecundity. *Proc. R. Soc. London*, Series B, **241**, 2217–21.

Cass, V.-C. (1983/84) Homosexual identity: a concept in need of defense. *J. Homo.*, **9**, 105–26.

Chapais, B. (1991) Primates and the origins of aggression, power, and politics among humans. In *Understanding Behavior: What Primate Studies Tell Us about Human Behaviour*, eds. J. D. Loy and C. B. Peters, pp. 190–228. New York: Oxford University Press.

Chevalier-Skolnikoff, S. (1976) Homosexual behaviour in a laboratory group of stumptail monkeys (*Macaca arctoides*): forms, context, and possible social functions. *Arch. Sex. Behav.*, **5**, 511–27.

Dagg, A. I. (1984) Homosexual behavior and female–male mounting in mammals – a first survey. *Mammal Rev.*, **14**, 155–85.

de Waal, F. B. M. (1987) Tension regulation and nonreproductive functions of sex in captive bonobos (*Pan paniscus*). *Nat. Geogr. Res.*, **3**, 318–38.

Diamond, J. M. (1989) Goslings of gay geese. *Nature*, **340**, 101.

Dickemann, M. (1993) Reproductive strategies and gender construction: an evolutionary view of homosexualities. *J. Homo.*, **24**, 55–71.

Dickemann, M. (1995) Wilson's panchreston: the inclusive fitness hypothesis of sociobiology reexamined. In *Sex, Cells and Same-Sex Desires: The Biology of Sexual Preference*, eds. J. P. Dececco and D. A. Parker, pp. 147–83. New York: Haworth Press.

Digby, L. (2000) Infanticide by female mammals: implications for the evolution of social systems. In *Infanticide by Males and its Implications*, eds. C. van Schaik and C. Janson, pp. 423–46. Cambridge: Cambridge University Press.

Dixson, A. F. (1977) Observations on the displays, menstrual cycles and sexual behavior of the 'Black ape' of Celebes (*Macaca nigra*). *J. Zool.*, **182**, 63–84.

Dixson, A. F. (1998) *Primate Sexuality*. Oxford: Oxford University Press.

Edwards, G. (1758–64) *Gleanings of Natural History: Exhibiting figures of quadrupeds, birds, insects, plants etc., many of which have not, till now, been either figured or described*, vol. III. London: Royal College of Physicians.

Erwin, J. and Maple, T. (1976) Ambisexual behavior with male–male anal penetration in male rhesus monkeys. *Arch. Sex. Behav.*, **5**, 9–14.

Fairbanks, L. A., McGuire, M. T. and Kerber, W. (1977) Sex and aggression during rhesus monkey group formation. *Aggress. Behav.*, **3**, 241–9.

Foucault, M. (1990) *The History of Sexuality: An Introduction*, vol. I. New York: Vintage.

Furuichi, T. (1989) Social interactions and the life history of female *Pan paniscus* in Wamba, Zaire. *Int. J. Primatol.*, **10**, 173–97.

Goy, R. W. and Wallen, K. (1979) Experimental variables influencing play, footclasp mounting, and adult sexual competence in male rhesus monkeys. *Psychoneuroendocrinol.*, **4**, 1–12.

Greenberg, D. F. (1988) *The Construction of Homosexuality*. Chicago: University of Chicago Press.

Guhl, A. M. (1948) Unisexual mating in a flock of white leghorn hens. *Transact. Kansas Acad. Scienc.*, **5**, 107–11.

Gywnne, D. T. (1991) Sexual competition among females: what causes courtship-role reversal? *Trends Ecol. Evol.*, **6**, 118–21.

Hale, E. B. (1955) Defects in sexual behavior as factors affecting fertility in turkeys. *Poultry Sci.*, **34**, 1059–67.

Hale, E. B. and Schein, M. W. (1962) The behavior of turkeys. In *The Behavior of Domestic Animals*, ed. E. S. E. Hafez, pp. 531–64. Baltimore: Williams & Wilkins.

Hamilton, W. D. (1964) The evolution of social behavior. *J. Theor. Biol.*, **7**, 1–52.

Hanby, J. P. (1974) Male–male mounting in Japanese monkeys (*Macaca fuscata*). *Anim. Behav.*, **22**, 836–49.

Haraway, D. J. (1989) *Primate Visions: Gender, Race and Nature in the World of Modern Science*. New York: Routledge.

Harvey, P. H. and Pagel, M. D. (1991) *The Comparative Method in Evolutionary Biology*. Oxford: Oxford University Press.

Herdt, G. (1997) *Same-Sex, Different Culture: Exploring Gay and Lesbian Lives*. Boulder, CO: Westview Press.

Hrdy, S. B. (1981) *The Woman That Never Evolved*. Cambridge, MA: Harvard University Press.

Hunt, G. L. and Hunt, W. M. (1977) Female–female pairings in Western gulls in Southern California. *Science*, **196**, 1466–7.

Hutchinson, G. E. (1959) A speculative consideration of certain possible forms of sexual selection in man. *Am. Nat.*, **93**, 81–91.

Idani, G. (1991) Social relationships between immigrant and resident bonobo (*Pan paniscus*) females at Wamba. *Folia Primatol.*, **57**, 83–95.

Kano, T. (1980) Social behavior of wild pygmy chimpanzees (*Pan paniscus*) of Wamba: a preliminary report. *J. Hum. Evol.*, **9**, 243–60.

Katz, J. N. (1995) *The Invention of Heterosexuality*. New York: Dulton.

Kirkpatrick, R. C. (2000) The evolution of human homosexual behavior. *Curr. Anthropol.*, **41**, 385–413.

Kirsch, J. A. W. and Rodman, J. E. (1982) Selection and sexuality: the Darwinian view of homosexuality. In *Homosexuality: Social, Psychological, and Biological Issues*, eds.

W. Paul, J. D. Weinrich, J. C. Gonsiorek and M. E. Hotvedt, pp. 183–95. Beverly Hills, CA: Sage.

Kirsch, J. A. W. and Weinrich, J. D. (1991) Homosexuality, nature, and biology: is homosexuality natural; does it matter? In *Homosexuality: Research Implications for Public Policy*, eds. J. C. Gonsiorek and J. D. Weinrich, pp. 13–31. Newbury Park: Sage.

Kitagawa, T. (1988) Ethosociological studies of the black-winged stilt, *Himantopus himantopus himantopus*: III. female–female pairing. *Jap. J. Ornithol.*, **37**, 63–7.

Kuroda, S. (1980) Social behavior of the pygmy chimpanzee. *Primates*, **21**, 181–97.

Kuroda, S. (1984) Interactions over food among pygmy chimpanzees. In *The Pygmy Chimpanzee*, ed. R. L. Susman, pp. 301–24. New York: Plenum.

LeVay, S. and Valente, S. M. (2003) *Human Sexuality*. 2nd edn, Sunderland, MA: Sinauer.

Lincoln, S. D. (1983) Etiology and control of buller syndrome. *Topics Vet. Med.*, **58**, 24–8.

MacRoberts, M. H. and MacRoberts, B. R. (1976) Social organization and behavior of the acorn woodpecker in central coastal California. *Ornitholog. Monogr. 21*. Washington, DC: American Ornithologists' Union.

McGrew, W. C. (1992) *Chimpanzee Material Culture: Implications for Human Evolution*. Cambridge: Cambridge University Press.

Moynihan, M. (1990) *Social, Sexual, and Pseudosexual Behavior of the Blue-Bellied Roller, Coracias Cyanogaster: The Consequences of Crowding or Concentration*. Smithsonian Contributions to Zoology 491. Washington, DC: Smithsonian Institution Press.

Murray, S. O. (2000) *Homosexualities*. Chicago: University of Chicago Press.

Myers, J. P. (1989) Making sense out of sexual nonsense. *Audubon*, **91**, 40–5.

Nanda, S. (2000) *Gender Diversity: Cross-Cultural Variations*. Prospect Heights, IL: Waveland.

Oi, T. (1991) Non-copulatory mounting in wild pig-tailed macaques (*Macaca nemestrina nemestrina*) in West Sumatra, Indonesia. In *Primatology Today*, eds. A. Ehara, T. Kimura, O. Takenaka and M. Iwamoto, pp. 147–50. Amsterdam: Elsevier.

Paredes, R. G. and Baum, M. J. (1995) Altered sexual partner preference in male ferrets given excitotoxic lesions of the preoptic area/anterior hypothalamus. *J. Neurosci.*, **15**, 6619–30.

Paredes, R. G. and Baum, M. J. (1997) Role of the medial preoptic area/anterior hypothalamus in the control of masculine sexual behavior. *Ann. Rev. Sex Res.*, **8**, 68–101.

Parish, A. R. (1994) Sex and food control in the 'uncommon chimpanzee': how bonobo females overcome a phylogenetic legacy of male dominance. *Ethol. & Sociobiol.*, **15**, 157–79.

Parker, G. A. and Pearson, R. G. (1976) A possible origin and adaptive significance of the mounting behavior shown by some female mammals in oestrous. *J. Nat. Hist.*, **10**, 241–5.

Phoenix, C. H., Goy, R. W., Gerall, A. A. and Young, W. C. (1959) Organizing action of prenatally administered testosterone propionate on the tissues mediating mating behavior in the female guinea pig. *Endocrinology*, **65**, 369–82.

Pomerantz, S. M., Goy, R. W. and Roy, M. M. (1986) Expression of male-typical behavior in adult female pseudohermaphroditic rhesus: comparisons with normal males and neonatally gonadectomized males and females. *Horm. & Behav.*, **20**, 483–500.

Quinn, T. W., Davies, J. C., Cooke, F. and White, B. N. (1989) Genetic analysis of offspring of a female–female pair in the lesser snow goose (*Chen c. caerulescens*). *Auk*, **106**, 177–84.

Reinhardt, V., Reinhardt, A., Bercovitch, F. B. and Goy, R. W. (1986) Does intermale mounting function as a dominance demonstration in rhesus monkeys? *Folia Primatol.*, **47**, 55–60.

Ritchie, J. P. (1926) Nesting of two male swans. *Scottish Naturalist*, **159**, 95.

Rodseth, L., Wrangham, R. W., Harrigan, A. M. and Smuts, B. B. (1991). The human community as a primate society. *Curr. Anthropol.*, **32**, 221–54.

Rose, N. A., Deutsch, C. J. and Le Boeuf, B. J. (1991) Sexual behaviour of the male northern elephant seals: III. the mounting of weaned pups. *Behaviour*, **119**, 171–92.

Ruse, M. (1981) Are there gay genes? Sociobiology and homosexuality. *J. Homosex.*, **6**, 5–34

Ruse, M. (1988) *Homosexuality: A Philosophical Inquiry*. Oxford: Basil Blackwell.

Smuts, B. B. and Smuts, R. W. (1991) Male aggression and sexual coercion of females in nonhuman primates and other mammals: evidence and theoretical implications. *Adv. Study Behav.*, **22**, 1–63.

Smuts, B. B. and Watanabe, J. M. (1990) Social relationships and ritualized greetings in adult male baboons (*Papio cynocephalus anubis*). *Int. J. Primatol.*, **11**, 147–72.

Sommer, V. (1989) Sexual harassment in langur monkeys (*Presbytis entellus*): competition for nature, eggs, and sperm? *Ethology*, **80**, 205–17.

Sommer, V. (1990) *Wider die Natur? Homosexualität und Evolution*. Munich: C. H. Beck.

Sommer, V., Denhan, A. and Little, K. (2002) Postconflict behaviour of wild Indian langur monkeys: avoidance of opponents but rarely affinity. *Anim. Behav.*, **63**, 637–48,

Srivastava, A., Borries, C. and Sommer, V. (1991) Homosexual mounting in free-ranging female Hanuman langurs (*Presbytis entellus*). *Arch. Sex. Behav.*, **20**, 487–512.

Stamp-Dawkins, M. (1995) *Unravelling Animal Behaviour*, 2nd edn. Harlow: Longman.

Talmage-Riggs, G. and Anschel, S. (1973) Homosexual behavior and dominance hierarchy in a group of captive female squirrel monkeys (*Saimiri sciureus*). *Folia Primatol.*, **19**, 61–72.

Taub, D. and Mehlman, P. (1991) Primate paternalistic investment: a cross-species view. In *Understanding Behavior: What Primate Studies Tell Us about Human Behavior*, eds. J. D. Loy and C. B. Peters, pp. 51–89. New York: Oxford University Press.

Tinbergen, N. (1963) On aims and methods of ethology. *Z. Tierpsychol.*, **20**, 410–33.

Tomasello, M. and Call, J. (1997) *Primate Cognition*. New York: Oxford University Press.

Travis, C. B. and Yeager, C. P. (1991) Sexual selection, parental investment, and sexism. *J. Soc. Iss.*, **47**, 117–29.

Trivers, R. L. (1971) The evolution of reciprocal altruism. *Quat. Rev. Biol.*, **46**, 35–57.

Trivers, R. L. (1972) Parental investment and sexual selection. In *Sexual Selection and the Descent of Man, 1871–1971*, ed. B. Campbell, pp. 136–79. Chicago: Aldine.

Trivers, R. L. (1985) *Social Evolution*. Menlo Park: Benjamin Cummings.

Tyler, P. A. (1984) Homosexual behaviour in animals. In *The Psychology of Sexual Diversity*, ed. K. Howells, pp. 42–62. Oxford: Basil Blackwell.

Vasey, P. L. (1995) Homosexual behavior in primates: a review of evidence and theory. *Int. J. Primatol.*, **16**, 173–204.

Vasey, P. L. (2002) Same-sex sexual partner preference in hormonally and neurologically unmanipulated animals. *Ann. Rev. Sex Res.*, **13**, 141–79.

Vasey, P. L., Foroud, A., Duckworth, N. and Kovacovsky, S. D. (2006). Male–female and female–female mounting in Japanese macaques: a comparative analysis of posture and movement. *Arch. Sex. Behav.*, **35**

Vasey, P. L. and Gauthier, C. (2000) Skewed sex ratios and female homosexual activity in Japanese macaques: an experimental analysis. *Primates*, **41**, 17–25.

Veenema, H. C. (2000) Methodological progress in postconflict research. In *Natural conflict resolution*, eds. F. Aureli and F. B. M. de Waal, pp. 21–3. Berkeley, CA: University of California Press.

Watanabe, J. M. and Smuts, B. B. (1999) Explaining religion without explaining it away: trust, truth, and the evolution of cooperation in Roy A. Rappaport's 'The Obvious Aspects of Ritual'. *Am. Anthropol.*, **101**, 98–112.

Wallen, K. and Parsons, W. A. (1997) Sexual behavior in same-sexed nonhuman primates: is it relevant to understanding human homosexuality? *Ann. Rev. Sex. Res.*, **8**, 195–223.

Weinrich, J. D. (1980) Homosexual behavior in animals: a new review of observations from the wild, and their relationship to human sexuality. In *Medical Sexology: The Third International Congress*, eds. R. Forleo and W. Pasini, pp. 288–95. Littleton, MA: PSG Publishing.

Werner, D. (1979) A cross-cultural perspective on theory and research on male homosexuality. *J. Homo.*, **1**, 345–62.

Whitman, C. O. (1919) *The Behavior of Pigeons*. Washington, DC: Carnegie Institution of Washington.

Wickler, W. (1967) Socio-sexual signals and their intra-specific imitation among primates. In *Primate Ethology*, ed. D. Morris, pp. 69–79. Chicago: Aldine.

Wilson, E. O. (1975) *Sociobiology: The New Synthesis*. Harvard, MA: Belknap.

Wolfe, L. D. (1991) Human evolution and the sexual behavior of female primates. In *Understanding Behavior: What Primate Studies Tell Us about Human Behavior*, eds. J. D. Loy and C. B. Peters, pp. 121–51. New York: Oxford University Press.

Zahavi, A. and Zahavi, A. (1997) *The Handicap Principle*. Oxford: Oxford University Press.

Zhang, S. D. and Odenwald, W. F. (1995) Misexpression of the white (w) gene triggers male–male courtship in Drosophila. *Proc. Nat. Acad. Sci. USA*, **92**, 5525–9.

PART II   NON-PRIMATES

2

Making the best of a bad situation: homosociality in male greylag geese

KURT KOTRSCHAL, JOSEF HEMETSBERGER
AND BRIGITTE M. WEIβ

Introduction

The issue of homosexuality has been a taboo and/or a matter of ideological controversy in many Western societies, particularly during the twentieth century. It is hard to judge whether this increased or decreased the attention to the common phenomenon of pair formation between same-sex partners in animals (Bagemihl, 1999; reviews in Meyer-Holzapfel, 1961; Morris, 1952; Schutz, 1966 and others; see Huber and Martys, 1993). Quite frequently, these reports dealt with birds (Allen, 1934; Lorenz, 1935, 1940; Sauer, 1972; Schutz, 1965a,b) and most were on captive or semi-captive animals. However, same-sex pairs were also reported from the wild (Hunt and Hunt, 1977; Schutz, 1965b and many more ever since).

Same-sex pairs seem to occur mainly in species with only moderate sexual dimorphism (Dilger, 1960). This gave rise to a number of mechanistic hypotheses to account for such behaviour (see Huber and Martys, 1993), including the failure to correctly identify the sex of a partner, the acceptance of an inappropriate partner when appropriate partners are short in supply and pseudo-female or pseudo-male behaviour in some of the partners. These mechanistic hypotheses still may have some merit in explaining how homosexual pair bonds are initiated (but see Bagemihl, 1999 for a different opinion). But they may also have supported the idea that homosexuality is generally dysfunctional (maladaptive), because it does not lead to offspring directly. This changed with kin selection theory (Hamilton, 1964), which opened the possibility that same-sex paired individuals may convey some reproductive advantage to close relatives.

It is well known that geese make good eating, but what, if anything, can be learned from them? Greylag geese (*Anser anser*) are one of the model species of 'classical' ethology. This was one of many species kept and observed by young Konrad Lorenz in preparation of one of his groundbreaking, comparative studies on the Anatinae (ducks, geese and swans; Lorenz, 1941). Greylag geese remained his major animal companions and research models throughout life (cf. Lorenz and Tinbergen, 1939; Lorenz, 1991). At the Konrad Lorenz Forschungsstelle (KLF) in Grünau/Austria, the 'Lorenzian' greylags still support behavioural research at the interface between mechanisms and functions (for example, Fritz and Kotrschal, 2000; Fritz *et al.*, 2000; Hirschenhauser *et al.*, 1999, 2000; Kotrschal *et al.*, 1993, 1998; Pfeffer *et al.*, 2002, this study). Geese also have been valuable research models elsewhere (see Black, 1996; Lamprecht, 1987; Loonen, 1997; Cooke *et al.*, 1995). Geese social organization can be quite complex. For example, there is now evidence for a female-bonded clan organization (Frigerio *et al.*, 2001), which is one of the facets where geese parallel primates. Also, male homosexual pairing in geese has long been described (Heinroth, 1911; see Huber and Martys, 1993).

The comparative approach has evident potential for explaining homosexuality in humans as well. In this regard, studies in monkeys and apes are certainly interesting, because of our close (phylo-)genetic relationships. However, due to a common ancestor, only a few million years ago, some psychological and social structures may be of common (homologous) origin, making explanations based on 'phylogenetic inertia' ('ghosts of selection past') hard to separate from present functions (or 'dys' functions). In contrast, birds and primates share a common reptilian ancestor some 350 million years ago. Phylogenetic distance between those two vertebrate clades, and between humans and geese in particular, makes it very unlikely that any similarity in the more complex psychological dispositions and social patterns are due to common ancestry. Although the basic structures of the body, brain and behaviour (even the basic emotional systems, see Panksepp, 1998) may be homologous, this is certainly not true for the complex phenomenon of homosexuality. Rather, similarities between geese and humans in such complex socio-sexual patterns may be due to convergent evolution leading in parallel to homosexual phenomena in geese and in humans.

The aim of this chapter is to describe patterns of homosexuality found in the semi-tame, but free-ranging Grünau flock of greylag geese during three decades of monitoring. We consider potential mechanisms leading into same-sex pairing and discuss functional aspects. Finally, we attempt to critically evaluate how representative these results may be for greylag geese in general and also what, if any, bearings our results may have for explaining homosexuality in other taxa, including humans.

Biology and social organization of greylag geese

Geese form flocks from summer into late winter, probably as a predator avoidance strategy related to foraging in open habitats. As facultative migrants, flocks may temporarily or permanently cease their migratory traditions when mild winters afford sufficient food supply (Kear, 1990; Rutschke, 1987, 1997). Because migration routes are passed on as family traditions, permanently resident flocks can be formed by interrupting this tradition. In February and March, after returning from their winter feeding areas, flocks disintegrate into pairs, mainly due to increasing agonism between males. At this time, greylag females nest in clusters, but individuals are visually isolated from each other and rely on individual camouflage. They form loose breeding colonies, usually in the reeds of flat wetlands. Greylags breed further to the South than any of the other Eurasian goose species. Only up to 20% of flock members are breeders (Rutschke, 1982). The non-breeders stay together or aggregate for moulting in May and June. After fledging of the young, breeders and non-breeders mingle again in big summer flocks.

Females lay up to seven eggs and tend to desert their clutches if egg number exceeds eight, due to egg dumping. Mainly young females tend to parasitise the nests of experienced geese by depositing their eggs there. Goslings hatch synchronously after 28 days. The gander stays in the vicinity, but does not approach the nest unless it needs to be defended against a predator. Upon hatching, the gander rejoins his mate and her goslings and the family leaves the nest site for an area where they raise their young in the proximity of other families. The herbivorous goslings usually hatch when the sprouting grasses and herbs provide maximum protein (Loonen, 1997). This allows them to grow from approximately 100 g at hatching to less than 3000 g at fledging, nine weeks later. Goslings forage for themselves from the beginning, but pay attention to the parental beak. Thereby, they quickly learn to take the same plants as their parents. In fact, feeding traditions form early via the social learning mechanism of local enhancement (Fritz and Kotrschal, 2000). Parental care is essentially restricted to staying close to the goslings. The gander may defend his female and goslings against other flock members or, occasionally, predators. After hatching, goslings are vulnerable to starvation and, hence, determine when and where to forage. Towards fledging, leadership within the family shifts to the parents (Lamprecht, 1991). Females brood the young until two weeks after hatching. The male is stimulated through the female and goslings to be vigilant and aggressive towards other geese (Lamprecht, 1986a,b). Within the first five weeks after hatching, aggressiveness decreases and parents start moulting their wing feathers. Parents regain the ability to fly again at approximately the time the young fledge.

Offspring stay with their parents until the end of their first winter and usually separate during the spring migration to the breeding grounds. Pairs, which show a high nest site fidelity, initiate a new breeding cycle, while their offspring from the previous year join non-breeding groups. During their first spring, individuals engage in courtship and may form a temporary bond. Permanent bonds, however, are not formed before the second year in the case of females and, often, the third year in the case of males. If the parents are unsuccessful with their new brood, offspring may remain with their parents for a second year (own, unpublished observations). Even when paired, females tend to stay close to their sisters, resulting in female-centred clans (Frigerio *et al.*, 2001), whereas paired males follow their females.

After fledging, the families form loose summer foraging flocks with the non-breeders. Subgroups of the flocks fly into different foraging areas from common night roosts at large bodies of water. Towards late autumn, they may form huge winter flocks. Geese from all over Europe, for example, aggregate at the south-eastern Neusiedlersee, a large lake at the border between Hungary and Austria. There, up to 20 000 greylag geese may stay put until a cold spell or snow inhibits access to food in the fields adjacent to the lake. At this time, the geese leave for their winter feeding areas on the Mediterranean coast of Tunisia, some 2000 km to the southwest of the Neusiedlersee (Rutschke, 1987, 1997). Most fly non-stop in altitudes of 4000–6000 m, where they utilize fast tail winds in jet streams for increasing their ground speed (Madsen *et al.*, 1999).

A goose flock may be seen as an aggregation of a number of social units, including heterosexual pairs, male–male pairs and families (that is pairs with their offspring) within female-centred clans (Frigerio *et al.*, 2001). Occasionally, clusters of unrelated individuals may form. Social units are characterized by the internal performance of the 'triumph ceremony' (Fischer, 1965; Lorenz, 1991; Radesäter, 1974). Members of the social unit provide each other with social support during agonistic interactions directed towards outside individuals. This support, in turn, provides high rank to members of the social unit (Frigerio *et al.*, 2003; Lamprecht, 1986a,b; Lamprecht 1987). Individuals within social units co-feed without agonistic interactions and behave in a relatively synchronized way.

The main benefit of flocking in geese seems to be predator avoidance (Lazarus and Inglis, 1978; Lazarus, 1978; Kotrschal *et al.*, 1992) via risk dilution and the many-eye effect (Krause and Ruxton, 2002). Alertness and flight by one individual usually provokes flight in all flock members. There is also active cooperation within the flock via mobbing potential predators. Mobbing calls communicate the presence of a potential source of danger and alert/teach naive individuals (Curio *et al.*, 1978). Alarm calls occur mainly in response to aerial predators. They

Figure 2.1. Greylag gander pair from the Grünau flock / Austria (photo: Brigitte M. Weiβ).

are faint and rare and, hence, may be directed towards the social unit, rather than towards the flock in general.

The Grünau flock of greylag geese

The Grünau flock (Figure 2.1) dates back to a group of geese established in the early 1950s at the first 'Max Planck Institut für Verhaltensphysiologie' in Northern German Westfalia. In the late 1950s, this early flock was transferred to the site of the second 'Max Planck Institut (MPI) für Verhaltensphysiologie' established for Konrad Lorenz in Southern German Seewiesen (near Starnberg). Upon his retirement from the MPI and after being awarded the Nobel price in 1973, Lorenz was invited to continue his studies in the Upper Austrian valley of the River Alm by the land owner, Ernst August, Duke of Hannover. Hence, the semi-tame, free-ranging flock of greylag geese was introduced into the valley by Konrad Lorenz and coworkers as described in (Kalas, 1977; Lorenz et al., 1982; Lorenz, 1991). Out of the 148 individuals moved to Grünau, only 74 remained at the site. Regular hand-raising stabilized the initially erratic spatio-temporal patterns. Flock size fluctuated between 130 and 160 over the years. From 1973 to 2003, the Grünau flock contained 794 fledged individuals (see Hemetsberger, 2001, 2002 for demography). All of the Grünau geese are individually marked with coloured leg rings and their whereabouts and social relationships are continuously monitored, usually every second day, yearlong.

The Grünau flock is non-migratory and relatively closed, that is only a limited number of sub-adults left the flock and very few outsiders ever joined. When such outsiders joined the Grünau flock, they did so mainly as new partners brought back by flock offspring that were off on a temporary excursion. The Grünau geese are supplemented with food yearlong, twice a day with pellets low in protein and with grain to keep their spatio-temporal patterns stable. Only a little food is provided from spring to fall, but geese are fully supplied in winter, when snow cover makes natural forage unavailable.

The flock usually spends the daylight hours in meadows surrounding the Konrad Lorenz Research station (E $= 13°57'$, N $= 47°49'$) and uses the Almsee, a lake 8 km south of the KLF as a night roost. Except for the lack of a migratory tradition, the seasonal patterns of our flock closely resemble that of wild greylag geese (see above; Hirschenhauser et al., 1999; Kotrschal, 1995; Rutschke, 1997). After break-up of the winter flock, in February–March, females nest at a number of sites in the valley, from the KLF to the Almsee. They either build relatively predator-safe nests in breeding boxes, which are provided by us, on ponds or they build their nests in reeds along the water. Predation pressure, particularly by red foxes, is high on the relatively 'unsafe' reed nests and per year two to seven females plus their clutches are lost that way. This leads to a male-biassed sex ratio in the flock (up to 70%) and fosters a tradition of breeding in safe boxes. Over the years, ravens at the lake have seemingly increased their skills, finding greylag nests and preying on the eggs as well. Approximately 80% of the hatched goslings do not survive until two weeks of age. They fall victim to predators (mainly red foxes and martens) and die of infections, probably as a result of inadequate care by inexperienced mothers. In fact, fledging success increases significantly with maternal age (Hemetsberger, 2001, 2002). In general, fledging success in the flock is low. Although the number of heterosexual pairs increased from 16 in 1973 to over 40 in 2000, the number of pairs fledging young remains remarkably constant at 4.44 ± 2.94 (SD) per year, with a total of 16.52 ± 11.44 geese per year fledging into the flock. This hardly balances the losses of fledged geese that fall victim to predators (Hemetsberger, 2001, 2002).

The flock only persists in the valley because we provide food in winter and we regularly introduce hand-raised geese. This results in a proportion of 20–40% of hand-raised individuals in the flock over the years (see below). Hand raising has provoked critique that this would be an 'unnatural' situation and our results would not be representative for greylag geese in general. This is unlikely, as the general patterns of our flock closely resemble that of greylag flocks in the wild (Loonen, 1997; Rutschke, 1997), but there are features that seem unique to the Grünau geese as well. For example, hand raising seems to play a role in the ontogeny of a certain type of same-sex gander pair (see below). Goslings will imprint immediately after hatching on the first creature they have contact with (usually their mother; Hess, 1973). Hand-raised goslings are hatched in an incubator and immediately after hatching contacted by the hand raiser. Rapid filial imprinting results and goslings will follow their human foster parent and initially avoid any close contact with other geese (Schutz, 1965a). When raised in a sibling group, these goslings will not sexually imprint on humans (Schutz, 1968).

Ever since Konrad Lorenz, hand raising follows a careful standard protocol to meet the social needs of goslings. Eggs are usually taken from abandoned clutches within the flock, but occasionally eggs were imported from breeding colonies, both from the East of Austria and from England, mainly to ensure genetic diversity in the flock. Humans that hand raise goslings care for 5–12 goslings, 24 hours a day, for a period of at least nine weeks, from hatching to fledging. This is usually done in an area of meadows and ponds 3 km South of the KLF, which is also used by the flock as the main breeding/raising site. At this site, small huts provide shelter during the night for the human caregiver and the goslings. During the day the herbivorous goslings forage in the company of their human caregiver as well as in close proximity to goose-raised goslings. Towards fledging, they increasingly socialize with the flock until complete integration in early fall. Hand-raised geese remain tame for life towards familiar humans in familiar contexts. The specific social bond with their substitute parent loosens during their first spring and even more so in their second year of life (Weiss, 2000), when females start to engage in long-term pair bonds with males.

Goslings are hand raised to utilize the resulting interspecific social bond for experiments, not possible with shy goose-raised offspring (for example Fritz and Kotrschal, 2000; Fritz et al., 2000; Pfeffer et al., 2002). Absence of the human caregiver causes distress calling and enhanced corticosterone excretion in the goslings (Pfeffer et al., 2002). It seems that the main function of the human caregiver is to provide social support for his/her geese even when these are already fledged. A familiar person close to the feeding flock, even if not interfering with other flock members, results in increased duration of feeding and enhanced success in agonistic interactions, as well as in decreased corticosterone excretion in hand-raised fledglings well into their second year of life (Frigerio *et al.*, 2003; Weiß and Kotrschal, 2004). The main effect of hand raising is that flock members are relatively tame and can be closely approached and managed. This also transmits to most non-hand-raised members of the flock. Also, some level of hand raising is necessary to maintain flock size despite considerable predation pressure (see above). Inadequate hand-raising conditions (that is non-permanent presence of hand raiser during the first weeks after fledging) results in socially incompetent or behaviourally deviant individuals.

Relevant behaviours involved in pair bond

The following account of behaviours is focused on the relevant elements for pairbond and is necessary, for example to appreciate the 'episodes' at the end of the Results section. For a relatively complete greylag ethogram see Lorenz (1991). Virtually all behavioural elements of geese are highly stereotyped (little variation within species with regards to pattern, but variation in intensity) and qualify as 'fixed (or modal) action patterns' (FAP, 'Erbkoordination', Lorenz, 1932, 1942). Hence, behavioural criteria are easily defined elements of the 'ethogram'. In the following, we use the classical Lorenzian terms for the relevant behaviours, although most of these are interpretative, suggesting partially untested or even untestable functions (for example 'triumph ceremony'). For a full discussion of these behaviours see Lorenz (1991).

Our behavioural criteria for *courtship*, for example, are *active maintenance of spatial proximity* by the courting individual, which is often, but not always, the male (young females may court gander pairs), *walking in parallel close* to the desired partner, or, generally, a high degree of *behavioural synchrony*. This is often accompanied by '*guarding*', when the rival courting an already paired individual (male or female) keeps between his desired partner and the pair partner of the latter, who may, in addition, frequently be chased off by the courting rival. '*Winkelhals*' (bent neck) is another courtship posture, when the neck forms a blunt angle and the peak points obliquely towards the ground. A new bond is only formed when the courted individual responds to the '*triumph ceremony*' of the courting rival (below). '*Triumphgeschrei*' (the '*triumph ceremony*'; Fischer, 1965; Radesäter, 1974) seems to serve as the establishing and reassuring of social bonds (not necessarily restricted to a dyad, see below), but this hypothesis was never vigorously tested. It consists of a complex, but fixed chain of action patterns performed mainly by males, which launch an initial attack towards another individual (not necessarily a rival), turn around and run towards his partner with high erect body, flapping wings and loud cackling vocalization. At short distance to the partner the male will continue with a gentle '*greeting behaviour*', with the neck stretched parallel to the ground, swinging left–right, the beak not directed directly at the partner and soft gan-gan-gan vocalization.

The criterion for interest between partners is close spatial proximity maintained for days and weeks; the criterion for a *pair bond* with mutual agreement of both partners is when the *triumph ceremony* (above) of the courting partner (usually a male), is responded to by '*greeting behaviour*' of the courted partner (the female or another male). One cannot help to see the analogy with a human wedding ceremony, where partners in front of witnesses confirm their bond to each other. However, the triumph ceremony is not restricted to dyads. It is performed within families (pair with offspring less that two years old), clans or even within groups of unrelated individuals, such as mixed-sex or male-only trios and quartets. The general Lorenzian term for this is '*Triumphgeschreigemeinschaft*' ('*triumph ceremony community*'; Fischer, 1965), which is characterized by the absence of agonistic interactions within the pair/family/group whose members perform the triumph ceremony towards each other. When a male-only triumph group breaks up, this may be accompanied by a fierce fight and may result in lasting hostility between former partners. The triumph ceremony may be performed within clans or other groups and indicates alliance formation.

Still, the socio-sexual pairbond in geese is a dyadic matter. Trios form by the secondary attachment of a third individual, male or female, to an existing pair, male–female or male–male. Such a triad always remains behaviourally asymmetrical. The triumph ceremony is mainly performed within the core dyad. Also, following a copulation, no matter whether within the core pair, the third partner, or extra pair, the '*postcopulatory display*' (Lorenz, 1991), a high-above-the-water display of the male towards his main partner, lasting for less than a second, is always performed towards the main partner. Hence, the primary partners form the central dyad and the secondary partner remains only loosely attached (Lamprecht, 1987).

In this text, the term '*fight*' stands for a '*Flügelbugkampf*': two ganders bite and hold each other where the wing is attached to the body and try to hit the opponent with heavy wing blows, using a horny knob at the wing front of the carpal area. Most of such fights last only for seconds, but occasionally may go on for several minutes, leaving geese exhausted or even injured. Series of fights between particular males may go on for days, occasionally even for months and are particularly intense in spring. Wing blows can also be delivered when running after the opponent or even in flight and are hard enough to break bones. The most common agonistic behaviour is the *threat display*, with the neck stretched towards the opponent (Fischer, 1965).

Results

Pair bond demography

In 30 years of continuous monitoring, the Grünau flock contained a total of 794 fledged individuals (352 males, 345 females). The sex of 97 individuals could not be unambiguously determined before they disappeared because of predation or dispersal. Of the males, 236 were paired at least once and 116 were never paired, most of them disappearing before reaching their third year of life. Pair formation has been observed between males and females and between males and males, but never between two females. Occasionally, trios form when a third individual, male or female, attaches to a male–female or a male–male pair. In that case, the dyadic core remains behaviourally distinct. Pair bonds in geese are defined by close spatial proximity, by the 'triumph ceremony' (see above), a lack of agonistic interactions within the dyad or triad, the direction of postcopulatory display and a duration of these interactions of at least one week. With these criteria, a total of 572 pair bonds were registered. As some 'affairs' may last for less than one week, this is a conservative estimate.

On average, the 218 ganders (100%; 18 males were excluded from analysis because of incomplete data) for which we have complete life history data, engaged in 2.53 pair bonds ± 1.78 (SD) throughout their lives (Figure 2.2). These pair bonds lasted on average, 2.69 ± 2.16 years. One hundred and six males (49%) only formed pair bonds with females. These males had, on average, 1.76 ± 1.07 partners, their pair bonds lasted 2.48 ± 2.35 years. The 80 'bisexual' ganders (37%) who pair bonded at least once with another male in addition to forming pair bonds with females had, on average, 3.81 ± 1.88 partners. The average pair bond duration for these 'bisexual' ganders was 2.52 ± 1.83 years. The 32 males who exclusively paired up with males (15%) had on average 1.84 ± 1.57 partners. Their male–male pair bonds lasted, on average, 2.42 ± 2.47 years.

At first pair bond, heterosexually paired males were on average 2.26 ± 1.09 years of age, whereas those which had their first pair bond with a male were significantly older (3.71 ± 2.77 years of age; Mann–Whitney U, $Z = -4.618$, $p < 0.0001$). This may indicate that males who had difficulties obtaining a female mate finally tended to take a male substitute. Successive male–male bonding was positively correlated with age. Hence, the older the ganders, the more likely they were paired with a male partner. Of 233 first pair bonds by 220 males, 185 were

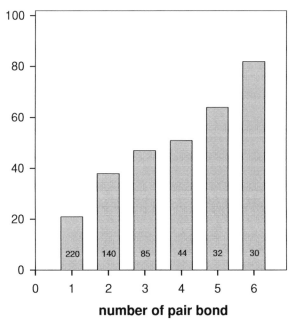

Figure 2.2. Male–male pair bonds as percentage of total pair bonds with increasing number of individual pair bonds. Numbers in bars show sample size.

with a female and 48 with a male (21% male–male). The reason why the number of bonds exceeds the number of males involved is explained by a certain number of first pair bonds which took the form of male–male–female trios. Of 160 second pair bonds by 140 males, 99 were with a female and 61 with a male (38% male–male). Of 94 third pair bonds by 85 males, 49 were with a female and 44 with a male (47% male–male). Of 49 fourth pair bonds by 44 males, 24 were with a female and 25 with a male (51% male–male). Of 33 fifth pair bonds by 32 males, 12 were with a female and 21 with a male (64% male–male). And of 33 sixth to tenth pair bonds by 30 males, only six were with a female, but 27 with a male (82% male–male).

Average life expectancy of fledged males was 6.6 years $\pm$ 5.9. Fledged females reached only 4.9 $\pm$ 4.01 years, mainly because they suffer more predation than males during incubation. This results in a male-biased sex ratio in the flock (below). Males that were exclusively heterosexually paired and older than five years of age had an average life expectancy of 10.7 $\pm$ 5.3 years, whereas ganders that were partially or exclusively homosexually paired, life expectancy was 12.3 $\pm$ 4.8 years (Mann–Whitney U, $Z = -2.151$, $p = 0.031$) once they reached five years of age. However, one has to be careful with a causal interpretation. Males may not get older simply *because* they are homosexually paired. Rather, it

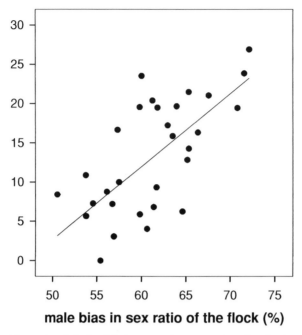

Figure 2.3. Relationship between the proportion of males in the Grünau flock of greylag geese and the relative number of males in gander pairs from 1973–2003 (R2 = 0.461; F = 24.854; p < 0.0001; y = −43.773 + 0.929x).

seems the other way round: elderly males having lost their female are increasingly likely to engage in a homosexual pair bond (above).

Influence of flock sex ratio on frequency of male–male pairs

The proportion of male–male pairs in the flock depends on sex ratios (Huber and Martys, 1993). When balanced (50–55% males, for example, 1973, 1974 and 1999), less than 12% of the pair bonds in the flock were male–male (Figure 2.3). When the sex ratio was heavily male biased (>65%, for example, 1980–1982 and 1989–1992), more than 20% of the pair bonds in the flock were male–male. Hence, as already found by Huber and Martys (1993) on a considerably smaller data base, the percentage of males in the flock can explain the percentage of gander pairs formed. But this explanation is probably not sufficient, as even at balanced sex ratios gander pairs occur.

In contrast to the Grünau flock of greylag geese, the Seewiesen (MPI) flock of barheaded geese (*Anser indicus*) was protected against red fox predation and because of male dispersal, developed a female-biased sex ratio (Buhrow and Lamprecht, pers.comm.). Female–female pairings were never observed even under these conditions, but, occasionally, male–male pairs occurred. Excess females formed trios by associating with a heterosexual pair as a secondary female. Such females copulated with the pair gander and raised their goslings close to the primary pair, but with little support from the gander; therefore, reproductive success of such secondary females was lower than of the primary female in the pair (Lamprecht, 1987).

Behaviour of male-paired males

As a base for hypothesizing about potential mechanisms of pair formation and the functionality of male–male pairing, it is relevant to compare the behaviour of heterosexually paired to male–male-paired ganders. In geese, the sexes show quantitative differences in behaviour. Males engage in agonistic interactions more often. In addition, they almost exclusively perform the 'triumph ceremony' and the dominance-related vigilance posture 'extreme head up' (Lazarus and Inglis, 1978). In addition, paired males are generally more vigilant and spend less time feeding than their female partner (Lamprecht, 1986a,b; Lorenz, 1991; Waldenberger and Kotrschal, 1993; Waldenberger, 1994).

If one of the partners in the male–male pair assumed pseudo-female behaviour, this might suggest that pair formation could be based on the inability of at least one of the partners to discriminate between sexes (Allen, 1934). This hypothesis was tested by Huber and Martys (1993) using quantitative behavioural observations of pairs from the Grünau flock. Both partners in homosexually paired males show male-typical behaviour in the agonistic categories approach and avoidance, as well as in vigilance and in social behaviour towards the partner. Ganders paired with males were found to be even more significantly vigilant and active in social bonding behaviour (mainly triumph ceremony) than males in heterosexual pair bonds. Hence, a gander in a male–male pair does not adopt a female role; rather both ganders are hyper-masculine in terms of their behaviour. These observations accord well with the generally loud, conspicuous and aggressive appearance male–male pairs make in the flock and with our own, quantitative data (Kotrschal et al., unpublished). However, not all male–male pairs fit that pattern. Occasionally, they may behave less conspicuously and resemble male–female pairs (see episodes, below). Despite such exceptions, in 30 years we have never observed one of the males assuming typical female behavioural patterns.

Temporal patterning of copulation and pair formation

As geese usually copulate in spring (January to April), but pair bonds are maintained year long, the importance of sexuality in such a social relationship

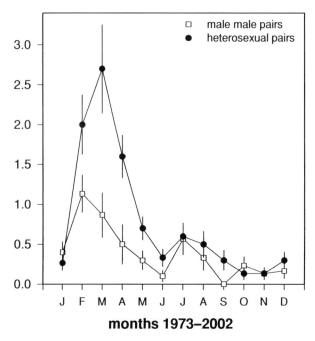

Figure 2.4. Pair bond frequencies for heterosexual and male–male pairs vary in parallel over the year.

is hard to judge. Two mutually compatible groups of hypotheses most plausibly explain this year-long and long-term monogamy: (1) Securing a suitable partner. If all other individuals are permanently paired, it may be hard to find a new and high-quality partner every spring. (2) The need for a social ally for maintaining high status within the flock (Lamprecht, 1986a,b). Allies facilitate an individual's access to resources (Kotrschal *et al.*, 1993) and the range of benefits of social support (Frigerio *et al.*, 2003; Weiß and Kotrschal, 2004). A male losing his female or male partner will immediately drop to low rank (Lorenz, 1991; Weiss unpublished). The social ally hypothesis would help explain male--male pairing. Although male–male pair formation mainly occurs during the sexual/reproductive phase of the year, pairs may form and break up throughout the year (Hemetsberger, 2002). This is true for heterosexual as well as male–male pairs (Figure 2.4)

Copulating may be important in pair formation. This is supported by a number of copulations that were observed in newly formed heterosexual pairs in fall, when fertile gametes are certainly lacking in both sexes. However, such 'untimely' copulations between partners of a newly formed pair were only seen in heterosexual pairings, which does not mean that they will never occur in male–male pairs.

Copulations occur only on water, following a distinct sequence of action patterns and, hence, are easily detected (Lorenz, 1991). This sequence is initiated by swimming in parallel, with the male or both partners assuming a position with the tail region high above water, the so-called 'frigate posture'. Then partners start dipping their necks into the water. This is usually initiated by the male partner. When neck dipping becomes synchronized, the male mounts the female and grabs her neck feathers. Thereby, the female may be completely submerged. After a few seconds, the male dismounts and performs a short postcopulatory display towards his partner, with the tips of the wings, the tail and also the beak highly elevated (Lorenz, 1991).

In most of the attempted male–male mountings, both partners tried to get on top, resulting in circling around each other and more often than not, this ended in a severe fight between the two. However, in quite a few of the observed cases, one of the partners ended up assuming the female position. Such copulations between males were indistinguishable from heterosexual copulations, with a male partner on top and the second male below, in the female position. However, it remains unclear whether or not intromissions occurred in these cases (geese, as all Anatidae have an intromittant organ) and whether sperm was released. If a male from a gander pair copulated with a (often young, unpaired) female, which happened quite frequently, the postcopulatory display was always directed towards the male pair partner and not towards the female he copulated with (e.g. Fischer, 1965; Heinroth, 1911; Huber and Martys, 1993).

Over the past three decades, it was generally not possible to collect quantitative data on copulations, because a majority of copulations on the ponds and lakes in the valley occur out of sight of the limited number of observers. However, quantitative data were taken in March 1996, at the peak of copulation activity (Kotrschal et al., unpublished). Due to an unusually severe late winter, most ponds in the area were still covered by ice and snow and, hence, were unsuitable for copulating. An exception was a pond, approx 10 m × 30 m in front of the research station, which receives a flow of well water and, therefore, is open all winter. The 126 geese in the flock at this time formed 33 heterosexual pairs (82.5% of pair bonds), seven homosexual pairs (17.5%); the rest were singles and other social categories. With 69 males and 57 females, the flock sex ratio was somewhat male biased (55%). One hundred and twenty one individuals were observed on this pond for a total of 130 hours from 6 to 20 March 1996 and a total of 297 copulations were recorded.

Inter-individual differences in male copulation frequencies were found. Whereas one male was seen copulating 26 times, 40% of the males were not seen copulating at all and only three males accounted for 25% of all copulations. Of the 40 ganders seen copulating, only three (7.5%) were male–male paired. As 20% of all males in the flock at that time were in a homosexual pair bond, they were underrepresented among the copulating males. Copulation frequencies were even more biased towards heterosexual males: only ten of a total of 297 observed copulations (3.4%) were by homosexually paired ganders. Hence, heterosexually paired males copulated significantly more often than male–male paired (Mann–Whitney U, $p = 0.0003$). Extra-pair copulations (EPC) of males were at 38%, of paired females at only 1.3%. Most copulations of homosexually paired ganders were with females, but exact numbers cannot be given in this case.

Dominance

The dominance rank of individuals in male–male pairs was as high as that of males in heterosexual pair bonds. If the percentage of agonistic interactions that individuals win in the flock is taken as a measure of individual dominance rank (Figure 2.5), males in male–male pairs and in male–female pairs won approximately 60% of their interactions (Mann–Whitney U, $Z = -0.112$, $p = 0.91$, Bonferroni-corrected). Family ganders (that is males in company of their females and of offspring) won significantly more than male–male-paired ganders (80%, Mann–Whitney U, $Z = -3.912$, $p < 0.001$, Bonferroni-corrected) and single ganders won significantly less than paired males (20%, Mann–Whitney U, $Z = -5.807$, $p < 0.001$, Bonferroni-corrected). Because this pattern was the same in all annual phases, sexual (February to May), summer flock (July to September) and winter flock (October to January), only the winter flock data are shown (Figure 2.5). Even though high-ranking heterosexual males copulated significantly more often than low-ranking males (Kruskal–Wallis, $p < 0.05$) and male-paired ganders did not differ in rank from heterosexually paired ganders, homosexual ganders still showed low copulation rates (above).

Reproductive success

Of a total of 352 males included in this study, 79 mainly heterosexually paired males reached more than five years of age, with a reproductive life span of two to three years and 53 of these also had at least one pair bond with a male in addition to bonds with females. Reproductive success (goslings raised to fledging per year of reproductive life span) in males older than five years tended to be higher in exclusively heterosexually bound males as compared with males who had at least one homosexual pairing, even though this difference was not significant (Mann–Whitney U, $Z = -1.155$, $p = 0.248$). We do not have data

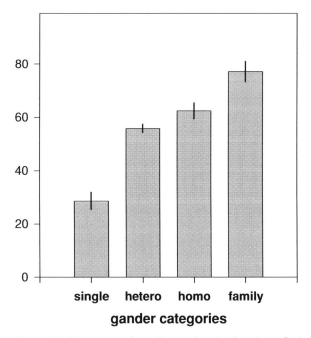

Figure 2.5. Percentage of won interactions in the winter flock (October–January). Pooled from one week winter flock observation periods per year, including the years from 1997 to 2001 and from 2002 to 2003. Males paired with males are not different from males paired with females (Mann–Whitney U, $Z = -0.112$, $p = 0.91$, Bonferroni-corrected), Singles won significantly less (Mann–Whitney U, $Z = -5.807$, $p < 0.001$, Bonferroni-corrected), family ganders more (Mann–Whitney U, $Z = -3.912$, $p < 0.001$, Bonferroni-corrected) interactions than males in either homosexual or heterosexual pair bonds.

on the potential reproductive success achieved by heterosexual and homosexual males via extra pair copulations. However, this should be low in male-paired males, considering their generally low copulation rates.

Mechanisms of male–male pair formation: episodes

The following anecdotes are not 'just so' stories. They result from regular and continuous qualitative observations of social relationships over many years (mainly by JH and BW). These case histories will exemplify the many different ways male same-sex pair bonds in geese form and break up, illustrating the mechanisms of such pair formation and hinting at potential functions. M = male, F = female.

Episode 1: 'Love and hate'

Subjects: Lorenz (M, goose raised, hatched 1998), Kiowa (M, goose raised, 1998), Lincoln (F, goose raised 1996, the older full sister of Lorenz) and Franzi (F, goose raised 1985).

The events: In February 2000, Lorenz-M and Lincoln-F paired up. It was their first pair bond. In March 2000, Kiowa-M started courting Lincoln-F. After a series of intense fights, Kiowa-M replaced Lorenz-M. This was Kiowa-M's first pair bond. However, Lorenz-M stayed close and was increasingly tolerated by Kiowa-M. In July 2000, Lorenz-M, Lincoln-F and Kiowa-M formed a trio. Lincoln-F disappeared in August 2001, but Lorenz-M and Kiowa-M remained paired. In February 2003, Lorenz-M left Kiowa-M and successfully fought for Franzi-F and Lorenz-M and Franzi-F formed a pair bond. A few weeks later, Kiowa-M disappeared.

Comment: This example shows common elements of the pair-bond dynamics in geese. The two males entered a same-sex pairing via a trio stage (attaching to the same female). After the disappearance of this female, the two males stayed together for nearly 1.5 years. When a female became available, the male–male bond disintegrated.

Episode 2: Dynamic changes

Subjects: Punki (M, hand raised 1992), Pirat (M, hand raised 1992, the social sibling of Punki), Sebastian (M, goose raised, 1992), Diana (F, goose-raised 1992), Fremder (M of unknown age and origin, one of the few cases an outsider ever integrated into the Grünau flock, arrived February 1993).

The events: In Spring 1994, the brothers Punki-M and Pirat-M, established a pair bond without female participation. In September 1994, Pirat-M disappeared. In the following winter, Punki-M and Sebastian-M had a short 'affair'. In February 1995, Sebastian-M paired up with Diana-F. At the beginning of April 1996, Sebastian-M disappeared, and a few days later Punki-M and Diana-F formed a pair. Their pair bond was characterized by social instability in spring 1997, but they successfully raised young together in spring 1998. Diana-F disappeared in April 2002, leaving Punki-M as a single. From Fall 2002 onwards, Punki-M was joined by Fremder-M and starting in December 2002 they performed triumph ceremonies together. They formed a relatively inconspicuous and 'asymmetric' male–male pair. Fremder-M was tolerated rather than desired as a partner by Punki-M.

Comment: Pair bond is not only a matter of individual choice (see Choudhury and Black, 1993). Chance events, such as the frequent disappearing of individuals have a great influence on the dynamics of pairing, and individuals seem to be relatively opportunistic with respect to their potential partners. For a male in a flock, the pairing threshold seems to be low when the choice is between no partner at all or accepting an available partner, which might not be first choice. As of spring 2003, Punki-M was 11 years old and had only marginal chances of ever engaging in a stable pair bond with a female again. Without a male partner, his chances would have been close to zero, because his status in the flock and his chances of survival would have been marginal. In fact, lone old ganders are low ranking, usually stay at the margin of the flock and tend to disappear soon.

Episode 3: Brothers 1

Subjects: Tian (M, hand raised 1999), Corrie (M, hand raised 1999, the social sibling of Tian), Jeanne (F, goose raised 1999).

The events: In spring 2001 Tian-M courted Jeanne-F, resulting in a loose pair bond (that is, the partners were not always close together; the triumph ceremonies were not intense), Tian-M's brother Corrie-M joined the loose pair bond and was tolerated. This was the first social bond for all of these individuals who were two years old at that time. In Spring 2002, Jeanne-F split off, leaving the brothers in a pair bond. In January 2003, Corrie-M left the valley and stayed two weeks at a lake, approximately 60 km to the South and, at this time, Tian-M courted Jeanne-F again. Two weeks later, Corrie-M returned and Tian-M immediately abandoned Jeanne-F. Corrie-M and Tian-M then formed a close and aggressive male–male pair and have remained together ever since.

Comment: One of the 'typical' elements in this story is the formation of a stable pair bond between hand-raised brothers. Females may have roles, but tend to remain marginal. Even if a brother copulates with a female, which is quite common in spring, the postcopulatory display is directed towards his brother, who is at the same time his male partner. Such male–male pairs are usually louder, more aggressive (Huber and Martys 1993), dominant and competitive when compared with heterosexual pairs.

Episode 4: Brothers 2

Subjects: Keiko (M, hand raised 1996), Skana (M, hand raised 1996 as a social sibling of Keiko), Jana (F, hand raised 1998).

The events: In January 1998, the hand-raised brothers Keiko-M and Skana-M paired up and formed the 'classical' pair for their social ontogeny background: tightly bonded, loud and aggressive (Huber and Martys, 1993). As often in such cases, they seemed attractive to young females, with whom they copulated and whom they tolerated close by for a while, so loose trios formed from time to time. However, the brothers never directed their (frequent) triumph ceremonies towards any of these females. Skana-M disappeared in February 2002, Keiko-M then loosely re-attached with the sibling sisters he was brought up with.

In February 2003, Keiko-M courted Jana-F and was accepted. A loose pair bond formed.

Comment: An episode similar to 3 (above). Again, a strong pair bond is formed between two hand-raised brothers and is only terminated when one of them disappears.

Episode 5: Brothers 3

Subjects: Iwan (M, hand raised 1993), Kobold (M, hand raised 1993 as a brother of Iwan), Nobody (F, goose raised 1994).

The events: In July 1996, Iwan-M lost his second female partner after having lost his first female partner just before. Both females had disappeared. Then he paired up with his brother Kobold-M, who had loosely paired with Nobody-F before. However, in March 1997, Iwan-M paired up with Nobody-F and Kobold-M paired with Lilleman-F. Kobold-M disappeared a few days later. Nobody-F successfully hatched, but never fledged offspring. She disappeared in April 2000. Iwan-M's female Lumpazi-F successfully fledged four offspring in 2003.

Comment: This episode shows that pair bonds between brothers may be followed by heterosexual pair bonds.

Episode 6: Opportunistic bonds

Subjects: Tsitika (M, hand raised 1999), James Bond (M, goose raised 1999), Jimmy (F, goose raised 1999), Noah (F, hand raised 2000), and many more . . .

The events: Tsitika-M and Jimmy-F paired up in fall 2000. It was the first pair bond for both individuals. At this time, Tsitika-M also showed increasing interest in James Bond-M. From Spring 2001, Tsitika-M's initiatives led to pairing with James Bond-M, with Jimmy-F remaining a loose third in the triad. From Fall 2002, Noah-F followed Tsitika-M and, from February 2003 onward, was integrated into the triumph ceremony group. Jimmy-F remained loosely attached to her party. Initially, Jimmy-F and Noah-F were mainly interested in Tsitika-M, who tolerated both females in close proximity and copulated frequently with Jimmy-F. Tsitika-M remained strongly attached to James Bond-M, however, who has formed a pair bond with Noah-F and has copulated frequently with her.

Comments: This episode shows that a relatively complex network of socio-sexual relationships between more than two individuals, with a male–male pair at the centre, may lead towards (non-exclusive) heterosexual pairs.

Episode 7: A series of social dramas

Subjects: Daniel (M, goose raised 1992), Lupo (M, goose raised 1994), Silva (M, goose raised 1990), Rockygrün (M, goose raised 1990), Lafayette (M, goose raised 1998), Alien (M of unknown age and origin, one of the very few cases an outsider ever integrated into the Grünau flock, arrived February 1998 together with a temporarily absent female from the Grünau flock), Tiger (F, hand raised 1995) and many more . . . None of the males involved were brothers.

The events: In March 1995, Daniel-M guarded Rockygrün-M, who himself was paired with Nina-F. This guarding behaviour on the part of Daniel-M triggered social chaos, because a number of males started courting and fighting over Nina-F. Her male partner Rockygrün-M could not defend her any more, because he himself was guarded by Daniel-M. During these events, Lupo-M, whose primary interest was in Daniel-M, fiercely attacked Rockygrün-M (probably as a rival) and also Daniel-M, who stopped guarding Rockygrün-M in April. In February 1997, Daniel-M paired up with Lupo-M. This was the first pair bond for both of them. In October 1998, a series of severe fights over Daniel-M occurred between Silva-M and Lupo-M. Lupo-M lost Daniel-M to Silva-M, but regained Daniel-M in November after another series of severe fights with Silva-M. Lupo-M disappeared in February 1999. Ever since, Daniel-M has been paired with Silva-M. In April 2003, Daniel-M caused another crisis simply by following Rockygrün-M, who, at that time, was paired with Tiger-F and had just left the nest with seven goslings. This caused Silva-M, Daniels-M mate, to attack Rockygrün-M fiercely for a number of days in a row. This situation was used by Lafayette-M, who was paired to Alien-M, after having been courted by Alien-M for a long time. Lafayette-M was at that time strongly courting Tiger-F, who was, in turn, permanently chased by Lafayette-M's male partner Alien-M. At the same time, Lafayette-M fiercely attacked Gandalf-M, another gander courting Tiger-F. Due to these events, six of the seven goslings of Tiger-F and Rockygrün-M were killed on their first day after hatching. The last gosling survived for nearly a week, as long as Gandalf-M remained on Tiger-F's side. This gosling finally died as a result of the social unrest connected with the change of Tiger-F back to Rockygrün-M again. After a few more days, things were calming down and Rockygrün-M was the unchallenged mate of Tiger-F once again.

Comment: An initial event may trigger a cascade of courtship and partner defence between different personalities with a spectrum of social relationships. The final result was the loss of all of the offspring of Rockygrün-M and Tiger-F in 2003. This episode shows particularly well the potential complexity of socio-sexual relationships. Dynamic events like this one are hardly done justice with categorial data and statistics alone. For example, although Lafayette-M did not respond to Alien-M's triumph ceremonies, and Alien-M rather than Lafayette-M actively maintained close spatial proximity, Lafayette-M was observed to copulate with Alien-M, the latter resuming the female position.

Often, brothers that are hand raised together are particularly close to each other, loud, aggressive, symmetric (that is showing the same amount and intensity of pair bond behaviours towards each other) and stable in their first mutual pair bond with each other. Their second pairing can be heterosexual. Since 1990, for example, we observed five pairings between brothers, four of which were first pair bonds, one pairing was a second pair bond of one and a third pair bond of the other brother and only lasted half a year (see above, Iwan & Co.). Three of the four brother pairs were stable until one of the brothers disappeared, the fourth still existed in spring 2003 as a trio with a female. In contrast, homosocial pairings between non-brothers are frequently characterized by asymmetry. One may take the initiative, while the other (paired with a female or interested in a female) hardly responds to the triumph-ceremony advances, but finally tolerates the persistent attention of the courting male.

Discussion

Based on our present evidence, 'sexual orientation' is an unlikely (mechanistic) explanation for the majority of male-male pairs in geese. Still, it cannot be excluded, at least for those 15% of males which were exclusively male paired. In most cases, male–male pairings may be seen as a tactic of securing an ally needed to maintain a high rank in the flock. High rank is important for access to resources, including female mates. Hence, if the number of available female partners is limited, males tend to resort to pairing with another male. This idea is supported by the fact that the frequency of male–male pairs is positively correlated with the male bias in the flock. Also, with successive pairings and increasing age of the ganders, the likelihood of pairing with a male increases, because virtually all suitable females of similar age (Choudhury and Black, 1993) are already engaged in pair bonds with other males.

Over the year, the social component of pair bonds, male–female or male–male, may be even more important than sexuality. Pair bonds are maintained year long and pair partners are crucial social allies, whereas copulations occur mainly from February to May. Even when a male-paired male copulates with a female, which is quite common, the postcopulatory display is directed towards his male partner. This indicates that irrespective of the female sexual partner, the social bond is between the two males. The same may happen when a male performs his postcopulatory display after an extra pair copulation (EPC) towards his long-term female mate, but not towards his EPC partner. Only members of 'triumph ceremony groups', either pairs or families, but not singles, are able to maintain a high rank within a flock (Lamprecht, 1986a; Weiß, 2001). Most males only facultatively engage in a pair bond with another male, either in parallel with, or in succession to bonds with females. Male–male bonds tend to disintegrate when a female becomes available. All this indicates that pair bonds in geese, including male–male, are mainly social alliances for most of the year. However, in a minority of ganders (15%), the first and all of the following pair bonds are with males only. This suggests that male same-sex pairing in geese may defy a simple, monocausal explanation.

Throughout the rest of the discussion, we will discuss some mechanistic aspects of pair formation and then tackle the difficult question of how functional these male–male pairs may be. We then continue with a critical evaluation of how representative the results of our semi-tame geese may be for geese in general and end with a discussion of what, if any, relevance our results may have for explaining homosexuality in general.

Mechanisms

Mechanistic explanations are not alternative to functional, ultimate-level explanations, but are complementary. Every natural feature or trait (that is any bodily or behavioural structure generated by evolution) needs to be explained at all 'four Tinbergian levels' (Tinbergen, 1963), ultimate functions, proximate mechanisms, ontogeny and evolutionary history. The adaptive benefits of the male–male pair bond, for example, will not explain how such a bond is at all possible. The following discussion of potential mechanisms also considers some of the more traditional ideas on homosexual pair bonds.

As outlined above, there are different ways by which pair formation between males can take place. Most commonly, after losing their female partner, ganders court males and finally engage in a same-sex pair bond, if no appropriate female is available. Other males form a same-sex pair when still young. Such male–male bonds tend to break up when females become available. Bonds between brothers may persist for life. It seems that the relatively monomorphic appearance and behaviour of the sexes produces only a low perceptional threshold for male same-sex pairings when compared with sexually dimorphic species, such as the peacock. Hence, the preferences for the looks of a partner choice may not be too different in both sexes. This, however, does not mean that such same-sex pairings are mechanistic mistakes in a sense, that individuals choose same-sex partners because they would be unable do discriminate between sexes (Allen, 1934), because same-sex-paired males do not show pseudo-female behaviour (Schutz, 1965a,b). To the contrary, male–male pairs exhibit an 'over-expression' of male-typical behaviours (Huber and Martys, 1993). This should allow for even easier sex discrimination by the potential partners. Also, if homosexual pairings were a matter of sex discrimination error, pairing between females should also form, which is never the case in the greylag geese.

Since agonistic behaviour and individual dominance are functions of motivational stimulation by the social partner (Lamprecht, 1986a,b; Lorenz, 1991), it is possible that male partners even provide 'supernormal' social stimulation for a bonding male, as compared with a behaviourally relatively inconspicuous (and, hence, behaviourally less stimulating) female. Males are behaviourally more active and conspicuous than females during the courting phase and are also more active and conspicuous than females once the pair bond is established and needs to be regularly re-affirmed, for example by triumph ceremonies and greeting following agonistic interactions with flock members. As such, males may be 'trapped' into pairing with another male, because the pair-bond related stimuli show a greater quantitative expression in males than in females. This mechanistic, 'supernormal stimulus' hypothesis is also supported by the fact that, even at a balanced sex ratio, the flock may contain a few male–male pairs (Figure 2.2), which is not in alignment with a simple functional–evolutionary explanation (below). Alternatively, this would also support the idea, that at least in a minority of male–male pairs, sexual orientation would be a relevant mechanism.

In agreement with this mechanistic explanation and with the 'low-threshold hypothesis' (above) are the cases of mainly, but not exclusively, hand-raised brothers, who develop a strong early bond, never overridden by bonds formed later with other partners. This shows how easily socio-sexual preference is affected by early social experience (Fischer-Mamblona, 2000; Schutz, 1968). Human caregivers, even if they do an optimal job, will unavoidably differ from a goose with respect to parental stimuli. Hence, the template of expectation in the brain of a gosling will necessarily be better matched by a goose than a human. This will also affect the social bond towards siblings and their role in early ontogeny.

How functional are male–male pairs?

As homosexually paired individuals do not produce offspring, the straight-forward evolutionary argument is that such pairings are dysfunctional. However, modern evolutionary theory allows for more sophisticated interpretations. In fact, the editors of this volume have produced a list of functional hypotheses (introduction). So, how adaptive/maladaptive are the patterns found in geese? A number of group selectionist arguments have been advanced, including: (a) gander pairs assuming the role of guardians for the flock, (b) gander pairs aggressively pushing unpaired males out of the flock, (c) gander pairs as a buffer system providing potential male partners when needed (Huber and Martys, 1993). For a number of reasons such arguments should be considered with caution, mainly because it is hard to explain how such selection pressures would make homosexuality an evolutionary stable strategy. Based on

Hamilton's theory of inclusive fitness (1964; Wilson, 1975), more plausible functional hypotheses postulate that individuals that engage in homosexual behaviour may behave altruistically towards others, when the recipients of altruistic acts are relatives.

In the case of greylag geese, more than 30 years of observation do not provide the slightest evidence that homosexually paired sons or brothers assume helping roles or functions for their mothers or sisters. Alternatively, as discussed above, a pair partner is a highly useful social ally. Hence, the majority of male–male pairings may be explained as a strategy for alliance formation, making the best of a bad situation. Our limited data and long-term observations suggest that copulation frequencies of male-paired males are lower than of heterosexually paired males. Hence, the reproductive success of heterosexually paired males may be higher due to intra-pair copulations (IPC) and, possibly, extra-pair copulations (EPC), than that of male-bonded males (via EPCs only). However, male–male pairings frequently break up when a female becomes available. Over their lifetime, facultatively male-paired males had an only insignificantly lower reproductive success than exclusively heterosexually paired males. This indicates that the tactic of pairing up with another male when no female is available is quite successful. 'Homosexual' pairing may be considered functional, because the alternative would be living in the flock as a low-ranking single (Lamprecht, 1986a,b), with no social support, resulting in low success rates in agonistic interactions, low feeding rates within the flock (Frigerio *et al.*, 2003; Weiß and Kotrschal, 2004) and virtually no chance in competing for an available female. Single ganders have an enhanced mortality risk, as they are pushed towards the margins of the flock, where their foraging efficiency is compromised and they are at greater risk of predation (Lazarus, 1978). This functional context of male–male bonds versus remaining single is supported by numerous cases recorded from the Grünau flock (unpublished).

For a minority of males (15%) who never engage in a heterosexual pair bond, this adaptive explanation may not apply. Some of these males may have simply not grown old enough to get a chance for pairing with a female, but others never seem to court females. These may be candidate individuals for 'true' homosexuality, involving sexual orientation. In general, male-paired males do copulate with females, however less frequently than heterosexually paired males. And it is unknown whether all of the 15% exclusively male-paired males ever copulated with a female. Their reproductive success, if any, is unknown.

How representative is the Grünau flock?

Critiques may challenge the relevance of our results, because the Grünau flock is non-migratory and semi-tame. Part of the flock is hand raised and geese may experience other ecological conditions than those that occur in the wild. The Grünau geese stay yearlong within a relatively small area of the narrow valley, extending 8 km N–S and a few hundred metres E–W. Food is provided year long, allowing much time for social interactions, resulting in a between-individual familiarity within the flock. This may not be the case in the wild, much bigger and dynamic flocks. Still, it has been shown that, even in such flocks, neighbours at the winter feeding grounds may remain stable over years (Black and Owen, 1995). The relatively small and closed Grünau flock may lead to enhanced opportunities for unusual social constellations unprecedented in large, migratory flocks, where only a limited number of familiar individuals surround each goose. In wild flocks, geese need to spend much more time and effort per day for feeding and travelling than in the Grünau flock, thus limiting their time for social interactions. Still, pairs have been observed in natural goose flocks which resemble male–male pairs in body size and behaviour (Hunt and Hunt, 1977; Schutz, 1965b). It is therefore unlikely that homosexually paired ganders only occur in semi-tame flocks.

Considerable predation pressure on females leads to a skewed sex ratio in the Grünau flock and this is the main cause of male–male pairing. However, in contrast to most wild flocks, the Grünau geese are not at all hunted. Despite the 'unnatural' elements in flock management, such as hand raising and food provision, the general social and annual patterns of our flock closely resemble greylag flocks in the wild (Loonen, 1997; Rutschke, 1997, 1982). Although not all aspects of the Grünau habitat may be representative for the species, geese in the study flock are able to have a full social and reproductive life, which would not be possible in a fenced situation. Hence, our semi-tame flock is probably as close to 'natural' as possible. Only through the tameness of the geese, were close and continuous long-term observations of individuals possible, allowing data collections such as this one.

Whether or not all the social details of our flock's social behaviour (for example, female bonding: Frigerio et al., 2001) are indeed representative for all greylag geese, is ultimately a futile question, because (a) there is hardly a way to test this and (b) there is some merit in the argument of Tooby and Cosmides (1990) that the psychological dispositions (such as emotions) within a species (in their case human) rather than behaviour itself are adapted. This is demonstrated by the fact that most social systems tend to be surprisingly flexible (Lott, 1991). Geese, as any other social animals, would adjust their social behaviour to local conditions based on their social dispositions and the tradition of the group in question within the general frame of the 'Reaktionsnorm' ('reaction norm'; see Lorenz and Tinbergen, 1939 for discussion) of the species. The patterns found in Grünau are certainly 'typical for greylag geese', in a way that the greylag genotype is the ultimate background for any social play greylag phenotypes may perform on the stage that is provided by the flock. The Grünau patterns may not exactly match wild greylag populations elsewhere, however wild greylag populations will also differ from each other with regard to their peculiar background of genes, ecology and social traditions.

Hand raising tends to be criticized as 'unnatural'. Obviously, hand raisers are not greylag geese. Even with utmost care, humans are not able to provide the same stimulus environment as greylag parents. Indeed, hand-raised goslings tend to be more vocally active (that is more 'wi-wi-wi-greeting') and less vigilant than same-aged goose-raised offspring; these minor differences may persist into adulthood (Sanders, 2002; own, unpublished data). Still, hand-raised individuals develop into socially 'normal' flock members, usually within the normal range of behavioural variation that goose-raised flock members show, except for some tendency to form long-term pairs between brothers. Also goose-raised individuals engage in male–male pair bonds, but in some cases the bond between hand-raised brothers seems to be more stable. There seems to be no barrier in mate choice between hand-raised and goose-raised individuals, because our flock contains many mixed pairs. Hand-raised individuals tend to be slightly bigger and heavier than goose raised, but they do not lay larger eggs than goose-raised females, nor are they distinguishable with regard to any of the behavioural parameters related to reproduction that we have analysed (Hemetsberger, unpublished). Nevertheless, hand raising needs to be done with care. Decades of experience have shown that inadequate hand-raising conditions (that is non-permanent presence of hand raiser) will result in socially incompetent or behaviourally deviant individuals (Fischer-Mamblona, 2000).

What does greylag geese behaviour imply for the phenomenon of homosexuality in general?

As is true for any biological result, generalizations must be made with caution. Results for one species are valuable for developing working hypotheses for another species, but cannot be directly transferred. So, what, if any, insights can be gained from geese? In contrast to primates, humans are only very distantly related to birds. A common ancestor of geese and humans probably lived some 350 million years ago. It is therefore highly unlikely that dispositions or behaviours related to homosexuality in birds and primates will be due to common ancestry. This allows us to compare the ecological and social conditions which may directly cause homosexuality. Gander homosexuality, as it regularly occurs in geese, seems to be particularly common in sexually relatively monomorphic species with long-term pair bond. Along such simple lines, humans are no exception. In greylags, sexuality *per se* seems to play only a minor role in the formation and maintenance of male same-sex greylag pairs. Far more important is that male pairing seems to be adaptive. It facilitates alliance formation among males that are unable to form pair bonds with females. Allies enjoy increased social status within the flock, which, in turn, may result in increased foraging efficiency and increased protection from predators. Thus, in a sense, most single males are making the best of a bad social job when they form pair bonds with other males. Hence, it seems justified to use the term 'homosocial' instead of 'homosexual' for most male–male pairings in geese. The social context in which a single greylag gander finds itself exerts severe stress on the individual, thereby promoting homosocial pairing if no adequate female partners are available. As the stress-reducing and socially integrating mechanisms of social support are very similar from the social birds to mammals (Frigerio *et al.*, 2003; von Holst, 1999; Weiß and Kotrschal, 2004), close attention should be paid to the 'social support' hypothesis, at least as a co-factor for also explaining human homosexuality. In a minority of male geese, early ontogeny may even have led to some degree of 'homosexual orientation'. The major take-home message from our study is that, even in geese, there may be no simple, mono-causal explanation for homosociality/sexuality, despite the fact that they are considerably less complex in their social interactions when compared with primates, and most notably humans.

Conclusions

Geese are classical research models of behavioural biology. It was soon recognized that at least in captive and semi-captive conditions these sexually monomorphic and monogamous birds may form same-sex pairs. Records of 794 fledged individuals, collected from 1973 to 2003 from our semi-tame, but free-roaming Grünau flock of Greylag geese, showed that only male–male pairs occurred, but never same-sex pairing between females. The 218 males for which we have complete life history records from the past 30 years engaged in a total of 572 pair bonds, resulting in an average of 2.5 pair bonds over the lifespan (range 1–10 pair bonds). At first pair bond, 21% of the males paired with a male, whereas 82% of the males did between pair bonds 6–10. One hundred and six males (49%) paired with females only, 32 males (15%) with males only and 80 (37%) paired sequentially or simultaneously (trios) with females and males. The number of male–male pairs in the flock was positively correlated with the male bias in the flock sex ratio. However, even at a balanced sex ratio, male–male pairs occur. Male-paired males typically show dominance rank positions similar to males paired with females. Lifetime reproductive success of males paired both with females and males was not significantly lower than that of males paired exclusively with females. Copulations between males were observed but at lower frequencies than those observed between males and females. The social backgrounds and the events leading to male–male pairing in geese are complex and extremely varied. For a majority of cases, the most likely functional explanation is alliance formation. Males pair with one another if no female is available, because unpaired males are at the bottom of the flock hierarchy. Hence, we refer to male–male pair bondings for alliance formation as 'homosociality'. However, 15% of males only engage in homosexual pair bonding, suggesting that sexual orientation may be an issue. Stable, long-term pair bonds may form between hand-raised brothers. Because of the near-natural conditions of the Grünau flock, we are confident that our results are representative of greylag geese in general. Certainly, male–male pair bonding is well within the species' behavioural repertoire.

Acknowledgements

We wish to thank the Verein der Förderer der Konrad Lorenz Forschungsstelle and the Herzog von Cumberland Stiftung; notably Ernst August; Prinz von Hannover for permanent and continuous support. We are also grateful to I. Scheiber for critically reading our manuscript.

References

Allen, A. A. (1934) Sex rhythm in the ruffed grouse (*Bonasa umbellus* Linn.) and other birds. *Auk*, **51**, 180–99.

Bagemihl, B. (1999) *Biological Exuberance: Animal Homosexuality and Natural Diversity*. New York: St. Martin's.

Black, J. (1996) *Partnership in Birds: The Study of Monogamy*. Oxford: Oxford University Press.

Black, J. and Owen, M. (1995) Reproductive performance and assortative pairing in relation to age in barnacle geese. *J. Anim. Ecol.*, **64**, 234–44.

Choudhury, S. and Black, J. M. (1993) Mate selection behaviour and sampling strategies in geese. *Anim. Behav.*, 46, 747–55.

Cooke, F., Rockwell, R. F. and Lank, D. B. (1995) *The Snow Geese of La Pérouse Bay*. Oxford: Oxford University Press.

Curio, E., Ernst, U. and Vieth, W. (1978) The adaptive significance of avian mobbing: II. Cultural transmission of enemy recognition in blackbirds: effectiveness and some constraints. *Z. Tierpsychol.*, **48**, 184–202.

Dilger, W. C. (1960) The comparative ethology of the African parrot genus Agapornis. *Z. Tierpsychol.*, **17**, 649–85.

Fischer, H. (1965) Das Triumphgeschrei der Graugans (*Anser anser*). *Z. Tierpsychol.*, **22**, 247–304.

Fischer-Mamblona, H. (2000) On the evolution of attachment-disordered behaviour. *Attachment & Hum. Devel.*, **2**, 8–21.

Frigerio, D., Weiβ, B. and Kotrschal, K. (2001) Spatial proximity among adult siblings in Greylag geese (*Anser anser*): evidence for female bonding? *Acta Ethol.*, **3**, 121–25.

Frigerio, D., Weiβ, B., Dittami, J. and Kotrschal, K. (2003) Social allies modulate corticosterone excretion and increase success in agonistic interactions in juvenile hand-raised greylag geese (*Anser anser*). *Can. J. Zool.*, **81**, 1746–54.

Fritz, J. and Kotrschal, K. (2000) On avian imitation: cognitive and ethological perspectives. In *Imitation in Animals and Artefacts*, eds. C. L. Nehaniv and K. Dauterhahn. Boston: MIT Press.

Fritz, J., Bisenberger, A. and Kotrschal, K. (2000) Stimulus enhancement in greylag geese: socially mediated learning of an operant task. *Anim. Behav.*, **59**, 1119–25.

Hamilton, W. D. (1964) The genetical evolution of social behaviour. *J. Theoret. Biol.*, **7**, 1–52.

Heinroth, O. (1911) Beiträge zur Biologie, namentlich zur Ethologie und Psychologie der Anatiden. *Verh. V. Int. Ornithol. Kongr. Berlin* 1910, 589–702.

Hemetsberger, J. (2001) The demographic development of Konrad Lorenz' flock of greylag geese in Grünau/Austria since 1973. In *Konrad Lorenz und seine verhaltensbiologischen Konzepte aus heutiger Sicht*, eds. K. Kotrschal, G. Müller and H. Winkler, pp. 249–60. Fürth: Filander.

Hemetsberger, J. (2002) Populationsbiologische Aspekte der Grünauer Graugansschar (*Anser anser*). Ph.D. thesis, University of Vienna.

Hess, E. H. (1973) *Imprinting Early Experience and the Psychobiology of Attachment*. New York: Van Nostrand Reinhold.

Hirschenhauser, K., Möstl, E. and Kotrschal, K. (1999) Seasonal patterns of sex steroids determined from feces in different social categories of Greylag geese (*Anser anser*). *Gen. Comp. Endocrinol.*, **114**, 67–79.

Hirschenhauser, K., Möstl, E., Wallner, B., Dittami, J. & Kotrschal, K. (2000) Endocrine and behavioural responses of male Greylag geese (*Anser anser*) to pair bond challenges during the reproductive season. *Ethology*, **106**, 63–77.

Huber, R. and Martys, M. (1993) Male–male pairs in greylag geese (*Anser anser*). *J. Ornithol.*, **134**, 155–64.

Hunt, G. L. and Hunt, W. M. (1977) Female–female pairings in Wester gulls in Southern California. *Science*, **196**, 1466–7.

Kalas, S. (1977) Ontogenie und Funktion der Rangordung innerhalb einer Geschwisterschar von Graugänsen (*Anser anser* L.) *Z. Tierpsychol.*, **45**, 174–98.

Kear, J. (1990) *Man and Wildfowl*. London: T. & A. D. Poyser.

Kotrschal, K. (1995) *Im Egoismus vereint?* Munich: Piper.

Kotrschal, K., Hemetsberger, J. and Dittami, J. (1992) Vigilance in a flock of semi-tame greylag geese (*Anser anser*) in response to approaching eagles (*Haliaeetus albicilla* and *Aquila chrysaetos*). *Wildfowl*, **43**, 215–19.

Kotrschal, K., Hemetsberger, J. and Dittami, J. (1993) Food exploitation by a winter flock of greylag geese: behavioral dynamics, strategies and social implications. *Behav. Ecol. Sociobiol.*, **33**, 289–95.

Kotrschal, K., Hirschenhauser, K. and Möstl, E. (1998) The relationship between social stress and dominance is seasonal in Greylag geese. *Anim. Behav.*, **55**, 171–6.

Krause, J. and Ruxton, G. D. (2002) *Living in Groups*. Oxford: Oxford University Press.

Lamprecht, J. (1986a) Structure und causation of the dominance hierarchy in a flock of barheaded geese (*Anser indicus*). *Behaviour*, **96**, 28–48.

Lamprecht, J. (1986b) Social dominance and reproductive success in a goose flock (*Anser indicus*). *Behaviour*, **97**, 50–65.

Lamprecht, J. (1987) Female reproductive strategies in barheaded geese (*Anser indicus*): why are geese monogamous? *Behav. Ecol. Sociobiol.*, **21**, 297–305.

Lamprecht, J. (1991) Factors influencing leadership: a study of goose families (*Anser indicus*). *Ethology*, **89**, 265–74.

Lazarus, J. (1978) Vigilance, flock size and domain of danger size in the white-fronted goose. *Wildfowl*, **29**, 135–45.

Lazarus, J. and Inglis, I. R. (1978) The breeding behaviour of the pink-footed goose: parental care and vigilant behaviour during the fledging period. *Behaviour*, **65**, 62–88.

Loonen, M. J. J. E. (1997). Goose breeding ecology: overcoming successive hurdles to raise goslings. Ph.D. thesis, University of Groningen.

Lorenz, K. (1932) Betrachtungen über das Erkennen der arteigenen Triebhandlungen der Vögel. *J. Ornithol.*, **80**, 50–98.

Lorenz, K. (1935) Der Kumpan in der Umwelt des Vogels. *J. Ornithol.*, **83**, 137–213.

Lorenz, K. (1940) Die Paarbildung beim Kolkraben. *Z. Tierpsychol.*, **3**, 278–92.

Lorenz, K. (1941) Vergleichende Bewegungsstudien an Anatinen. Suppl. *J. Ornithol.*, **89**, 194–294.

Lorenz, K. (1942) Die angeborenen Formen möglicher Erfahrung. *Z. Tierpsychol.*, **5**, 16–409.

Lorenz, K. (1991) *Here I am – Where Are You? The Behaviour of the Greylag Goose*. London: Harper Collins.

Lorenz, K. and Tinbergen, N. (1939) Taxis und Instinkthandlung in der Eirollbewegung der Graugans 1. *Z. Tierpsychol.*, **2**, 1–29.

Lott, D. F. (1991) *Intraspecific Variation in the Social Systems of Wild Vertebrates*. Cambridge: Cambridge University Press.

Madsen, J., Cracknell, G. and Fox, T. (eds.) (1999) *Goose Populations of the Western Palearctic*. Denmark: Wetlands International. National Environmental Research Institute.

Meyer-Holzapfel, M. (1961) Homosexualität bei Tieren. *Praxis*, **50**, 1266–72.

Morris, D. (1952) Homosexuality in the ten-spined stickleback (*Pygosteus pungitius* L.). *Behaviour*, **4**, 233–61.

Panksepp, J. (1998) *Affective Neuroscience. The Foundations of Human and Animal Emotions*. Oxford, New York: Oxford University Press.

Pfeffer, K., Fritz, J. and Kotrschal, K. (2002) Hormonal correlates of being an innovative greylag goose. *Anim. Behav.*, **63**, 687–95.

Radesäter, T. (1974) On the ontogeny of orienting movements in the triumph ceremony of two species of geese, *Anser anser* L. *and Branta canadensis* L. *Behaviour*, **50**, 1–15.

Rutschke, E. (1982) Stability and dynamics in the social structure of the greylag goose (*Anser anser*) L. *Aquila*, **89**, 39–55.

Rutschke, E. (1987) *Die Wildgänse Europas. Biologie. Okologie. Verhalten*. Wiesbaden: Aula.

Rutschke, E. (1997) *Wildgänse: Lebensweise, Schutz, Nutzung*. Berlin: Parey.

Sanders, B. (2002) Verhaltensentwicklung bei Graugänsen. Masters thesis, Julius Maximilians Universität Würzburg & Konrad Lorenz Forschungsstelle Grünau.

Sauer, E. G. F. (1972) Aberrant sexual behaviour in the South African ostrich. *Auk*, **89**, 717–37.

Schutz, F. (1965a) Sexuelle Prägung bei Anatiden. *Z. Tierpsychol.*, **22**, 50–103.

Schutz, F. (1965b) Homosexualität und Prägung. *Psychol. Forsch.*, **28**, 439–63.

Schutz, F. (1966) Homosexualität bei Tieren. *Studium Generale*, **19**, 273–85.

Schutz, F. (1968) Sexuelle Prägungserscheinungen bei Tieren. In *Die Sexualität des Menschen. Handbuch der medizinischen Sexualforschung*, ed. H. Giese, pp. 284–317. Stuttgart: Ferdinand Enke.

Tinbergen, N. (1963) On aims and methods of ethology. *Z. Tierpsychol.*, **20**, 410–33.

Tooby, J. and Cosmides, L. (1990) The past explains the present: emotional adaptations and the structure of ancestral environments. *Ethol. Sociobiol.*, **11**, 375–424.

Waldenberger, F. and Kotrschal, K. (1993) Individual vigilance in male greylag geese (*Anser anser*) depends on flock density and social rank. *Ecol. Birds*, **15**, 193–9.

Waldenberger, F. (1994). Social contexts of vigilance behaviour in greylag geese (*Anser anser*). Diploma thesis, University of Salzburg.

Weiß, B. M. (2000). Social support in juvenile greylag geese (*Anser anser*). Diplom thesis, University of Vienna.

Weiß, B. M. and Kotrschal, K. (2004) Effects of passive social support in juvenile greylag geese (*Anser anser*): a study from fledging to adulthood. *Can. J. Zool.*

Wilson, E. O. (1975) *Sociobiology: The New Synthesis*. Harvard, MA: Belknap.

3

Pink flamingos: atypical partnerships and sexual activity in colonially breeding birds

CATHERINE E. KING

Introduction

Flamingos are highly gregarious wading birds. They nest in colonies that may number thousands of individuals, which form pair bonds during the breeding season. Such bonds do not only develop between males and females, but may include same-sex pairs and trios. The current paper will detail the occurance of such 'atypical' relationships and investigate their potential causes.

Reports about same-sex sexual behaviour of wild flamingos are lacking. Homosexual behaviour is not even described for the well-studied colony of greater flamingos in the Camargue, France. However, this lack might be due to constraints imposed by field studies because it is difficult to detail behaviour and to confirm the sex of individual birds from a distance. More information may soon become available because chicks caught for ringing in the Camargue and at Fuente de Piedre, Spain have been sexed and will soon reach breeding age (A. Johnson, personal communication (pers. comm.)).

Nevertheless, the formation of same-sex pairs and trios is well known from captivity (for example Studer-Thiersch, 1967, 1975; Wilkinson, 1989; Blythe, 1994; King, 1994; Peters, 1996; Reinertsen, 1999; Shannon, 2000), although descriptions of the sexual behaviour in these relationships are all but absent.

Flamingos share characteristics with groups such as Anseriformes (especially geese), Cathartidae (New World vultures), Pelecanidae (pelicans) and families within the Ciconiiformes, including Ardeidae (herons), Ciconiidae (storks) and Threskiornithidae (ibises; del Hoyo et al., 1992).

These flamingo relatives do also form same-sex pairs and trios, which may interact sexually (review in Bagemihl, 1999). Among herons, male–male mounts constitute a sizeable proportion of extra-pair copulations, that is 5–6% in cattle egrets, 3–5% in little blue herons, 5–6% in little egrets and 8% in gray herons. Most of these mounts were by males who had a female partner, and pair bonding between same-sex individuals was not observed. However, same-sex mounting as well as pair bonding occurs in both sexes of semi-wild and captive black-crowned night herons. Same-sex pair bonding, despite availability of individuals of the opposite sex, has also been observed among free-ranging white storks. Such pairs have successfully incubated and reared foster young (K. Vos and M. Bloesch, pers. comms.). A female–female pair of scarlet ibis at the Bronx Zoo (USA) also produced fertile eggs (Elbin and Lyles, 1994). Male–male pairings of greylag geese in semi-captivity may include extended ritualized courtship and pair behaviours. Pairings are relatively fluid. After the death of a partner, males of same-sex pairs may pair with a female, or males in female–male pairs may pair with a male. Trios can be formed by a two-male pair with a female, or by two females with one male (Huber and Martys, 1993).

Flamingos also demonstrate versatility in the formation of partnerships. The causes for combinations other than male–female pairs can be best understood if viewed within the context of a colony as a whole. I will therefore relate descriptions of atypical relationships to colony characteristics, including size, sex-ratio and age structure. Moreover, I will compare the length of pair bonds, the number of partnerships and copulation activities in same-sex pairs and/or trios with individuals that engaged exclusively in male–female pairs. These data – largely unpublished and supplemented by personal communications from numerous institutions that keep flamingo colonies – will be used to judge functional explanations of same-sex behaviour, such as the practice and alliance hypothesis.

Methods

Flamingo taxonomy and socio-ecology

Flamingo taxa are closely related to each other but only distantly related to other birds (del Hoyo et al., 1992). Therefore, most taxonomists award them their own order, Phoenicopteriformes. The three Phoenicopterus taxa, the Chilean flamingo (Phoenicopterus chilensis) of South America, the Caribbean flamingo (Phoenicopterus ruber ruber) of the Carribean region and the Greater flamingo (Phoenicopterus ruber roseus) of Africa, Asia and Europe share many similarities.

Males are larger than females in all three taxa (Richter and Bourne, 1990; Richter et al., 1991). Annual survival rate of the greater flamingo is 94% (Cezilly and Johnson, 1995). Despite the harsh saline environments that most wild flamingos live in, they can be quite long-lived; a wild male greater flamingo died at the age of 40 years (Johnson, 1997). Captive flamingos have survived and even successfully bred for more than 60 years (A. Studer-Thiersch, pers. comm.).

Figure 3.1. Nest mounds in a breeding colony of Greater flamingos and Carribean flamingos at the Rotterdam Zoo (photo: Peter Huijskes).

All flamingos use their bills to filter-feed in shallow waters, primarily consuming small invertebrates, but possibly also plant seeds and fish (Jenkin, 1957; Zweers *et al.*, 1995). The birds feed in flocks where aggressive encounters may occur (for example Bildstein *et al.*, 1991; Schmitz and Baldasarre, 1992).

Flamingos are considered nomadic rather than migratory, and movements are thought to be mostly related to food availability. While flamingos are almost exclusively found in groups in all stages of the life cycle, movements of individuals appear to be independent, that is flamingos do not remain in a given group in the wild (del Hoyo *et al.*, 1992; A. Johnson, pers. comm.).

Flamingos often nest in huge colonies numbering thousands of individuals, but sometimes breed in smaller colonies numbering hundreds or just tens of individuals, for example in the Galapagos, Turkey and Italy (Kirwan, 1992; Bacetti *et al.*, 1994; Albanese *et al.*, 1997). Nesting is typically associated with the formation of heterosexual pairs. Both males and females build nests and alternate in incubation and chick rearing. Nest mounds built of mud and other nearby materials (for example pebbles, feathers) must be constantly defended against conspecifics (Figure 3.1). Squabbles over nests and between flamingos occupying

Figure 3.2. A flamingo chick is being fed 'crop milk' by an adult bird (photo: Peter Huijskes).

adjacent nests are particularly common at the beginning of the breeding season. These conflicts can result directly in egg breakage or increased vulnerability of the eggs to predation.

Older birds with more experience have a greater chance of accessing favourable nest sites. Greater flamingos nesting in the dense Camargue colony are often not successful until seven years of age (Cezilly *et al.*, 1997; A. Johnson, pers. comm.). Newly formed colonies of greater flamingos in the Mediterranean have a younger age structure at first breeding than in the Camargue, and flamingos in captivity may breed successfully as early as two years of age (pers. obs.). Flamingos nesting in the Camargue were found to be mostly seasonally monogamous. However, if a first nesting attempt is not successful in a given season, the pair members will generally re-pair with others if they make a second nesting attempt (Cezilly and Johnson, 1995). Greater flamingos also have an age-assortative mating pattern, as partners tend to be from the same age cohort (Cezilly *et al.*, 1997).

Generally only one egg is laid, and both parents feed the young a 'crop milk' (Figure 3.2) secreted from glands found in the upper digestive tract (Studer-Thiersch, 1975). The chick is initially guarded and brooded by a parent on the nest, and at approximately one to two weeks of age joins a 'creche' of chicks. The creche is normally supervised and led by a few adults, who may possibly be failed breeders. Parents and young appear to recognize each other primarily by vocalizations, and this recognition seems to be already established when the chick hatches (de Hoyo *et al.*, 1994; Mathevon, 1996, 1997).

Flamingo courtship displays and breeding are performed in groups consisting of both sexes. The most famous display is the frequently filmed 'marching' in which thousands of flamingos may move together in a 'synchronized quick-step, going first in one direction and then in the other' (Kahl, 1975). The ritualized displays of 'head-flagging', 'wing-salute', 'inverted wing salute', 'twist-preen' and 'wing-leg stretch' all seem to be derived from comfort movements (Kahl, 1975). Such displays are performed in sequences, generally among smaller subgroups within larger congregations and may function to synchronize potential breeders and aid pair formation (Studer-Thiersch, 2000). In temperate regions where breeding is seasonal, frequency of ritualized displays increases greatly as the nesting period approaches. Periods of high-intensity display may again occur months after nesting activities have ceased (Kahl, 1975). Under the favourable conditions of the tropics, both displays and breeding may occur year round.

Copulation behaviours generally begin weeks and even months before nesting starts, both in the wild and in captivity. Frequency is initially low, and many approaches or attempts break off. Copulation intensity increases as egg laying approaches (Brown, 1958; Studer-Thiersch, 2000). Copulations occur any time in the 24-hour daily cycle, but levels are clearly increased around sunrise and sunset (A. Studer-Thiersch, pers. comm.). Copulation rates are highly individual. For example, the number of copulations varied from 1 to 49 among 16 partnerships (ten pairs, two trios) during 520 hours of observation at Rotterdam Zoo (King, 1994).

Nothing is known about extra-pair copulations in the wild, but rates in captivity vary across individuals and with respect to numbers of extra-pair partners (for example King, 1994; Peters, 1996: de Bruin and de Haan, 2001). The pair partner of the copulating bird tries to interrupt an extra-pair copulation, but this becomes more difficult once the partner has to remain on a nest to defend it.

Other behaviours that could be termed 'sexual' are absent among flamingos. They do not engage in tactile social behaviours, for example courtship feeding or allo-preening, seen among many other avian groups.

Terminology

Descriptions of flamingo behaviour provided in the current paper rely on explicit definitions of central terms.

Partnership: Flamingos that sleep side by side, simultaneously carry out the same activities, call together and form coalitions during aggressive encounters (Studer-Thiersch, 1975).

Pair: A partnership of two flamingos. Pairs can be male–male, female–female or male–female. This definition does not encompass sexual activity, as pairs can be clearly identifiable in colonies that are not reproductively active.

Trio: A partnership of three birds (male–female–female, male–male–female, male–male–male, female–female–female). Male–female–female and male–male–female trios are defined as 'heterosexual' if the members only engage in sexual activity with a member of the opposite sex and 'bisexual' if two (or three) same-sexed members engage in sexual activities together and also with the member of the other sex (Bagemihl, 1999). Trios can be 'associative' when all three birds share the same nest and 'disassociative' when two nests are maintained next to each other (Shannon, 2000).

Quartet: A partnership of four birds (any combination of the two sexes).

Active pair (breeding season): A pair that manages to hold a nest for at least a period of three days.

Inactive pair (breeding season): A pair that nest builds but are not successful in holding a nest for even a period of three days.

Breeding: Holding a nest, producing and incubating an egg. The egg does not necessarily have to hatch for the attempt to be considered a breeding event.

Copulation approach (typical): The male and female walk, usually to the edge of a flock, with the male behind the female. The male maintains a short distance between himself and the female, and looks from side to side, possibly watching for males that may potentially disrupt copulation, as this regularly occurs. This stage is sometimes referred to as 'driving', despite the fact that the female may actually be leading. Copulation activities may proceed no further than the approach, particularly at the beginning of the breeding season.

Copulation attempt (typical): If the female is receptive for copulation she lowers her head towards the ground or to the water, adapting a 'false-feeding posture' (Studer-Thiersch, 1975). If the male wants to proceed with the copulation he stretches his neck diagonally to touch her back with his bill or breast, stimulating her to stand still and open her wings slightly away from the body. The male then jumps on the female's back, lowering his tarsometatarsi in between the female's wings and body, with his toes resting on her 'shoulders'. If for some reason a copulation effort breaks off before cloacal contact is made, it is referred to as a copulation attempt.

Copulation (typical): The male dips the distal body region down, and the female tips the distal part of the body up, to make cloacal contact. Semen often drips from the female's cloaca when cloacal contact is broken. The male then jumps off the female, usually over her head. Often directly after the male's feet are again on a solid substrate the male, and sometimes also the female, fleetingly perform 'hooking', a posture in which the lower neck is held in a rigid line, with the upper neck and head forming a hook. Hooking seems to be primarily a threat display (Kahl, 1975).

Copulation (atypical): Because copulation is defined as cloacal contact among birds such as flamingos that do not have a sexual organ (for example, a penis), copulations may occur between same-sex birds. A male may assume the female role and a female may assume the male role in copulation activities.

Within-pair copulation: A copulation involving two individuals in the same pair or other partnership.

Extra-pair copulation: A copulation involving an individual that is in a pair (or trio or quartet) with another bird not in the same partnership.

Lone-nester: A flamingo that builds a nest and incubates an egg alone.

Populations: Animal numbers are denoted as 'males.females.unknown sex'. For example, six of each would be '6.6.6', whereas six males would be denoted as '6.0.0'.

Identifiers: Code on individual leg bands.

Data collection

Survey of institutions that keep flamingo colonies. To get some idea about how prevalent atypical partnerships are among flamingo colonies in captivity, I put a small survey on two zoo bird-oriented listservers, asking for records of reproductive relationships of individually identifiable flamingos of known sex and whether same-sex pairs and trios were observed. No effort was made to discover the frequency of specific partnerships within one colony or how long they endured.

Basle Zoo, Switzerland. A breeding population of greater flamingos consisted of 47.51.0 flamingos on 1 May 1996, and changed little throughout the study. Video recordings of individually identifiable birds were made to study the mechanics of copulation behaviour and to compare full-flighted and flight-restricted birds. Flight was restricted through pinioning (surgical removal of the distal wing portion) and tenonectomy (cutting and cauterization of the main wing tendon). Copulation activities were recorded three to four times per year for five to six consecutive days during April and May in 1996, 1997 and 1998, for a total of 58 days. Recordings focused on the beginning of the breeding season, prior to most of the colony incubating eggs. Observations were spread within a season to capture copulation activity of more individuals, as different individuals had different peaks. The 'all occurrence' sampling method (Lehner, 1979) was used.

Rotterdam Zoo, The Netherlands (Table 3.1). A colony of *Phoenicopterus ruber* grew steadily from 21.19.2 in 1992 to 52.46.0 in 2002. In 1992 the colony consisted of all wild-caught birds with the exception of 5.0.1 flamingos hatched at the zoo between 1989 and 1991. The effective breeding size of the population (birds two years or older) varied between 19.18.0 in 1992 and 41.39.0 in 2002. Ninety-three flamingos were hatched at Rotterdam, 18 wild-caught specimens were imported, 36 died, and 14 flamingos were exported to other zoos. Some manipulation of eggs has occurred since 1993. As the colony consists of both subspecies of *P. ruber* (*P. r. ruber* and *P. r. roseus*), eggs produced by pairs consisting of a member of each subspecies are discarded. Some males are not successful in copulation attempts, perhaps because of pinioning of one wing, and their partners produce infertile eggs unless the egg is fertilized by an extra-pair copulation. Infertile eggs of such pairs have often been switched for a fertile egg so that all pairs have the opportunity to rear young. The birds have been fitted with plastic coloured rings engraved with individual codes since 1991. The colony has been observed daily during egg laying and chick hatching since 1992, for a minimum of 200 hours and a maximum 520 hours over a 124 day period. Records on dominance, parental and other social behaviours cover 1994–2002 (or 2003, in some instances). They are collected by volunteers that have been with the project since 1993. Dominance behaviours are sampled year-round using the 'focal-animal' method (Lehner, 1979). Nest occupation is recorded during the breeding season using a scan sampling (Lehner, 1979) with the nest occupant recorded every ten minutes. Courtship (including copulation), feeding of chicks and aggressive behaviours are recorded *ad libitum* between scan and focal samples.

Durrell Wildlife Conservation Trust, Jersey, UK (Table 3.1). Records on *Phoenicopterus chilensis* breeding in trios and same-sex pairs, as well as information on colony size, cover the period 1990–2002 (G. Young, pers. comm.). Multiple flamingos were moved between the breeding enclosure and another enclosure several times. The colony did not breed in 1991 and 2001, as well as during regroupings in 1997 and 1998 (to form an even sex-ratio of 19.19.0 birds) and 1999 (resulting in 17.24.0 birds).

Amersfoort Zoo, The Netherlands (Table 3.1). The colony of Chilean flamingos was very stable, mostly consisting of 37.52.0 individuals from 1995 to 2001. One male died and one male hatched in 1996. Copulation activity and partnerships formed were recorded in 1996 (Peters, 1996) and 2001 (de Bruin and de Haan, 2001), and partnerships formed were also identified in 1995 (Kerkman, 1995) and 1997 (Bolhuis and Lamers, 1997). A number of rings were lost between 1997 and 2001, making it impossible to identify some individuals and their sex during the 2001 observations.

Table 3.1 *Effective breeding populations and atypical partnerships in three flamingo colonies: Rotterdam Zoo, Greater flamingo and Carribean flamingo, 1992–2002; Durrell Wildlife Conservation Trust, Chilean flamingos, 1990–2002 (raw data G. Young, pers. comm.); Amersfoort Zoo, Chilean flamingo, 1995–1997 (data from Kerkman, 1995; Peters, 1996; Bolhuis and Lamers, 1997)*

Location	Year	Census (a)	Extra (b)		Same-sex pairs		Trios			Atypical relationships (c)		
			m	f	m-m	f-f	f-f-f	m-f-f	m-m-f	m	f	%
Rotterdam	1992	19.18.0	1	–	1	–	–	2	1	6	5	29.7
	1993	23.17.0	6	–	2	–	–	2	1	8	5	32.5
	1994	25.20.0	5	–	1	–	–	1	–	3	2	11.1
	1995	20.20.0	–	–	–	1	–	–	1	2	3	12.5
	1996	20.20.0	–	–	–	1	–	1	–	1	4	12.5
	1997	23.23.0	–	–	–	–	–	2	1	4	5	19.6
	1998	25.27.0	–	2	–	–	–	2	–	2	4	11.5
	1999	29.29.0	–	–	–	–	–	2	–	2	4	10.3
	2000	30.31.0	–	1	–	–	–	2	–	2	4	9.8
	2001	37.35.0	2	–	–	–	–	2	–	2	4	8.3
	2002	41.39.0	2	–	1	–	–	2	–	4	4	10.0
Durrell	1990	10.15.0	–	5	–	1	1	1	–	1	7	32.0
	1991	9.15.0	–	6	–	–	–	–	–	–	–	–
	1992	9.14.0	–	5	–	1	1	1	–	1	7	34.8
	1993	14.21.0	–	7	–	2	–	–	–	–	4	11.4
	1994	15.21.0	–	6	–	2	1	2	–	2	11	36.1
	1995	18.27.0	–	9	–	4	–	2	–	2	12	31.1
	1996	20.25.0	–	5	–	1	–	2	–	2	6	17.8
	1997	19.19.0	–	–	–	–	–	–	–	–	–	–
	1998	19.19.0	–	–	–	–	–	–	–	–	–	–
	1999	17.24.0	–	7	–	–	–	–	–	–	–	–
	2000	17.25.0	–	8	–	2	1	1	–	–	9	21.4
	2001	17.24.0	–	7	–	–	–	–	–	–	–	–
	2002	20.23.0	–	3	–	6	–	–	–	–	12	27.9
Amersfoort	1995	37.52.0	–	15	4	7	–	1	–	9	16	28.1
	1996	37.52.0	–	15	3	6	–	1	–	7	14	23.6
	1997	37.52.0	–	15	2	4	–	3	–	5	10	17.1

Notes: (a) m.f.unknown sex.

(b) Extra = individuals exceeding an even sex ratio.

(c) m / f / % = number of males / number of females / % individuals in the colony involved in atypical relationships.

Table 3.2 *Number of flamingo colonies surveyed (n = 25) with reports of atypical relationships. Data from: Birmingham Zoo (USA), Bristol Zoo (UK), Cincinnati Zoo (USA), Dallas Zoo (USA), Durrell (UK), Honolulu Zoo (USA), Lowry Park Zoo (USA), Minnesota (USA), National Zoo (USA), North Carolina Zoo (USA), Paignton Zoo (UK), Poland Zoo), Riverbanks Zoo (USA), Rotterdam Zoo (The Netherlands), San Diego Wild Animal Park (USA), Tautphaus Park (USA), Vogelpark Walsrode (Germany), Zoo Planckendael (The Netherlands)*

	None	Same-sex pairs			Trios				
		m-m	f-f	m-m, f-f	f-f-f	m-f-f	m-m-f	m-f-f, m-m-f	m-m-m
Total colonies	5	5	5	9	1	3	1	3	1
Sex-ratio f-biased		4	4	2					1
Sex-ratio m-biased				2	1	3		3	
Sex-ratio bias changing or 1:1				4					
No sex ratio data		1	1	1			1		
m-f before (a)		X	X						
m-f after (a)		X	X						
m-f before or after (a)		X	X		X	X		X	

Note: (a) X = individuals in a given category of atypical relationships were in a male–female relationship before / after / before or after; based on data from 12 colonies only.

Statistics

Tests (Wilcoxon, binomial; Siegel and Castellan, 1988) were one-tailed and significance was accepted at $p < 0.05$.

Results

Atypical relationships at 23 flamingo-keeping institutions

Information on relationships of individually identified flamingos of known sex was provided for 25 colonies in 23 zoos. The number of flamingos per colony ranged from 16 birds to 98. Atypical relationships are detailed in Table 3.2. Neither same-sex pairs nor trios had been observed in 5/25 colonies. However, records covered only three years or less in three of the zoos with no such records, and the fourth (with a colony of 16 flamingos) had only sporadic breeding.

Twelve zoos with trios or same-sex pairs were able to indicate whether flamingos in atypical relationships had bred in male–female pairs either before or after being in a trio or same-sex pair. In only two colonies had the birds definitely not bred in a male–female pair. One of these cases was a female–female pair, and in the other case only males were held.

A quartet formed at the Dallas Zoo (USA; S. Maltsbarger, pers. comm.) and another quartet at the Audubon Zoo (USA) when one member of two pairs formed a new pair together and the other members of the previous pairs followed along (P. Shannon, pers. comm.).

The percentage of individuals engaged in atypical relationships could be detailed for three institutions (see Table 3.1) and turned out to be suprisingly similar. At Rotterdam, they averaged 15.3%, with a maximum of 32.5%. At Durrell, they averaged 26.6%, with a maximum of 36.1%, and at Amersfoort 22.9% with a maximum of 28.1%. Thus, 18.2% of all flamingos for which we have detailed data from captivity live in an atypical relationship at any given time, that is every fifth bird.

Same-sex copulation activity

At Basle Zoo, 4057 copulation activities of greater flamingos performed by 47 males and 51 females were recorded over three years. None of these activities involved two males. Copulation approaches and attempts were observed among four female–female pairs, and they proceeded in the same way as a male–female copulation. Roles of mounter and mountee did not change in any of the four pairs. One female ('Gr5') attempted to copulate with a male ('Rr3') on three occasions, but her female partner ('gg') always intervened. The female partner was once also observed in a copulation approach with the same male. Copulation behaviour of a male–female–female–female quartet was also recorded. The male typically made a copulation approach with one female, and, if she was not receptive, he then approached another one and then the third. 'Reverse' copulation attempts were observed on several occasions in consecutive years between one male–female pair. The copulation approach proceeded normally, but with the female in the male role and the male in the female role. The male assumed the female copulation posture but the pinioned female never attempted to jump on his back. This pair also copulated normally.

At Rotterdam Zoo, observations of male–male copulation activities are restricted to two years. In 1993, 326 copulation approaches and attempts were recorded. Of these, 302 were among 18 partnerships and 24 were extra-pair activities. One copulation approach involved two males which were both in male–female partnerships, and seven others (2.1%) involved a male–male pair ('113' and '116'). Male '116' always assumed the female role, and neither bird engaged in other copulation activities. In 1994, 2413 copulation approaches and attempts were observed among 17 pairs, two trios, three birds nesting alone as well as (other) extra-partnership copulation activities. Ten of these (0.4%) were copulation approaches involving male '116' and male '113', but this time with male '113' always in the female role. Male '116' also engaged in copulation approaches with females (six with female '114', four with female '152', 30 with female '150') as well 27 copulation attempts (with female '150'). Male '113' was not observed in any other copulation activities.

Female–female copulation activity at Rotterdam Zoo was restricted to a single female–female pair ('033' and 'CBZ'). Of 1458 copulation activities recorded in 1995, 59 (4.0%) involved this female–female pair, including 13 copulation approaches, and 46 copulation attempts and successful copulations. Six copulation approaches and 12 copulation attempts and successful copulations were recorded in 1996, constituting 2.2% of all 814 observations. Of 700 copulation activities recorded in 1997, 18 (2.6%) involved this female–female pair, and female '033' was always the mountee. Female '033' also copulated with males (male '158': twice in 1995, seven times in 1997, 21 times in 1997; male '022': twice in 1997). Female 'CBZ' was observed in one copulation approach in 1997 with male '026' who formed a trio with these females in 1998.

Extra-pair copulation associations at Rotterdam Zoo tended to be long term (over several breeding seasons) and/or with previous or future breeding partners (examples, see below). In male–male relationships, no extra-pair copulation activity with another male was observed. A forced copulation took place in 2003 when male 'EAM' (hatched at Rotterdam Zoo in 1996) tried to take possession of a nest occupied by male '109' (unknown age, arrived at Rotterdam Zoo in 1988) and his female partner '155' (unknown age, arrived at Rotterdam Zoo in 1984). 'EAM' lost the battle, and then jumped on the back of '109', who turned his head to try to pick at the mounter. '109' stood still, apparently to keep his balance, and 'EAM' was able to achieve cloaca contact. This occurred twice in succession.

At Amersfoort Zoo, copulation attempts or approaches were observed among all three male–male pairs and among 3/6 female–female pairs in 1996. Copulation activity with at least one female was recorded for 5/6 males in male–male pairs. No extra-pair male–male copulation activity was observed. Copulation activity with at least one male was observed among 8/12 females in female–female pairs. No extra-pair female–female copulation activity was recorded. Similarly, all 74 extra-pair copulation activities recorded in 2001 involved a male and female together.

Longevity of male–female and atypical partnerships

The Rotterdam Zoo colony contained 26 adult flamingos, which were present during all ten breeding seasons from 1993 to 2002. All of these had partnerships at least some years, but 14/26 were without a partner for at least one nesting season. One female '150' bred in seven different partnerships, including one trio. The mean number of different partnerships was 3.1 ± 1.4 (range 1–7). Two males ('116' and '102') were each in one same-sex partnership (with males not present all ten years). Male '116' was involved in three male–female partnerships and was active in 9/10 years. Male '102' was also involved in three male–female partnerships and active in 8/10 years. Ten flamingos were involved in trios, and one female participated in three different trios.

The mean number of partners for the 12 flamingos in at least one atypical relationship was 3.1 ± 1.4 (range 2–7), whereas the 14 flamingos not in any atypical relationships had 2.8 ± 1.3 partners (range 1–6). This difference was not significant ($Z = 1.26$; ns, Wilcoxon test). The mean number of breeding seasons that the 12 flamingos in at least one atypical relationship were without a partner was 0.8 ± 0.8 (range 0–2), whereas the 14 flamingos not ever in any atypical relationship were without a partner for 1.2 ± 1.5 (range 0–4) breeding seasons. Again, this difference was not significant ($Z = 0.28$; ns, Wilcoxon test).

At Durrell Wildlife Conservation Trust individuals that were in at least one atypical relationship had an average of 4.4 ± 1.4 partnership changes over eight years. This was significantly more common than the 3.1 ± 1.4 changes recorded at Rotterdam over ten years ($Z = 1.68\,p < 0.0465$, Wilcoxon test).

Age-assortive pair selection

Flamingos are reproductively mature when they are at least two to four years old. Flamingos at Rotterdam Zoo would attempt their first breeding significantly more often with birds in the same or next age cohort than with flamingos in cohorts more than one year different ($Z = -2.92$, $p < 0.0018$, binomial test, Table 3.3), even though the number of potential partners in cohorts which are two or more years different in age is far larger.

Examples from the Rotterdam Zoo illustrate this pattern. Of the ten flamingos involved in same-sex pairs between 1992 and 2003, two males ('113', '116') arrived together at Rotterdam and were a pair for their first breeding season. Three flamingos (female '026', male 'ECT', male 'ECX') formed their first (same-sex) relationship with an individual in the next age cohort. Male '102' (hatched in 1980) and male '024' (hatched in 1982) were already a pair when observations began. As the only other flamingo (a female, hatched between 1980 and 1982) was paired to a male (unknown age, received in 1964) these were the most similarly aged birds remaining from those years. Female '033' bred first

Table 3.3 *Proportion of flamingos at Rotterdam Zoo pairing with different age cohort categories during the first breeding attempt. Based on data for 57 birds hatched between 1992 and 2000*

Age difference	n	%
Same age	17	29.8
±1 year	17	29.8
2 years or more	13	22.8
Not paired	10	17.5

with a male of unknown age, but her second partner was a female ('026') one year younger than her. Male 'CAD' bred in a male–female pair before forming a same-sex pair, although the age of both his female and male partner is unknown.

Influence of sex-ratio on frequency of atypical relationships

Colonies at zoos for which detailed data are available had consistently uneven sex ratios (see Table 3.1). The imbalance at Rotterdam varied between zero (four years) and six birds (one year). The population was skewed towards males in five years and females in two years, with a mean deviation of 1.7 ± 2.1 birds. The flock at Durrell was female-biased in all but two years, with the number of extra females ranging from 3 to 9 (mean 6.1 ± 1.6), whereas the flock at Amersfoort had 15 extra females in all years.

There was no difference in the percentages of individuals engaged in atypical relationships each year between the two female-biased institutions (ns, $p = 0.5000$, Wilcoxon test). However, the percentages of individuals engaged in atypical behaviours at Durrell was significantly greater than at Rotterdam ($Z = 2.52$, $p < 0.006$, Wilcoxon test) where the sex ratio was mostly close to even.

Six males and 13 females at Durrell, all but two of which were wild-caught flamingos of unknown age received between 1975 and 1985, were involved in atypical relationships in the eight breeding years (eight female–female pairs, seven male–female–female trios, four female–female–female trios). While each partnership was different, some of the members were the same. For example, two females were in both female–female pairs in some years and female–female–female trios in others. Two females formed a trio with one male one year and another male another year. Seven male–female pairs formed that involved females that had been in atypical partnerships and two males that had been in male–female–female trios.

Several female–female pairs at Durrell produced fertile eggs, but in many cases the eggs were predated upon by gulls or perhaps colony members. Infertile eggs were also produced. However this does not necessarily indicate that copulation was not occurring, as fertility of captive flamingos is often compromised if the flamingos are pinioned.

Examples of atypical partnerships

The fluidity of flamingo partnerships is illustrated by the following case studies from Rotterdam Zoo.

Trio A (male '107', female '140', female '150')

Male '107' (unknown age, arrived in 1960) nested with a first female '140' (unknown age, arrived in 1973). A second female, '150' (unknown age, arrived in 1989), built a nest next to theirs in 1992 and produced an egg. The second female had no partner to help her defend the nest, and the egg was quickly destroyed by male '024', a member of a male–male pair. Male '107' did not guard the nest that he shared with the first female well, and their egg was also lost. Male '107' and the second female promptly nested, and an egg was produced less than a week later. Male '107' shared incubation duties and protected this nest better, and the egg was successfully incubated full term, but was infertile. Male '107' was often sighted with the first female, when not incubating the egg in the nest shared with the second female.

Male '107' copulated with both females in 1993, but more approaches and attempts were observed with his first mate (28 with female '140', seven with female '150'). The two females built nests next to each other, and produced eggs one day apart. While male '107' would stand on both nests he only incubated the first female's egg. He again proved to be a poor nest guard, and both eggs were broken by male '024', 12 and 13 days after they were laid. Male '107' and the first female incubated three more eggs that season, all of which were broken. The second female did not hold another nest, but may have laid one or more eggs of the seven eggs that were either rolled from a nest or were laid without a nest and for which parentage could not be ascribed. Male '107' died at the end of 1993.

The second female bred with seven different males each year between 1994 and 2002. The first female also bred with four different males during those years, except in 1998 when she was loosely associated with a male–female–female trio. She also copulated with male '101' from this trio throughout the years.

> Trio B (male '105', female '152', female '153'), trio C (male '101', female '152', female '153'), trio D (male '101', female '152', female '114')

Male '105' (unknown age, arrived in 1960) bred with a first female ('152', unknown age, arrived in 1973) in 1992. This pair formed trio B with a second female ('153', unknown age, arrived in 1973) in 1993. The females shared a nest and the second female was the first to lay an egg. This egg was broken (how was not observed) five days later and the first female laid an egg the next day. This infertile egg was removed by zoo staff three days after being laid. The trio occupied a different nest site together. The second female laid a second egg ten days after removal of the first female's egg. The first female rolled this new egg out of the nest the same afternoon and proceeded to lay her own. The egg of the first female was incubated by all three adults, but also proved infertile. A fertile egg was given to the still incubating trio 58 days later and the trio reared the chick together.

Male '105' bred with another female ('154', unknown age, arrived in 1960) in a male–female pair in 1994, but both birds died at the end of that year. The two females of trio B ('152', '153') formed trio C in 1994 with a different male ('101', unknown age, arrived in 1960). Four eggs were laid in the same nest, the first of which was broken and the second and third were removed by staff after a week. All four eggs were believed to be laid by the first female '152'. The females did most of the incubation; the male mainly nest guarded and regularly engaged in copulation activities with a third female ('154', his male–female pair partner in 1992, and the mate of male '105' of trio B in 1994). He also copulated with a fourth female ('114', unknown age, arrived in 1993) who nested alone, as well as with the two females '152' and '153'. No copulation activity between the two females '152' and '153' together was observed. All trio-members reared the chick resulting from the fourth egg laid.

The second female '153' of trios B and C went on to form a pair with male '016' (hatched in Cuba 1976, arrived in 1977) in 1995. This pair bred together every year through to 2002. Male '101' copulated with the first female '152' and fourth female '114' in 1995 and the three formed trio D. They shared one nest, and although only female '114' appeared to lay in 1995, the first female helped rear the chick. The trio remained intact and both females laid eggs in all years from 1996 to 2002. One (or more) of these eggs was generally given to a pair that produced infertile eggs and one was left in the nest. If a chick resulted, it was reared by all trio members. Two eggs, including an egg laid by each female, was left in the nest in 2000 to see if the trio could incubate two eggs and rear two chicks. One egg hatched after a slightly prolonged incubation period (34 days instead of the normally 27–31 days). The chick was weak and died after two days. The other egg was infertile.

Male–male pair A (male '024', male '102')

Male '102' (hatched in 1980) and male '024' (hatched in 1982) had a partnership in 1992 and 1993. Male '102' initiated a few copulation approaches with females in 1992, but was never successful. Both males attempted to sit on the head of incubating flamingos. Copulation activity between the two males was not observed. This pair was quite disruptive, breaking eggs of solitary females or of poorly guarded male–female nests. The males usurped a newly laid egg and nest from a male–female pair in 1992. Incubation behaviour of this male–male pair was similar to that of male–female pairs, and the two birds ceased disturbing the colony. The egg was infertile, but they incubated it for the rest of the breeding season. Primarily to deter this male–male pair from disturbing the colony, it was given an infertile egg once it began nest building in 1993. The pair eventually received a fertile egg, and after a total incubation period of 65 days (for both eggs) both males produced crop milk when the chick hatched. Parental behaviour was normal, and the pair had no problems rearing the chick.

Male '024' was sent to another zoo in April 1994. There are no data on subsequent breeding, although there was little breeding activity in the flock that he joined. He died in May 2000.

Male '102' quickly formed a pair-bond with female '156' (unknown age, arrived in 1988). The pair built nests and raised some chicks during a six year period (1994–1997, 1999–2000). The male bred with female '150' (member of trio A) in 2001.

Male–male pair B (male '113' and male '116')

Wild-caught males '113' and '116' (ages unknown) arrived in 1993. They maintained a partnership in 1993 and 1994 and were generally disruptive, breaking eggs of lone nesters and other nests that were not well defended. Male '113' was sent to another zoo in March 1995 where he promptly paired with a female who produced an egg, and he has since continued to be reproductively active with females (W. Schoo, pers. comm.).

Male '116' bred with female '150' (of trio A) from 1995 to 1998, and with two more females thereafter.

Male–male pair C (male '116', male 'CAD')

Female '156' was paired to male '116' in 2002, but formed a pair in 2003 with male '160', her partner in 1998. Male '116' banded together in 2003 with male 'CAD' (hatched in 1998). These males held a nest that was highly desirable to other flamingos, as when they left the site their nest was generally occupied upon their return. As both birds were quite large they won the battles to regain the nest. However, the situation created considerable disturbance to neighbours, resulting in egg loss, even though the male–male pair did neither actively disturb nests of others nor try to gain possession of an egg. No copulation activity between the two was observed.

Male–male pair D (male 'ECT', male 'ECX')

Male 'ECT' (hatched in 1999) and male 'ECX' (hatched in 2000) formed a pair bond in 2002 and maintained it during the breeding season of 2003. They possessed a nest but did not disturb others. Neither male was observed engaging in copulation activities with each other or with other birds of either sex.

Female–female pair A (female '033', female 'CBZ') and trio E (male '029', female '033', female 'CBZ')

Female '033' (hatched in 1991) formed a male–female pair with male '139' (unknown age, arrived in 1993) in 1994. Male '139' was sent to another zoo in March 1995. Female '033' quickly formed a female–female pair with 'CBZ' (hatched in 1992). Successful copulations were regularly witnessed in 1995, with female 'CBZ' always assuming the mounter role. Female '033' copulated with male '158' (unknown age, arrived in 1994) in all three years. Three of five eggs laid in 1995, 1996 and 1997 were fertile. These eggs were given to foster parents because female–female pairs have difficulties protecting a nest from larger males.

In 1998 these two females formed a trio with male '029' (hatched in 1993, departed to another zoo in 1993, returned in March 1997), and this trio remained intact throughout 2003. The male copulated with both females and guarded both equally against copulation approaches by other males. No copulations between female '033' with male '158' or with any other male were observed after formation of the trio. The two females regularly copulated, too, with female 'CBZ' always acting as mounter. These female–female copulations took place when male '029' was on the nest and unable to interfere. The trio has successfully reared a chick, with all three flamingos sharing incubation and parent-rearing tasks. Eggs in addition to the first one laid each year are removed from the nest when produced.

Discussion

The current paper provides the most detailed information about same-sex pairs and trios amongst flamingos till date, and will attempt to identify some main factors that influence the formation of such 'atypical' partnerships.

Atypical partnerships are not that atypical

Surveys of institutions that keep flamingos indicate that same-sex and trio partnerships are fairly common in captivity, as they were observed in 80% of all colonies (see Table 3.2). This proportion will almost certainly rise further, since most colonies without reports of atypical partnerships have not been studied for very long. It is quite striking to note that 12 (46.2%) of 26 flamingos formed atypical partnerships during at least one breeding season in ten years at Rotterdam Zoo, while this zoo had a significantly lower percentage of individuals engaging in atypical partnerships than observed at Durrell Wildlife Conservation Trust and Amersfoort Zoo.

Bagemihl's (1999) review of homosexual behaviour among animals states for flamingos that 'in most captive populations with same-sex couples about 5–6% of pairs are homosexual, although some populations may have over a quarter same-sex couples' (p. 526). This general statement, as well as earlier, more specific reports are supported by the data presented here, since same-sex pair partnerships do occur regularly, although individual proportions per colony are quite variable. Same-sex pairs (not including trios) constituted 3–5% during 11 study years at Rotterdam Zoo, 28–33% during 8/13 years at Durrell Wildlife Conservation Trust, and 14–25% during three years at Amersfoort Zoo.

The proportion of partnerships considered to be atypical is quite sensitive to the criteria employed. For example, criteria for Rotterdam, Amersfoort and Atlanta Zoos (see below) were similar (sleeping side by side, simultanous and same activities, calling together, coalitions during aggressive encounters). Thus, if two males engaged with a female, but had an agonistic relationship with each other, they were not considered to form a trio.

Contrarily, studies on flamingo relationships have generally concentrated on nesting behaviour, with partners (usually referred to as pairs) defined as individuals sharing a nest and nesting behaviours, for example nest building and incubation (Bennett, 1987; Wilkinson, 1989; Pickering, 1992; Farrell et al., 2000; Stevens et al., 1992). However this approach fails to identify partnerships, either same-sex or heterosexual, that do not possess a nest. Such partnerships do occur and are of social significance within the colony. For example a male–female pair of Caribbean flamingos (male '022', female '036') at Rotterdam has a pair bond extending at least from 1992 to present. These birds copulate each year but are not able to hold a nest, even though they are usually one of the first pairs to begin nest building. It is therefore noteworthy that partnership criteria for Durrell were based on nesting behaviour, as this was all that was available. An even higher percentage of individuals in atypical partnerships may have been identified if nesting had not been the only criterion.

The number of individual flamingos that engaged in same-sex partnerships during a given year at the three best-studied zoos (see Table 3.1) ranged from 2.5 to 5.0% at Rotterdam, 27.9 to 33.3% at Durrell and 13.6 to 24.7% at Amersfoort. At the Atlanta Zoo in 1997, 17.5% (10/57) of wild-caught adult Chilean flamingos were in same-sex pairs (Reinertsen, 1999). Thus, atypical partnerships occur regularly, but their frequency is situation dependent.

Formation of atypical partnerships in flamingos may be facilitated by the group courtship strategy employed, particularly if the function of the displays is to hormonally synchronize the birds for breeding and the flamingos select their mates in displaying groups, as described in the introduction. At Rotterdam Zoo, reshuffling takes place a couple of months before nesting commences. It may be that flamingos select a bird who is closer synchronized in its reproductive physiology than the previous partner (Studer-Thiersch, pers. comm.), and this new partner may well be an individual of the same sex.

Definition problems for trios

Bagemihl (1999) states for trios that 'all three birds share incubation and chick-raising duties (although no same-sex bonding occurs)' (p. 526). This may hold true if 'bond' is defined sexually. However, all trio members are considered to be bonded if the criteria used in this chapter are employed. It may be that the bond is stronger between two of the three birds, and these two birds may be the same sex. Same-sex sexual activities may indeed occur, as demonstrated by trio E in which females 'CBZ' and '033' copulated with each other as well as male '029'.

P. Shannon (pers. comm.) interprets the formation of trios as a situation in which one bird in a pair forms a bond with a different bird, while the previous mate refuses to leave the partnership. However, using the definition applied here, the bird that refuses to leave has to be accepted by the newly formed pair in order to constitute a trio, as the three individuals sleep side by side, carry out the same activities simultaneously, call together and form coalitions during aggressive encounters.

The descriptions of trios B, C and D above demonstrate that trios may also form in other situations. Similarly, twice in five years the same two greater flamingo females began the breeding season at Disney's Animal Kingdom (USA) in a male–female–female trio. However, they evicted the male from the nest once an egg had been produced, incubated the egg and reared the chick as a female–female pair (S. Maher, pers. comm.). Likewise, two males imported together to Basle Zoo began a breeding season in a male–male–female trio, but once the female had laid an egg the males did not let her on the nest much, and mostly kept her away from the chick (A. Studer-Thiersch, pers. comm.).

Thus, same-sex pairs may sometimes form trios because of 'necessity': a female–female pair needs the male to fertilize the egg and help defend the nest, whereas a male–male pair needs the female to provide the needed egg.

Same-sex partnerships are not necessarily sexual

Despite thousands of copulation activities recorded at Basle Zoo, none involved two males (but neither were male–male pairs observed). All three male–male pairs at Amersfoort Zoo in 1996 and one of four at Rotterdam Zoo did participate together in low levels of copulation activity but no extra-pair male–male copulation activity was ever witnessed. Numerous copulation approaches were performed at Rotterdam Zoo between males but only a single copulation approach between two males paired to females was observed. The single 'successful' male–male copulation constituted a 'forced-mating', possibly a dominance display. P. Shannon (pers. comm.) noted that males in same-sex pairs never open their wings for the other male to mount. Interestingly, a male observed in reverse copulation attempts with his female partner at Basle Zoo did in fact open his wings.

On the other hand, 'successful' copulation attempts were recorded among female–female pairs at both Rotterdam and Basle Zoos. However, none of the females in the eight female–female pairs collectively recorded at Rotterdam, Basle and Amersfoort Zoos engaged in extra-pair copulation activity with another female.

All 74 extra-pair copulation activities recorded at Amersfoort Zoo in 2001 were by a male and female together (de Bruin and de Haan, 2001). The lack of extra-pair copulation activity by same-sex pair members with flamingos of the same sex at Amersfoort, Rotterdam and Basle Zoos is particularly interesting, given that at least some male–female pairs tend to engage in extra-pair copulations (for example Bennett, 1987; Peters 1996; Farrell *et al.*, 2000; King, 1994; de Bruin and de Haan, 2001).

Bagemihl (1999) stated that 'single males seeking male partners have been known to harass heterosexual pairs, following them around and disrupting their copulations and incubation shifts in an attempt to gain access to the male' (p. 525). Given data presented here, it is doubtful that this occurs with any regularity. Although males do disturb copulation attempts, it may be for reasons other than to gain access to the male. Several males routinely disturbed copulation behaviour of both male–female and female–female pairs at Basle Zoo. The interfering bird generally waited until an attempt was in progress, and then threw his body against the mounting bird. In at least one case the disrupting male took over the copulation (pers. obs.).

Thus, copulations might be a more important bonding behaviour for females in same-sex pairs than for males in same-sex pairs, although this assumption is based on a small sample.

Sexual interest in the opposite sex does not cease in same-sex partnerships

Males in male–male pairs were more frequently engaged in copulation activities or extra-pair copulations with females than with males (3/7 at Rotterdam; 5/6 at Amersfoort, unpublished data, Peters, 1996).

At Amersfoort Zoo in 1996, 3/6 female–female pairs engaged in copulation activities with each other, while 8/12 females engaged in extra-pair copulation activity with males. In fact, females living in same-sex partnerships copulated with males at all four zoos for which data are reported here, and fertile eggs were produced within female–female pairs at Basle, Durrell and Rotterdam Zoos.

Copulation activities occur each year among all male–female pairs at Rotterdam, and were observed in 28/30 male–female pairs at Amersfoort in 1996 (Peters, 1996). Thus, copulation activity is higher amongst male–female pairs than amongst same-sex pairs.

Nesting, incubation and chick rearing in atypical relationships

Our data illustrate that the frequency of partner change is variable among individuals and among colonies. At Rotterdam Zoo, reshuffling takes place a couple of months before nesting commences. It may be that flamingos select a bird who is closer synchronized in its reproductive physiology than the previous partner (Studer-Thiersch, pers. comm.) – and this new partner may well be an individual of the same sex.

Trios and same sex pairs nested as often as heterosexual pairs at Amersfoort Zoo (Peters, 1996). Nest building behaviour and parenting of a closely observed male–male pair rearing a chick at Rotterdam Zoo were similar to those of male–female pairs (pers. obs.). Two respondents from the zoo survey did likewise report successful chick rearing by male–male pairs.

Male–male pairs in a male-biased colony at Honolulu Zoo (USA) are given infertile eggs to incubate, because otherwise they are disruptive to the colony, pushing male–female pairs off their nests. The same-sex pairs incubate the eggs throughout the entire breeding season if allowed to, and from fertile eggs they rear chicks (L. Santos, pers. comm.). A male–male pair in another zoo usurped an egg and nest from a male–female pair and proceeded to incubate it, although the egg did not hatch. Similarly, at Rotterdam, both a male–female pair and male–male pair took over a nest and proceeded to incubate the already laid egg. In both cases the egg was infertile.

Females are smaller than males and therefore at a disadvantage in confrontations. Nevertheless, female–female pairs have managed to rear chicks in the Basle colony (A. Studer-Thiersch, pers. comm.), but not at the Durrell Wildlife Conservation Trust (G. Young, pers. comm). However, the survey revealed successful rearing of chicks by female–female pairs in three zoos.

Sex ratio and mate choice

The pool of potential mates is influenced by the colony's sex ratio. Our long-term data demonstrate that two colonies with a rather biased sex ratio had relatively more atypical partnerships than a colony with a relatively even sex ratio (see Table 3.1). The likelihood that the formation of same-sex and trio partnerships increases if sex ratios are biased is also illustrated by results from the survey of institutions that breed flamingos (see Table 3.2). The formation of female–female pairs at Chester Zoo (Wilkinson, 1989) and Basle Zoo has likewise been attributed to a female biased sex ratio.

Nonetheless, the picture is often more complicated. Both Shannon (2000) and Reinertsen (1999) deny a clear-cut correlation between atypical partnerships and biased sex ratios. Similarly, the survey revealed the existence of two male–male pairs in a group of 7.15.0 Chilean flamingos (S. Maltsbarger, pers. comm.), and male–male pairs were present in the strongly female-biased colony at Amersfoort Zoo (Table 3.1), while as many as eight females were unpaired (Peters, 1996).

However, such data need careful interpretation because flamingo partnerships tend to span years and sex ratios are not static. Experience gained in previous partnerships may be more important than current sex ratios. As a consequence, flamingos may actually be selecting partners from a sub-population with a different sex ratio than the whole population.

Colony movements and longevity of partnerships

Similar to the findings for Rotterdam Zoo over a period of ten years, no difference was found between mate fidelity in the formation of pair-bonds of same-sex and male–female pairs studied for two years at Atlanta (Reinertsen,

1999). Mate fidelity of both same-sex and male–female pairs was lower in pairs separated from their last partner for several weeks prior to the breeding season than in flamingos that were not, perhaps due to their inability to perform displays associated with pair bond maintenance (Reinertsen, 1999), creating an asynchrony with respect to the previous mate. At Durrell Wildlife Conservation Trust, flamingos were often moved back and forth between the breeding flock and another flock and changes in atypical partnership were more common than at Rotterdam.

Nevertheless, moving captive flamingos around simulates the situation in the wild more closely, as these birds are nomadic and move independently. Such movements may at least partly explain why flamingos of the huge Camargue colony are serially monogamous (Cezilly and Johnson, 1995), whereas captive flamingos that move less often maintain longer-term partnerships (for example Studer-Thiersch, 1967; Bennett, 1987; Wilkinson, 1989; Pickering, 1992; Stevens et al., 1992; King, 1994, Reinertsen, 1999; Shannon, 2000). Similarly, mallards are considered seasonally monogamous (for example Losito and Baldassarre, 1996) yet most ringed mallards in a small, sedentary population, were found to be monogamous in consecutive years (Mjelstad and Sataersdal, 1990).

Switches between male–female and atypical partnerships

Results reported here indicate that flamingos in same-sex pairs and in trios have bred in male–female pairs before and may do so later on. A male at National Zoo (USA) left his eight-year long male–male partnership, formed a pair with a female and reared a chick (S. Hallager, pers.comm.). Members of a male–male pair found in Amersfoort in 1995 and beginning of 1996 both bred with females in 1996 (Peters, 1996). A male that remains if his partner from a same-sex pair is removed may go on breeding with a female, particularly if the male has previously been in a male–female pair.

Thus, mate fidelity in flamingos – whether heterosexual or same-sexual – seems to be a very plastic trait, largely depending on environmental conditions.

Age cohorts, arrival cohorts and mate choice

Greater flamingos in the Camargue generally nest with partners from the same age cohort (Cezilly et al., 1997). This age-assortative mating pattern is not due to mate fidelity or time of arrival at the colony but rather is proposed to be from a directional pairing preference for older and more experienced individuals.

Flamingos reared at Rotterdam Zoo did likewise form the first pair bond most often with a bird of similar age (see Table 3.3), including those that formed same-sex partnerships. At least five but possibly seven out of nine birds at Rotterdam Zoo formed such partnerships either with a similar-aged bird and/or with one they arrived with.

It seems to be a general pattern that flamingos will at least initially form partnerships within a group which has been transferred to another zoo – even if the available partners have the same sex (A. Studer-Thiersch, pers. comm.; Wilkinson, 1989). Acquired birds are usually wild caught, thus whether such arrivals are indeed of similar age is difficult to ascertain, but mutual experiences could likewise function to mould mate choice preferences.

Therefore, age or arrival cohorts do in all likelihood pose yet another constraint on the pool of potential partners, both in the wild and in captivity.

The 'alliance formation' hypothesis

Aggressive interactions related to the acquisition and defence of resources such as loafing, roosting and feeding sites are frequent amongst flamingos. For example, Caribbean flamingos in Venezuela quarreled during 26% of 160 feeding observations, which significantly reduced the time spent feeding (Bildstein et al., 1991). Therefore, individual fitness may be greatly enhanced if a flamingo can make it more likely that an aggressive encounter will be won. For this, wild flamingos seem to form alliances. For example, 38% of 371 aggressive interactions between foraging Caribbean flamingos in Mexico involved more than two birds (Schmitz and Baldassarre, 1992). Most of these conflicts were initiated and won by male–female pairs. Although coalitions must be temporary, due to the nomadic lifestyle, they can last as long as the partners are both in the same area. Low-level agonistic interactions occur among flamingos at Rotterdam Zoo year round (King, unpubl.), and partners usually support each other (for example Studer-Thiersch, 1975; Peters, 1996; King, unpublished). Support may indeed be the primary reason why captive flamingos often maintain partnerships year round.

Same-sex pairs and trios can confer the same advantages as male–female alliances. Male–male pairs and trios possibly even offer additional advantages through the male's larger size (Schmitz and Baldassarre, 1992) and increased numbers.

The female can maintain a nest, incubate eggs and rear young without assistance of a male partner in at least 500 avian species (Skutch, 1976). This is clearly not a possibility for flamingos, where nests are highly conspicuous and are vulnerable to destruction by conspecifics and predation, e.g. by herons, corvids, gulls and marabou storks (del Hoyo *et al.*, 1992; Cezilly, 1993; pers. obs.). Some lone individuals, both males and females, have attempted to incubate an egg at Rotterdam Zoo, and also at Basle Zoo (A. Studer-Thiersch pers. comm.). If the egg is not broken and the nest not taken over by conspecifics, the incubating bird has to leave the egg defenceless while it eats and drinks. At this point the nest is often abandoned and quickly marauded by conspecifics (Cezilly, 1993). Thus, lone flamingo breeders are never successful, and reproduction is entirely dependent on team work.

Therefore, same-sex pairs or trios will have a greater chance of rearing a chick successfully than a lone nester. However, atypical relationships should provide an additional advantage since chick rearing does not take place in most cases. Moreover, in the various potential same-sex partnerships and trios, at least one partner is by definition excluded from direct reproduction, even if fertile eggs are acquired. Such additional benefits may be conferred upon by the continued practice and experience associated with a partnership.

The 'practice' hypothesis

Flamingo parenting skills are quite hard-wired compared with other wading birds, such as storks. At the Rotterdam Zoo no chicks have been lost due to causes related to insufficient parenting (King, 1990). Therefore, the need to practice parenting skills seems to be an unlikely motivation in the formation of same-sex pairs and other atypical partnerships.

Nevertheless, practice and experience might be important with respect to accessing, defending and holding a good nest site. Data from the Camargue indicate that older birds are better at this (Cezilly *et al.*, 1997). Younger flamingos in the Camargue sometimes build nests on a less favourable site – a dyke – where predation is higher, or wait until the older birds have vacated the breeding island (A. Johnson, pers. comm.). That thousands of younger Camargue flamingos nest each year with little chance of success demonstrates how important experience might be. Therefore, same-sex pairs who nest – and even those who do not rear chicks or rear chicks that are not their own – may still gain experience that could enhance their long-term reproductive success.

Data presented here indicate that at least some individuals can move with fluidity between different types of partnerships. Even if some individuals have more of a tendency to do so (for example male '116' lived in two different male–male relationships and female '152' in three different trios), this does not necessarily indicate a specific 'sexual orientation'. Rather, the same-sex or trio partnership might represent the 'best bet' at that time.

The option to form different partnerships could be an advantage in colonies where availability of breeding partners is limited. Indeed, flamingos at

Rotterdam who had engaged in at least one atypical relationship were less often without a partner (8% of all breeding seasons) than those who did not engage in atypical relationships (12% of all breeding seasons), although this difference was not significant. In any case, this social fluidity allows flamingos that might otherwise be without a partner in a given season to stay in the game.

Nonetheless, flamingos do sometimes skip breeding years, both in the wild (A. Johnson, pers. comm.) and in captivity. At least a few adults of both sexes are not paired each year in most captive colonies (for example Bennett, 1987; Wilkinson, 1989; Peters, 1996; Stevens *et al.*, 1992; King, 1994, Reinertsen, 1999; Shannon, 2000). Approximately half (14/26) flamingos at Rotterdam Zoo skipped breeding years between 1993 and 2002. A partner from the previous year would often breed with another partner that year. However, it is difficult to say whether the previous partner did not want to breed or if the new pairing was the result of competition. It also occurred that both previous partners did not breed. Reproduction is physically quite demanding, and flamingos may well not be in sufficiently good condition to breed every year. In fact, female greater flamingos may desert nests in the Camargue because of the risk of decreased fitness through further parental investment when the flamingo is already physically compromised (Cezilly, 1993).

If same-sex partnerships become more likely when breeding is more difficult, it is as yet not borne out by the data, but it is possible.

Conclusion

Partnerships other than male–female are not uncommon in *Phoenicopterus* flamingos and include same-sex pairs as well as trios. The probability of atypical partnerships is higher in colonies with a skewed sex ratio. Other factors such as cohort (age) structure and physiological synchrony may place additional constraints on the suitability of available opposite-sex mates. Atypical partnerships may include sexual interactions but the basis for these atypical partnerships seem to be primarily social, at least in captivity. Extra-pair and extra-trio copulation activity is directed towards the opposite sex in both male–female pairs and atypical partnerships. Atypical partnerships offer possibilities to form coalitions in aggressive encounters and to gain experience in holding a nest against conspecifics, just as male–female partnerships do.

Acknowledgements

I would like to thank Wim van der Horst, Han Kemp, Marja Los, and most of all Lisette Kuijsters for patient and careful data collection in the Rotterdam Zoo flamingo colony, and the zoo keepers and veterinarians for contributing their knowledge of the flamingos as well. Adelheid Studer-Thiersch and Peter

Studer made it possible for me to study the flamingos at Basle Zoo under very enjoyable and stimulating circumstances, and Saskia Ober kindly helped with data collection.

The following are thanked for responding to the flamingo quick survey: Bill Aragon (Albuquerque Zoo, USA), James Ballance (Atlanta Zoo, Baltimore Zoo USA), Colin Bath (Paignton Zoo, UK), Sara Hallager (National Zoo, USA), Marleen Huyghe (Zoo Planckendael, The Netherlands), Andrzej Kruszewicz (Poland) Sherry Maltsbarger (Dallas Zoo, USA), Joyce Nickley (San Diego Wild Animal Park, USA), David Oehler (Cincinnati Zoo, USA), Dieter Rinke (Vogelpark Walsrode, Germany), Jay Robinson (Riverbanks Zoo, USA), Sara Rosenbloom (Lowry Park Zoo, USA), Linda Santos (Honolulu Zoo, USA), Wineke Schoo (Arnhem Zoo, The Netherlands), Nigel Simpson (Bristol Zoo, UK), Tim Snyder (Birmingham Zoo, USA), Adelheid Studer-Thiersch (Basle Zoo, Switzerland), Dana Urbanski (North Carolina Zoo, USA), Glyn Young (Durrell, UK) and anonymous respondents from the zoos of Minnesota and Tautphaus Park (USA).

References

Albanese, G., Bacetti, N., Magnani, A., Serra, L. and Zantello, M. (1997) Breeding of the greater flamingo *Phoenicopterus ruber roseus* in Apulia, southeastern Italy. *Alauda*, **65**, 202–4.

Bacetti, N., Cianchi, F., Dall'Antonia, P., De Faveri, A. and Serra, L. (1994) Nidificazione de fenicottero, *Phoenicopterus ruber*, nella Laguna di Orbetello. *Rivista Italiana di Ornitologia*, **64**, 86–7.

Bagemihl, B. (1999) *Biological Exuberance: Animal Homosexuality and Natural Diversity*. New York: St. Martin's.

Bennett, C. (1987) Breeding and reproduction of the Chilean flamingo (*Phoenicopterus chilensis*) at the Santa Barbara Zoo. AAZPA 1987, *Reg. Conf. Proc.*, pp. 313–17.

Bildstein, K. L., Frederick, P. C. and Spalding, M. G. (1991) Feeding patterns and aggressive behavior in juvenile and adult American flamingos. *Condor*, **93**, 916–25.

Blythe, E. (1994) *The History of the Lesser Flamingos (Phoeniconaias minor) at the Baltimore Zoo*. Baltimore: Baltimore Zoo.

Bolhuis, A. and Lamers, R. (1997) *Chileense flamingo's om Dierenpark Amersfoort: koppelvorming & paargedrag in 1997*. Leeuwarden, NL: Van Hall Institute.

Brown, L. H. (1958) The breeding of the greater flamingo *Phoenicopterus ruber* at Lake Elementeita, Kenya Colony. *Ibis*, **100**, 388–420.

Bruin, A. de and Haan, N. de (2001) *Koppelvorming en paargedraag van Chileense flamingo's in Dierenpark Amersfoort*. Leeuwarden: Van Hall Institute.

Cezilly, F. (1993) Nest desertion in the greater flamingo, *Phoenicopterus ruber roseus*. *Anim. Behav.*, **45**, 1038–40.

Cezilly, F. and Johnson, A. R. (1995) Re-mating between and within breeding seasons in the greater flamingo *Phoenicopterus ruber roseus*. *Ibis*, **137**, 543–6.

Cezilly, F., Boy, V., Tourenq, C. J. and Johnson, A. R. (1997) Age-assortive pairing in the greater flamingo *Phoenicopterus ruber*. *Ibis*, **139**, 331–6.

Elbin, S. B. and Lyles, A. M. (1994) Managing colonial waterbirds: the scarlet ibis *Eudocimus ruber* as a model species. *Int. Zoo Yb.*, **33**, 85–94.

Farrell, M. A., Barry, E. and Marples, N. (2000) Breeding biology in a flock of Chilean flamingos (*Phoenicopterus chilensis*) at the Dublin Zoo. *Zoo Biol.*, **19**, 227–37.

Hoyo, J. del, Elliot, A. and Sargatal, J. (1992) *Handbook of the Birds of the World*. Vol. 1. Barcelona: Lynx Edicions.

Huber, R. and Martys, M. (1993) Male–male pairs in greylag geese (*Anser anser*). *J. Ornithol.*, **134**, 155–64.

Jenkin, P. M. (1957) The filter-feeding and food of flamingos (Phoenicopteri). *Philosoph. Transact. Roy. Soc. London (B)*, **240**, 401–93.

Johnson, A. R. (1997) A noteworthy recovery. *Flamingo Specialist Group Newsl.*, 9, 12. La Sambuc: Station Biologique de la Tout du Valat.

Kahl, M. P. (1975) Ritualized displays. In *Flamingos*, eds. J. Kear and N. Duplaix-Hall, pp. 142–9. Berkhamsted: T. & A. D. Poyser.

Kerkman, M. (1995) *Diep-verslag over koppelvorming en verblijfsgebruik door de Chileense flamingo iin Dierenpark Amersfoort*. Amersfoort: Amersfoort Zoo.

King, C. (1990) Reproductive management of the Oriental white stork *Ciconia boyciana* in captivity. *Int. Zoo Yb.*, **29**, 85–90.

King, C. E. (1994) Management and research implications of selected behaviours in a mixed colony of flamingos at Rotterdam Zoo. *Int. Zoo Yb.*, **33**, 103–13.

Kirwan, G. (1992) A freshwater breeding record of greater flamingo *Phoenicopterus ruber* in Turkey. *Sandgrouse*, **14**, 56–7.

Lehner, P. N. (1979) *Handbook of Ethological Methods*. New York and London: Garland STPM Press.

Losito, M. P. and Baldassarre, G. A. (1996) Pair-bond dissolution in mallards. *Auk*, **113**, 692–5.

Mathevon, N. (1996) What parameters can be used for individual acoustic recognition by the greater flamingo? *Comptes Rendus de l'Academie des Science Serie Sciences de la Vie*, **319**, 29–32.

Mathevon, N. (1997) Individuality of contact calls in the greater flamingo *Phoenicopterus ruber* and the problem of background noise in a colony. *Ibis*, **139**, 513–17.

Merriam-Webster, Inc. (1988) *Webster's Ninth New Collegiate Dictionary*. Springfield: Merriam-Webster Inc.

Mjelstad, N. and Saetersdal, M. (1990) Reforming of resident mallard pairs *Anas platyrhynchos*, rule rather than exception? *Wildfowl*, **41**, 130–1.

Peters, L. (1996) *Voortplantingssucces Chileense flamingo's (Phoenicopterus chilensis) in Dierenpark Amersfoort*. Utrecht: University of Utrecht.

Pickering, S. P. C. (1992) The comparative breeding biology of flamingos Phoenicopteridae at the Wildlife and Wetlands Trust Centre, Slimbridge. *Int. Zoo Yb.*, **31**, 139–46.

Reinertsen, M. (1999) *Pairing in Captive Chilean Flamingos as a Function of Social Separation Methods*. Atlanta: Georgia Institute of Technology.

Richter, N. A. and Bourne, G. R. (1990) Sexing greater flamingos by weight and linear measurements. *Zoo Biol.*, **9**, 317–23.

Richter, N. A., Bourne, G. R. and Diebold, E. N. (1991) Gender determination by body weight and linear measurements in American and Chilean flamingos, previously surgically sexed: within-sex comparison to greater flamingo measurements. *Zoo Biol.*, **10**, 425–31.

Schmitz, R. A. and Baldassarre, G. A. (1992) Contest asymmetry and multiple bird conflicts during foraging among nonbreeding American flamingos in Yucatan, Mexico. *Condor*, **94**, 254–9.

Shannon, P. W. (2000) Social and reproductive relationships of captive Caribbean flamingos. *Waterbirds (Special Publication 1)*, **23**, 173–8.

Siegel, S. and Castellan, N. J. (1988) *Non-parametric Statistics for the Behavioral Sciences.* 2nd edn. New York: McGraw-Hill.

Skutch, A. F. (1976) *Parent Birds and Their Young.* Austin: University of Texas Press.

Stevens, E. F., Beaumont, J. H., Cusson, E. W. and Fowler, J. (1992) Nesting behavior in a flock of Chilean flamingos. *Zoo Biol.*, **11**, 209–14.

Studer-Thiersch, A. (1967) Beiträge zur Brutbiologie der Flamingos (Gattung *Phoenicopterus*). *Zool. Garten*, **34**, 159–229.

Studer-Thiersch, A. (1975) Group display in *Phoenicopterus*. In *Flamingos*, eds. J. Kear and N. Duplaix-Hall, pp. 1450–8. Berkhamsted: T & A. D. Poyser.

Studer-Thiersch, A. (2000) What 19 years of observations on greater flamingos suggests about adaptations to breeding under irregular conditions. *Waterbirds (Special Publication 1)*, **23**, 150–9.

Wilkinson, R. (1989) Breeding and management of flamingos at Chester Zoo. *Avicultural Mag.*, **95**, 51–61.

Zweers, G., de Jong, F., Berkhoudt, H. and Van Den Berge, J. C. (1995) Filter feeding in flamingos. *Condor*, **97**, 297–324.

4

Establishing trust: socio-sexual
behaviour and the development of
male–male bonds among Indian Ocean
bottlenose dolphins

JANET MANN

Introduction

In the popular press, bottlenose dolphins have been characterized as
'sexual' animals who frequently engage in non-reproductive sexual behaviour
(Kluger, 1999; CNN, 2002; Begos, 1999; Fahy, 2003; Kyodo News International,
2001), including homosexual encounters. However, actual accounts of same-sex
activity in bottlenose dolphins include only a few descriptive studies from captivity (for example, Caldwell and Caldwell, 1972; McBride and Hebb, 1948) and
the wild (for example, Herzing and Johnson, 1997); only one quantitative study,
which focused on just two captive individuals (Östman, 1991), has been published. Systematic research on homosexual behaviour in other cetaceans is limited to a killer whale dissertation study (Rose, 1992). Even though bottlenose
dolphins are one of the best-studied cetaceans, their sexual behaviour has not
been quantified in field settings. Given the difficulty in studying marine mammals, and cetaceans in particular, the dearth of research in this area is not
surprising. The present study of Indian Ocean bottlenose dolphin calves, followed closely since 1988, is the first to quantify homosexual behaviour in wild
cetaceans using focal animal sampling.

Since the start of this research, it became evident that all age–sex classes
participate in socio-sexual behaviour, which includes genital contact between
opposite-sex and same-sex individuals. However, calves and juveniles engage
in higher rates than adults, including homosexual activity. Relative to adults,
calf and juvenile socio-sexual behaviour is typically seen in a playful context,
and it seems likely that these behaviours fulfil some social function(s). But,
with so little previous research on play, development and sexual behaviour in

dolphins, specific hypotheses have not been developed other than Östman (1991), who characterized adult male homosexual behaviour among captive dolphins as dominance related.

The study of sexual play or socio-sexual behaviour in calves is of interest because elements of play often reflect components of adult behaviours (for a review of primates see Pereira and Fairbanks, 1993) and informs how sexual behaviour develops. Pre-pubertal sexual behaviour and especially male–male interactions are common in other species, for example, domestic pigs (Berry and Signoret, 1984) or primates (for example, Brown and Dixson, 2000). A fundamental question in the play literature is whether the relative contribution of play behaviours (including socio-sexual play) primarily has current or future utility (for example, Fairbanks, 1993). For example, socio-sexual play may allow calves to practice skills for future courtship, promote bonds of current or future value, have organizational effects on development, or be a by-product of hormonal activity during early development (that is, postnatal surge in testosterone found in some primates, Brown and Dixson, 2000). A profound difficulty in discriminating these functions is that, even in captivity, one cannot manipulate one variable, such as time spent in socio-sexual behaviour, without affecting other important behaviours, such as physical contact and social interaction. Developmental outcomes are multi-determined and it would be extremely challenging to assess the long-term effects of infantile socio-sexual behaviour.

Of particular interest in bottlenose dolphin research is the relationship, if any, between male homosexual behaviour and alliance formation, a crucial part of male mating strategies (Connor et al., 1992a,b, 1996, 1999). Males form first-order alliances (pairs and trios) that cooperate to sequester and maintain exclusive access to a single female for up to six weeks (although typically less than one week), an event known as a consortship. Some first-order alliances appear to remain highly stable for 15–20 years (Connor, unpublished data). They typically pair with one or two other alliances to form second-order alliances. Second-order alliances cooperate by helping each alliance keep their respective females during consortships. Although popular accounts occasionally infer that males coerce copulations on the female, such behaviour has never been observed. Males may also form a super-alliance of up to 14 individuals. Pairings and trios within the super-alliance are labile, with no more than three males consorting with a female at any time. However, if the pair or trio is challenged by an outside alliance, the entire super-alliance may help the pair or trio defend the female. Although adult male pairs (but not trios) have been noted at several research sites (for example, Owen et al., 2002), and not at other bottlenose dolphin research sites, only in Shark Bay, Western Australia are males known to form multi-level alliances (see review by Connor et al., 2000), a pattern otherwise seen only in humans.

We understand little concerning how such intense, prolonged male–male bonds are formed and maintained. The current study from Shark Bay may offer some insights into the mechanisms.

I begin with quantitative description of the patterns and frequencies of socio-sexual and homosexual behaviour in bottlenose dolphins, including sex-differences, preferential partnerships and the role of actor and recipient. In the discussion, hypotheses typically proposed to explain homosexual behaviour in other non-human animals will be examined against the dolphin data.

Methods

Bottlenose dolphin society

Three major long-term studies of bottlenose dolphins have been ongoing for more than 15 years, in the Moray Firth, Scotland; Sarasota, Florida, USA; and Shark Bay, Australia. Everywhere they have been studied, bottlenose dolphins (*Tursiops* sp.) live in coastal fission–fusion societies, characterized by sex-segregation and frequent changes in group membership (reviewed in Connor *et al.*, 2000). Adult females tend to associate with other adult females, juveniles and calves. Adult males tend to associate with other males or be solitary. Preferential male–male associations are particularly strong at two long-term study sites, Sarasota, Florida and Shark Bay, Australia (reviewed in Connor *et al.*, 2000). Females form loose networks, with weaker, but consistent associations with other females (Smolker *et al.*, 1992). In Shark Bay, the age at first birth is 12 and females nurse calves for three to six years (Mann *et al.*, 2000) and daughters continue to associate with their mothers after weaning; sons rarely do (Connor *et al.*, 2000). Some females are solitary, and are almost never sighted with other dolphins except their own calves, and some are highly social, almost never sighted without other dolphins (Mann *et al.*, 2000; Gibson and Mann, 2003). Their diet consists mainly of fish and squid (Connor *et al.*, 2000); lactating females spend on average 30% of the day foraging and <2% socializing (Mann and Watson-Capps, 2005; Mann and Sargeant, 2003). Calves tend to socialize for 10–15% of their time (Mann and Watson-Capps, 2005). Activity budgets for juveniles and adult males have not been reported. Both sexes are philopatric and the communities do not appear to be closed (Connor *et al.*, 2000), but males tend to disperse or range more widely than females (Krützen *et al.*, 2004a).

Field site

Shark Bay is located at 25°47′S, 113°43′E in Western Australia. After an initial visit in 1982, a long-term study of the Shark Bay dolphins was established in 1984 off of a fishing camp (now resort) called Monkey Mia (Connor and Smolker, 1985). The habitat consists mostly of embayment plains (5–13 m in depth) and shallow seagrass beds (0.5–4 m) bisected by deeper channels (7–13 m). Observations are in 6 m depth on average, making near continuous follows possible. The study area currently extends, 250 km^2 off the east side of the Peron Peninsula and includes over 600 animals that are monitored annually. Over 250 are observed in several long-term studies using focal follow methods. Most of the dolphins are well habituated to small boats (4–5 m), allowing us to follow individuals for many hours (Smolker et al., 1993; Mann and Smuts, 1998). Based on genetic haplotypes, the Shark Bay bottlenose dolphin species classification remains unresolved and the animals are hence referred to as *Tursiops* sp. (M. Krützen, unpublished).

Age–sex class determination

Calves are defined as still nursing, with an average weaning age of four years (Mann et al., 2000). Juveniles are weaned but still pre-reproductive (for females up to age 12, for males up to age 14). Adult females are age 12 or older and adult males age 14 or older (when males begin to form stable long-term alliances). Physiological data from captivity and the field suggest that males are capable of fathering offspring as early as eight to ten years of age (Schroeder, 1990; Wells et al., 1987), but in Shark Bay their access to reproductive females is limited until they have formed stable alliances (Krützen et al., 2004b). Although birth years are not known for most dolphins born prior to 1982, all dolphins could be classified into an age class based on ventral speckling (Smolker et al., 1992), age of first reproduction (for females, see Mann et al., 2000) or body size.

Subjects and dataset

A long-term study of bottlenose dolphin mothers and calves was initiated in 1988 (Mann and Smuts, 1998; Mann and Smuts, 1999; Mann et al., 2000; Mann and Sargeant, 2003; Mann and Watson-Capps, 2005). Between 1989 and 1999, 46 focal calves born to 26 mothers were observed for 1349 hours during the first four years of life (27 females for 725 hours 19 males for 624 hours). Focal mothers and calves were observed from 1 to 9 hours per day depending on weather conditions. The median and average follow duration is three hours. Most calves were observed for 10–15 hours per year or age class.

The population is residential and female home ranges are well known, minimizing the search effort required to find specific individuals. Group size and membership was determined by scan sampling the number of animals within a 10 m chain (see Smolker *et al.*, 1992) at either five or one minute intervals. For example, every minute we scan every individual that surfaces within 10 m of any individual in the group. Any individual who surfaces more than 10 m outside of any group member is considered to have left the group. Average and median group size for adult females is 4–5 individuals (Mann *et al.*, 2000). Maternal and calf behaviours were observed continuously but scored by either point sampling (1 min intervals) or predominant activity sampling (2.5 min intervals, Mann, 1999). Social events were noted continuously (frequency sampling) when possible for the focal dyad and on an *ad libitum* basis for non-focal individuals. For the latter, the direction of interactions (actor–recipient) were used, but rates of socio-sexual interactions were not calculated. For focal mothers and calves, the type of social behaviour was indicated every minute (sexual behaviour, various types of play, petting, rubbing, etc.), while event frequencies (mounts, body parts involved in petting) were recorded continuously or when observed. Activity budgets are fairly accurate, but event rates are an underestimate since some sub-surface activities were not observed or because participants could not be readily identified during some polyadic interactions.

Socio-sexual behaviour

Vasey (1995) defines homosexual behaviour as genital contact and/or manipulation involving same-sex individuals. The current study defines four behaviours – 'mounting', 'goosing', 'push-ups', 'petting' – as 'socio-sexual' when genital contact is involved. These activities are considered to be 'homosexual' when between same sex partners.

- *Mounting* is observed in three basic forms: dorso–ventral (Figure 4.1a), lateral–ventral and ventral–ventral. During lateral–ventral mounts, the mountee turns on-side, facilitating the mount. During ventral–ventral mounts, the mounter swims belly-up under the mountee. Mounts were mostly dorso–ventral.
- *Goosing* occurs when the actor brings his or her beak into contact (gently or not so gently) with the genital area of the recipient (Figure 4.1b). This has also been called 'beak to genital propulsion'.
- *Push-ups* occur when one dolphin pushes up the genital area of another with his or her head, usually so it clears the water.
- *Socio-sexual petting* is defined by pectoral fin-to-genital contact, when one dolphin either strokes the genital area of another dolphin with his or her pectoral fin (Figure 4.1c), or inserts the pectoral fin into the genital slit of another dolphin. Genital contact is one type of petting,

(a)

(b)

Figure 4.1. Socio-sexual behaviour in bottle-nosed dolphins. (a) Typical dorsal–ventral mount position. (b) Juvenile female with her beak to the genital area of a female calf (gentle goosing). (c) Pectoral fin petting of the genital area. (d) Interactions in a group of 11 immature males. As illustrated by the photograph, the actors and recipients are difficult to identify. At least four males were involved in the current interaction. The owner of the erection was not identified.

(c)

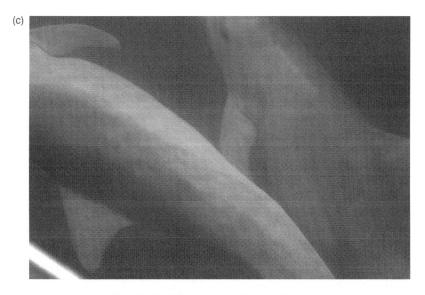

(d)

Figure 4.1. (*cont.*)

although other body parts are frequently involved during petting interactions. We can often observe petting, but not always specify the body parts involved. Thus, only those including specific contact with the genital area are included in this study.

Rubbing of the genital area on the body part of another dolphin is also a form of sexual contact, but this was infrequently observed except during mounts, push-ups or gooses and is therefore not considered a separate act. Because both male and female dolphins have genital slits, males can easily achieve penile intromission with other males. However, this was so rarely observed that it was not included in our analysis. Intromission is generally difficult to confirm because it lasts for a few seconds only, is underwater, and the body of the mounter is pressed against the mountee. Genital inspections, when the inspector brings his/her beak close to the genital area of another dolphin without touching, often occur, but are not considered sexual by this definition.

Each mount, petting of genital area, goose or push-up was considered one *socio-sexual event*. Events were considered part of a *socio-sexual bout* when they occurred within 5 min of the last event, involved at least one of the same participants and were not interrupted by non-social behaviour (for example, forage). Rates, calculated only for focal individuals, are underestimates because not all events were recorded, especially during long bouts of socio-sexual behaviour, which typically involved all or mostly male participants.

Several factors lead to potential underestimates. During long bouts, the dolphins tend to change direction often, making it difficult for observers to identify the actor and recipient (Figure 4.1d). Another bias concerns petting, which was classified as socio-sexual only if the observer was able to determine contact with the genital area. However, this was not always possible because of the difficulty in viewing which body parts are involved. Since females tended to engage in more petting interactions than males, the rate of female–female socio-sexual behaviour might also be underestimated.

Partner availability

Fission–fusion social systems such as in Shark Bay are difficult to quantify because the group composition is fluid and ever-changing. Since individuals who interact are likely to be together in social groups in the first place, preferential grouping is correlated with preferential relationships. For example, during or just prior to socio-sexual interactions, the participants often segregate themselves into a separate group for varying periods of time. Calves are not like most mammals with extensive parental care in that they venture hundreds of meters from their mothers and join groups without them. Compared with terrestrial mammals, the costs of locomotion are low for dolphins (Williams *et al.*, 1993), facilitating these separations and associations with others. Theoretically, a calf has numerous social options, for at least brief periods of time when away from the mother.

I computed the bias in interactions between individuals of different sex–age classes, the proportion of female and male calves, juveniles and adults in the entire population of frequently sighted individuals (>20 times) whose sexes were known for a nine-year period (1991–1999). The entire pool of 305 individuals varied depending on births and deaths, ranging from 253 to 197 in a given year. The proportion of each sex class varied little from year to year, no more than 2%. On average, the population consisted of 6.7% (SD = 0.2) female calves, 8.8% male calves (SD = 0.5), 17.9% juvenile females (SD = 0.4), 16.0% juvenile males (SD = 0.9), 27.0% adult females (SD = 0.8) and 23.5% adult males (SD = 0.6).

These proportions were used as the pool of available partners for focal calves and were contrasted with the sex difference for each age class. Using the average number of animals across all years, this meant that each year a focal calf had, on average, 19 potential male calf partners, 35 potential juvenile male partners, 52 potential adult male partners, 15 potential female calf partners, 39 potential juvenile female partners and 60 potential adult female partners. The mother was excluded from the analysis.

A likelihood score for each bout was then computed, taking into account the number of partners in each age and sex class involved for each bout, and the number of individuals in each age and sex class for the population. Basically, the expected likelihood of an interaction given the group composition is subtracted from the observed. The resulting value is between 1 and −1, where a positive value means that a male was more likely to be the partner than expected and a negative value means that a female was more likely to be the partner. This was contrasted within age class but between sexes to see if male or female calves preferred male or female calves, juveniles or adults as partners. These were averaged for each focal calf to compute an overall score. The sample size refers to the number of focal calves that interacted with either a male or female in each age class. Values could not be computed for calves who did not interact with individuals in a given age class. For example, if a calf did not interact with any juveniles, then a preference score for male vs. female juveniles could not be calculated. The Mann–Whitney U test was used to examine sex differences. The average value was also converted to binary scores for chi-square analysis of overall preference. Even though calves associate with far fewer individuals (averaging 34, range = 0–82; Gibson and Mann, 2003) than the total population available to them, the difference between observed and expected would not be affected by total number of associates, only by relative number of males and females within an age class.

Results

Events and bouts of socio-sexual behaviour

Data for focal calves and mothers included 1597 events during 245 bouts of socio-sexual behaviour. Focal calves or their mothers were involved in all but 11 bouts which were *ad libitum* and not used to calculate rates. Of the 1545

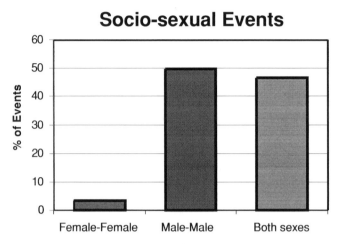

Figure 4.2. Proportion of socio-sexual events ($n = 1597$) that were homosexual or bisexual.

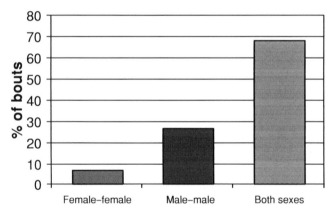

Figure 4.3. Proportion of bouts that were exclusively female, exclusively male or bisexual. Bout = events clustered within 5 min intervals ($n = 245$).

events involving males, at least one of the male participants was observed with an erection 15.5% of the time; this is an underestimate since erections are brief and difficult to see when the genital area is underwater or pressed against another individual. Most socio-sexual events were mounts (66.5%) and gooses (25.3%). Push-ups (3.7%) and petting (4.6%) were least common.

Half of all socio-sexual events were homosexual (Figure 4.2). This can largely be attributed to the fact that male–male interactions were more common. Out of 245 bouts (Figure 4.3), the majority (67.4%) involved both males and females; 26.1% involved males only and 6.5% involved females only. The types of socio-sexual interactions differed between the sexes. Nearly half of female–female

Table 4.1 *Sex-ratio percentages of actors to recipients for each socio-sexual event (n = 1597). M = male, F = female*

| Actor | Recipient | | | | |
	F	FF	M	MM	Events (*n*)
F	47.6	1.0	51.4	0.0	105
FF	0.0	0.0	100.0	0.0	3
MF	68.8	0.0	31.2	0.0	16
M	47.6	0.0	52.1	0.3	1037
MM	28.1	0.0	71.7	0.2	424
MMM	60.0	0.0	40.0	0.0	10
MMMM	100.0	0.0	0.0	0.0	2

bouts involved socio-sexual petting (43.8%). In contrast, none of the male–male bouts involved socio-sexual petting. Of all petting interactions involving the genital area (21 events), seven were female–female (33.3%) and 14 were male–female (67.7%). Six of the seven petting interactions between females were between four mother–daughter pairs. Five of the 14 male–female interactions involved a single male calf with his mother.

Although females participated in socio-sexual bouts, males were likely to continue the interaction after the female left and be more active. Sex ratios of actors versus recipients did also reveal a greater participation of males (Table 4.1). In a total of 1597 events, there were only five where two individuals were the recipients at the same time (all same-sex interactions), but males commonly acted together against a single individual. Females were the recipients in 683 of all events (42.8%), whereas male recipients constituted 914 events (57.2%). Females were actors in 8.1% of events and males in 94.4%. Males were both actors and recipients in the majority of events. In all cases, male–male interactions were more common than male–female for both dyadic and triadic interactions. Synchronized acts with more than one actor were recorded in 455 events (28.5%).

The more active role of males is also evident, if events are broken down by age and sex class (Table 4.2). The 42 events that involved both a calf and juvenile as actors simultaneously were added to the juvenile–juvenile category, whereas five cases of more than one recipient in an event were excluded.

Median partner number for heterosexual and male same-sex bouts was three, but all cases of female same-sex bouts involved two individuals. Male–male sociosexual bouts never involved more than four participants, but dyadic and triadic interactions were equally common. Heterosexual bouts were typically dyadic, but groups of 3–4 were also common (Table 4.3).

Table 4.2 *Proportion of socio-sexual events by age and sex class (divided by the total for a specific age-sex class of actor) (n = 1572). M = male F = female*

| | Recipient | | | | | | |
Actor	F calf	F juvenile	F adult	M calf	M juvenile	M adult	Events (n)
F calf	22.9	2.1	31.3	41.7	2.1	0.0	48
F juvenile	19.2	7.7	11.5	53.9	7.7	0.0	26
F adult	36.7	6.7	0.0	56.7	0.0	0.0	30
M calf	17.8	10.0	20.1	42.3	4.8	5.0	926
M juvenile	23.6	9.0	16.9	49.4	1.1	0.0	89
M adult	10.5	10.5	10.5	57. 9	0.0	10.5	19
MM calves	7.7	8.0	8.4	65.0	10.5	0.3	323
FF calves	18.6	6.9	21.6	34.3	18.6	0.0	102
MM adult	5.9	11.8	11.8	70.6	0.0	0.0	17

Table 4.3 *Percentage of socio-sexual bouts involving females (F) only, males (M) only, and both sexes*

| | | Number of participants | | | | | | | |
Sex	Socio-sexual bouts	2	3	4	5	6	7	8	Overall
FF	n	16							16
	Mean duration (min)	2.0							2.0
	SD duration (min)	1.6							1.6
MM	n	30	32	3					65
	Mean duration (min)	9.0	17.1	20.3					13.6
	SD duration (min)	9.2	10.2	19.6					10.9
MF	n	78	45	25	11	3	1	1	164
	Mean duration (min)	3.1	9.7	17. 8	18.6	39.0	75.0	8.8	9.3
	SD duration (min)	4.5	10.1	12.0	18.2	20.7	–	–	12.7
Total	n	124	77	28	11	3	1	1	245
	Mean duration (min)	4.3	12.8	18.1	18.6	39.0	75.0	8.8	9.9
	SD duration (min)	6.3	10.7	12.5	18.2	20.7	–	–	12.1

Bouts involving one male and one female were typically short, but with more participants they were longer (usually because more males were involved). The average bout length was 9.9 mins (SD = 12.1; median = 5.0; range = 0.1–75.0). These bouts often involved multiple dolphins and partner exchanges. Up to eight individuals participated in socio-sexual bouts and up to five in a single event. The maximum for one event was four juvenile males simultaneously mounting or attempting to mount an adult female.

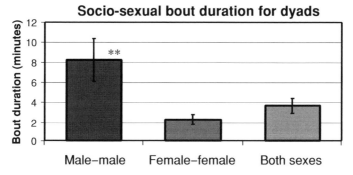

Figure 4.4. Mean bout durations and SE for dyads: male–male, female–female, male–female.

To increase statistical independence of bouts for comparison of bout length, only one bout per day that involved the same participants was used. If there were two such bouts, the one with longer duration was chosen. This method reduced the sample to 187. If we compare dyadic bouts only (thus controlling for the number of partners; Figure 4.4), male homosexual bouts lasted significantly longer than female homosexual bouts ($t = 2.06$, df $= 34$, $p = 0.047$) and heterosexual bouts ($t = 2.53$, df $= 72$, $p = 0.014$). However, female homosexual bouts were not significantly longer than heterosexual bouts ($t = 0.95$, df $= 62$, $p = 0.34$). For bouts that involved three individuals, male homosexual bouts (17.6 min, SD 2.1) were significantly longer than heterosexual bouts (10.9 min, SD $= 1.6$; $t = 2.31$, df $= 54$, $p = 0.025$).

Focal calves

Socio-sexual behaviour ('sex play') was often observed but significantly more common among male calves (mean $= 2.4$, *SD* $= 0.7$ events per hour, median $= 1.1$, range $= 0.0$–10.1) than female calves (mean $= 0.1$, SD $= 0.1$ events per hour; median $= 0.0$, range $= 0$–1.3; Mann–Whitney U $= 398.5$, $p < 0.001$, $n = 46$ calves).

The focal male was the actor in 79.8% of the events (SD $= 5.1$%, median $= 80.4$%). The focal female was the actor in 38.2% of events (SD $= 12.6$%, median $= 29.9$%), but the event rate was extremely low for most females. The three females who were actors more than recipients had only three or fewer events each. Even so, the difference in ratios between actor and recipient was significant between sexes (Mann–Whitney U $= 111.5$, $p = 0.015$, $n = 24$). Of 46 focal calves, 30.4% were not seen engaging in any socio-sexual behaviour. Nine of the 14 abstainers were females (34.6% of females), and five were males (26.3% of males).

Focal mothers and kin

Focal mothers were much less sexually active than their calves, but if they did engage in socio-sexual behaviour, their calf was almost always the partner. Average adult female (lactating) rate of socio-sexual interactions was 0.06 per hour (SD = 0.10, range = 0–0.34). In other words, one would have to observe a lactating (non-cycling female) for 17 hours, on average, before observing a single socio-sexual interaction. Fourteen of 26 mothers were not seen engaging in any socio-sexual behaviour. Of the 126 events (63 bouts) that involved a focal mother, nearly all, 91.3% of events and 92.0% of bouts, involved her offspring exclusively as either the actor or recipient. Only 0.8%, one event (petting of the genital area) involved an adult male with the focal adult female and ten socio-sexual events (7.9%) were between the focal female and unrelated juveniles or calves.

Fifty-eight socio-sexual bouts involved mothers with their calves only. Five focal female calves engaged in ten bouts with their mothers. All 13 male calves observed engaging in any form of sexual behaviour also mounted their mothers (44 bouts). The remaining four bouts also involved mother–calf pairs, but not during focal observations for that dyad. On average, female calves had socio-sexual interactions with their mothers once every 72.5 hours. Male calves had socio-sexual interactions with their mothers once every 14.2 hours.

Socio-sexual behaviour also occurred with other kin. For example, one male calf mounted his grandmother 67 times and another mounted his maternal sister five times. Eight maternal brother–sister pairs were never observed engaging in sexual behaviour. One pair of maternal brothers mounted each other 27 times, but five maternal brother dyads were not observed mounting each other. One female had a daughter by her own father, although in-breeding is generally quite low in the population (Krützen et al., 2004b).

Paternity is known for 16 offspring in the population (Krützen et al., 2004b). Adult males rarely mount calves, but one recently weaned juvenile male was mounted frequently by his father during one bout of socio-sexual behaviour. Several juveniles had been socializing and the juvenile was on the periphery of the group and not directly involved. Two adult males joined the group and the juvenile's father mounted his son (achieving intromission) repeatedly for several minutes, while his son lay passively on his side.

Socio-sexual interactions between same-sex partners

Only 3.7% of all socio-sexual events (n = 51 events, 16 bouts) involved only females (cf. Tables 4.1, 4.2). Nearly all events were dyadic, but one event involved three females. In eight bouts, adult females were the actors (either mounting, goosing or petting the genital area of another female). In all cases, a juvenile or calf female was the recipient. No homosexual interactions involving two adult females were observed in this study. In four of the eight bouts, the recipient was also the young daughter of the actor. Calves also directed socio-sexual behaviour towards immature and adult females in 12 bouts. Five of those involved daughter–mother pairs. Roughly half of all female homosexual interactions involved mothers and daughters, but such interactions were infrequent.

Male homosexual interactions typically involved more than two individuals (cf. Tables 4.1, 4.2). Another striking aspect of these interactions is that dolphins of all age and sex classes direct most of their socio-sexual interactions towards male calves. Male calves were the most common actors and recipients in socio-sexual events.

Synchronous socio-sexual behaviour

In 28.9% of 1572 events, two or more dolphins simultaneously acted on a third (cf. Tables 4.1, 4.2). Nearly all of the actors were male (95.8% of 455 events), with only three cases where females acted synchronously (all involved juvenile females mounting or goosing a male calf, 0.7%) and 16 where a male and female synchronously acting on a third (3.5%). Of the 436 events in which two males acted simultaneously in mounting, goosing or pushing-up (but never petting) a third, a male, was the recipient in 70.4% of the events. Most synchronized behaviour involved pairs of male (93.2%) with 12 cases (2.6%) when three or four males attempted syncronized matings, gooses or push-up of a single individual. Thus, most synchronized socio-sexual interactions involved males exclusively.

Partner preferences

Preferences within each age class were examined to determine if male or female calves biased their interactions towards calves, juveniles or adults by sex, given their availability in the population. For an individual to be included in the analysis, s/he had to interact with at least one member of the age class. First, mean scores for observed minus expected (see Methods) were ranked to test for calf sex differences in partner preferences. Male and female calves did not significantly differ in their preferences (Mann–Whitney U, all NS). Second, these scores (positive score = male preference and negative score = female preference) were converted to binary scores to examine whether calves interacted significantly more with males than females within an age class. Calves of both sexes showed a significant preference for interacting with male calves than with female calves (chi-square $= 8.9$, df $= 1$, $p = 0.003$, $n = 22$). For 20 calves that interacted with either a male or female juvenile, there were no significant preferences (chi-square $= 1.8$, df $= 1$, $p = 0.18$). Similarly, for ten calves that interacted with adults (excluding the mother), there was no significant sex preference (chi-square $= 1.6$, df $= 1$, $p = 0.21$).

Symmetry of male–male socio-sexual relationships

Most male–male interactions among calves were symmetrical (2.7 dyads, SD $= 1.5$), with regular role exchanges between the pair in terms of actor and recipient (Table 4.4). All calves had at least one symmetrical relationship. Male calves averaged 1.7 asymmetrical relationships (SD $= 1.3$). Individual differences were obvious; for example, COO had six symmetrical relationships and no asymmetrical, whereas SMO and SRY had more asymmetrical relationships than symmetrical. Overall, male calves who interact regularly with other males tend to form socio-sexual relationships (defined as males they mounted with more than five times) with up to seven males (average 4.2 males, SD $= 1.8$).

Discussion

In bottlenose dolphins, male calves engage in higher rates of both socio-sexual and homosexual behaviour than female calves and adults. Male homosexual bouts were also significantly longer than either female–female or male–female socio-sexual bouts and male calves were more often actors than recipients. Female homosexual interactions were infrequent, characterized by petting and were typically dyadic. Male socio-sexual behaviour among immatures (calves and juveniles) was characterized by mounting, goosing, synchrony and multiple participants (typically involving three or more individuals), not unlike adult male alliance behaviour. This suggests that these behaviours help mediate the development of male–male bonds. In addition, polyadic, particularly triadic interactions, are likely to help males practice adult courtship and sexual behaviour.

Bottlenose dolphin calves engage in very high rates of socio-sexual contact. The event rate for males (2.38/h) is nearly 40 times that for wild female bonobos (0.06–0.03 /h, Hohmann and Fruth, 2000), a species already characterized as hypersexual. Female calves engage in lower rates of socio-sexual contact (0.15/h), but still more than twice as often as that reported for bonobo females. In addition, the rate of bottlenose dolphin calf socio-sexual behaviour is higher than that reported for wild and captive primates during comparable developmental periods (for example, Brown and Dixson, 2000).

Table 4.4 Symmetry and asymmetry patterns of mounts for ten male calves and their primary male sexual partners. Only individuals with 10 or more mounts are shown. For each pair with at least five interactions, the proportion that were in one direction determined a symmetry score. For example, COO mounted SMO 110 times which represents 59% of the mounts involving SMO. Symmetrical dyads = scores of 25–75% (bold); assymetrical dyads = scores of 0–24% or >75% (underlined and italicized)

		Recipients																		Relationships (n) that are	
		M calves										M juveniles						M adults		Symmetrical dyads	Asymmetrical dyads
Actors		COO	DIN	JBO	JSE	NOM	QUA	RAB	SMO	SRY	URC	ARC	BOO	DAG	EDG	LAN	LIN	SHA	WAV		
M calves	COO				13			35	**110**			3		8				11	_5_	6	0
	DIN					16				_22_										1	1
	JBO		1			19				**42**										2	1
	JSE			27			67		**44**		_24_	14		_7_	_1_					3	3
	NOM	1		23					**43**							_11_				2	1
	QUA					12			**14**											2	0
	RAB	52			37				**44**	**44**		13	2	**13**			6		_29_	4	3
	SMO	75								_0_			17				1			2	3
	SRY			26				_7_	_0_		_9_			_0_	_11_					2	3
	URC	_5_			_3_	19			_9_											1	2
M juveniles	ARC							28			5					3				1	0
	BOO		2		7				**11**											2	0
	DAG						9			_0_					**14**	1				1	2
	EDG				5				**14**							_0_				0	2
	LAN							2	_0_											0	1
	LIN						4		1											1	0
M adults	SHA	**12**							1											1	0
	WAV	**11**						_2_	1									_3_		1	1

A number of hypotheses have been proposed to explain homosexual behaviour in animals (review in Vasey, 1995), such as group stability, tension reduction, reconciliation, dominance assertion, bond/alliance formation and mating practice. Little is known about hormonal or other proximate factors that might correlate with the hypersexual activities of male dolphin calves. But the patterning of male behaviour, including partner selection and synchronicity with other males, suggests that male homosexual behaviour is more than a hormonal by-product.

One functional explanation maintains that socio-sexual behaviour could enhance *group stability* or solidarity. However, with a fission–fusion social structure, groups are typically unstable, except between members of a stable adult male alliance. Homosexual behaviour is therefore unlikely to enhance group stability or solidarity but – as will be discussed below – may help foster and strengthen long-term alliance formation and maintenance between certain individuals.

Tension-reduction around food, other resources or during social conflict, is also an unlikely explanation, since socio-sexual behaviour rarely occurs in connection with hunting activities, which in Shark Bay is a solitary endeavour (though more than one dolphin may be attracted to large schools of fish) and dolphins do not share prey.

Similarly, *reconciliation* does not seem to be a likely function explanation. Although there seem to be competitive elements to socio-sexual interactions, most occur in a playful context and agonistic interactions involving calves are rare (Scott *et al.*, 2005).

Subtle formation of *dominance relationships* between males may occur during this early play period, but such relationships would be difficult to detect. Physical asymmetries would be reinforced during socio-sexual play and male socio-sexual behaviour may acquire more pronounced rank-related functions at later stages of development. Whether those mounted are generally subordinate to the mounter or actor is unknown and needs further study. Several male calves were mounted more often than other male calves. The receiver of mounts, especially when more than two males were involved, frequently displays with slaps of various body parts on the water, and belly-ups to avoid being mounted. The actors frequently chase the receiver who attempts to reverse roles by swimming behind the others. These attempts sometimes result in circle swimming, with two trying to remain behind a third, and the third trying to swim behind the others. Being the receiver appears to be the less desirable position for males, although females are sometimes observed swimming in front of males, apparently inviting males to chase and mount. Gooses and push-ups on the genital area can be forceful and calves clearly compete for the 'behind' position. The sequence of these interactions needs to be more systematically quantified. This is difficult given that dolphins move very quickly in the water, go in and out of view, and identification of dorsal fins needs to be rapid for determining who is the actor and who is the receiver (see Figure 4.1d). In non-human primates, the literature is equivocal regarding the relationships between mounting and rank, with some studies finding a correlation, others not (reviewed by Vasey, 1995). In the single quantitative study of captive adult male bottlenose dolphins, the more dominant male mounted the subordinate more often than reverse (Östman, 1991).

Similar to the view presented by Smuts and Watanabe (1990) for savanna baboons, *Papio anubis*, bottlenose dolphin male–male socio-sexual interactions appear to be more like negotiation rather than dominance assertion, especially given the duration and frequent role exchanges. Although rank within eventual alliances may be important, rank between alliances is likely to be equally important. The development of alliances would be facilitated by establishing reciprocity or 'trust' (cf. Zahavi, 1977) through repeated interactions and by providing an opportunity to assess the manoeuvrability and social skills of potential partners. For example, male partnerships in socio-sexual activities and *long-term bonds* could be established through taking turns as actor and recipient (symmetrical relationships, as above) and practicing synchronous movement in chasing, mounting, displaying and goosing other males or females. Trust is crucial to these interactions, because the recipient of socio-sexual behaviour is vulnerable by exposing the belly and genital area to one or more males in the advantaged rear position. Role exchanges may therefore be important for establishing trusted allies. A more detailed longitudinal study of male–male relationships across the lifespan is needed to understand how same-sex interactions relate to *bond* or *alliance formation*. Relationship negotiation, mediated by socio-sexual behaviour, may be particularly significant when there is multi-level alliance formation (alliances of alliances), a pattern found only in humans and Shark Bay bottlenose dolphins (Connor *et al.*, 1992a).

Most of the bonds that develop during infancy remain strong post-weaning, up to 16 years of age, when alliances begin to stabilize (Connor, unpublished data). Males may change their associations throughout development, but clearly male social relationships begin to form at a very early age, long before the formation of stable alliance partnerships. The number of symmetrical socio-sexual bonds formed in infancy averaged 2.7, approximating the typical size of first-order alliances (Connor *et al.*, 1992a).

A longitudinal study of alliance development in males with varying degrees of social experience as calves would provide insights concerning the importance of early male–male interactions. Early social experience is influenced by the sociality of the mother and the local cohort available, particularly male calves of the same age. All four calves whose mothers were sponge-carriers (a specific foraging strategy where the mother carries a marine sponge on her beak and uses it to ferret fish from the seafloor; Mann and Sargeant, 2003) were fairly solitary, spending over 80% of their time alone with their mothers and had no socio-sexual interactions. Solitary females, such as the sponge-carriers, may afford few social opportunities for their calves. Although calves have associates independent from their mothers, the pattern of calf sociality (number of associates and proportion of time in social groups away from the mother) is predicted by the mother's sociality (Gibson and Mann, 2003) However, calves vary in their tendency to separate from the mother (Mann and Watson-Capps, 2005), even though potential social partners may be a few hundred meters away. This could have long-term consequences for calves. As Crews (1998) points out, animals age as they gain socio-sexual experience, but do not necessarily gain such experience with age. Without experimental manipulation, it would be difficult to assess the effects of age and socio-sexual experience separately, or what determines male alliance size, stability, kinship and structure.

Bottlenose dolphin homosexual behaviour differs from that of most other mammals. Few species have homosexual interactions as often as heterosexual (Vasey, 1995). Bottlenose dolphin male calves have higher rates of same-sex interactions than opposite sex interactions. In contrast to primates and virtually all reports of mammalian sexual behaviour, a high proportion of sexual interactions included multiple partners and synchronous mountings, in which two individuals mount a third. Synchronous gooses and push-ups were also observed, as were leaps and displays. Male trios or two males with a female as the mountee, were the most common combinations.

The fact that males were typically acting in pairs or trios is consistent with the patterning of adult male consortships, in which the males mate with and defend a female for a period of time (Connor *et al.*, 1992a,b, 1996). Four or more males have never been known to consort with an individual female. Males do, however, form second-order alliance relationships with more than three males or form more labile first-order relationships with larger groups of males (Connor *et al.*, 1999).

Socio-sexual interactions are also likely to benefit males by providing opportunities for practice mating, which may be critical to male reproductive success. Practice may be more important in dolphins than terrestrial mammals because cetaceans are constantly in motion and females can easily turn belly-up or away from males during mating attempts. Despite the fact that adult males clearly coerce females to stay with them during some consortships and despite hundreds of hours of intensive observation of females in consortships by my colleagues Richard Connor and Jana Watson-Capps, a successful copulation (intromission), forced or cooperative, has not been observed. During mounts, it is nearly impossible to see if intromission occurs. Since synchronous behaviour and mounts are characteristic of adult courtship, synchronous socio-sexual practice may therefore be important.

Although relations between members of an alliance are clearly cooperative, with males cooperating to capture and retain cycling females, while keeping competing alliances away from the female, males are also in direct competition over fertilizations within an alliance. Thus, Shark Bay dolphin males are cooperating and competing at multiple levels in a fluid, three-dimensional environment, placing additional demands on socio-sexual practice not found in terrestrial animals. Females may need to practice avoiding unwanted matings, but this would favour more heterosexual than homosexual interactions. Most female homosexual interactions were between mothers and daughters, were dyadic and involved petting – possibly analogous to primate grooming. This would suggest that female same-sex interactions were less 'sexual' in nature and more affiliative.

Thus, homosexual interactions in bottlenose dolphins are expected to be much more common amongst males. They seem to serve multiple functions, although the exact fitness consequences, if any, are unknown. Our understanding of the social structure and relationships in a larger context would suggest that male–male socio-sexual interactions are significant for the development of close bonds or alliance formation, negotiating dominance relations within and between eventual alliances, and practicing courtship behaviours for adulthood.

References

Begos, K. (1999) Behind the smile. *The News Herald*, Panama City, Florida, 29 August.

Berry, M. and Signoret, J. P. (1984) Sex play and behavioural sexualization in the pig. *Reproduct. Nutrit. Develop.*, **24**, 507–13.

Brown, G. R. and Dixson, A. F. (2000) The development of behavioural sex differences in infant rhesus macaques (*Macaca mulatta*). *Primates*, **41**, 63–77.

Caldwell, D. K. and Caldwell, M. C. (1972) Behavior of marine mammals. In *Mammals of the Sea, Biology and Medicine*, ed. S. H. Ridgway, pp. 419–65. Springfield, IL: Charles Thomas.

CNN.com. (2002) Amorous dolphin targeting swimmers, 4 June.

Connor, R. C. and Smolker, R. A. (1985) Habituated dolphins (*Tursiops sp.*) in Western Australia. *J. Mammalogy*, **66**, 398–400.

Connor, R. C. Smolker, R. A. and Richards, A. F. (1992a) Dolphin alliances and coalitions. In *Coalitions and Alliances in Humans and Other Animals*, eds. A. H. Harcourt and F. B. M. de Waal, pp. 415–43. Oxford: Oxford University Press.

Connor, R. C. Smolker, R. A. and Richards, A. F. (1992b) Two levels of alliance formation among bottlenose dolphins (*Tursiops sp.*). *Proc. Nat. Acad. Sci. USA*, **89**, 987–90.

Connor, R. C., Richard, A. F., Smolker, R. A. and Mann, J. (1996). Patterns of female attractiveness in Indian Ocean bottlenose dolphins. *Behaviour*, **133**, 37–69.

Connor, R. C., Heithaus, M. R. and Barre, L. M. (1999) Superalliance of bottlenose dolphins. *Nature*, **397**, 571–72.

Connor, R. C., Wells, R. S., Mann, J. and Read, A. J. (2000) The bottlenose dolphin: social relationships in a fission–fusion society. In *Cetacean Societies: Field Studies of Dolphins and Whales*, eds. J. Mann, R. C. Connor, P. L. Tyack and H. Whitehead, pp. 91–126. Chicago: University of Chicago Press.

Crews, D. (1998) On the organization of individual differences in sexual behavior. *Am. Zool.*, **38**, 118–32.

Fahy, D. (2003) Is fungi gay? *The Mirror* (London), 5 July.

Fairbanks, L. A. (2002) Juvenile Veruet Monkeys: establishing relationships and practicing skills for the future. In *Juvenile Primates: Life History, Development, and Behavior*, eds. M. E. Pereira and L. A. Fairbanks, pp. 211–27. Chicago: University of Chicago Press.

Gibson, Q. A. and Mann, J. (2003) Individual variability in wild bottlenose dolphin calf social relationships in Shark Bay, Western Australia. Abstracts, 15th Biennial Conference on the Biology of Marine Mammals. Greensboro, NC. December.

Herzing, D. L. and Johnson, C. M. (1997) Interspecific interactions between Atlantic spotted dolphins (*Stenella frontalis*) and bottlenose dolphins (*Tursiops truncatus*) in the Bahamas, 1985–1995. *Aquatic Mammals*, **23**, 85–99.

Hohmann, G. and Fruth, B. (2000) Use and function of genital contacts among female bonobos. *Anim. Behav.*, **60**, 107–20.

Holecková, J., Bartos, L. and Tománek, M. (2000) Inter-male mounting in fallow deer (*Dama dama*) – its seasonal pattern and social meaning. *Folia Zool.*, **49**, 175–81.

Holekamp, K. E. and Smale, L. (1998) Behavioral development in the spotted hyena. *Bioscience*, **48**, 997–1005.

Kirkpatrick, R. C. (2000) The evolution of human homosexual behavior. *Curr. Anthropol.*, **41**, 385–413.

Kluger, J. (1999) The gay side of nature. *Time Magazine*, 26 April.

Krützen, M., Sherwin, W. B., Berggren, P. and Gales, N. (2004a) Population structure in an inshore cetacean revealed by microsatellite and MtDna analysis: bottlenose dolphins (*Tursiops sp.*) in Shark Bay, Western Australia. *Marine Mammal Sci.*, **20**, 28–47.

Krützen, M., Barre, L. M., Connor, R. C., Mann, J. and Sherwin, W. B. (2004b) O father: where art thou? Paternity assessment in an open fission–fusion society of wild bottlenose dolphins (*Tursiops sp.*) in Shark Bay, Western Australia. *Molecular Ecol.*, **13**, 1975–90.

Kyodo News International (2001) Wild dolphins off Brazil engage in homosexual behavior. *Kyodo News Int.*, 9 July.

Mann, J. (1999) Behavioral sampling methods for cetaceans: a review and critique. *Marine Mammal Sci.*, **15**, 102–22.

Mann, J. and Sargeant, B. (2003) Like mother, like calf: the ontogeny of foraging traditions in wild bottlenose dolphins (*Tursiops sp.*). In *The Biology of Traditions*, eds. D. Fragaszy and S. Perry, pp. 236–66. Cambridge: Cambridge University Press.

Mann, J. and Smuts, B. B. (1998) Natal attraction: allomaternal care and mother–infant separations in wild bottlenose dolphins. *Anim. Behav.*, **55**, 1097–113.

Mann, J. and Smuts, B. B. (1999) Behavioral development in wild bottlenose dolphin newborns (*Tursiops sp.*). *Behaviour*, **136**, 529–66.

Mann, J. and Watson-Capps, J. J. (2005) Surviving at sea: ecological and behavioral predictors of calf mortality in Indian Ocean bottlenose dolphins (*Tursiops sp.*). *Anim. Behav.*, **69**, 899–909.

Mann, J., Smolker, R. A. and Smuts, B. B. (1995) Responses to calf entanglement in free-ranging bottlenose dolphins. *Marine Mammal Sci.*, **11**, 168–75.

Mann, J., Connor, R. C., Barre, L. M. and Heithaus, M. R. (2000) Female reproductive success in bottlenose dolphins (*Tursiops sp.*): life history, habitat, provisioning, and group size effects. *Behav. Ecol.*, **11**, 210–19.

McBride, A. F. and Hebb, D. O. (1948) Behavior of the captive bottlenose dolphin *Tursiops truncatus*. *J. Comp. Physiol. Psychol.*, **41**, 111–23.

Östman, J. (1991) Changes in aggressive and sexual behavior between two male bottlenose dolphins (*Tursiops truncatus*) in a captive colony. In *Dolphin Societies: Discoveries and Puzzles*, eds. K. Pryor and K. S. Norris, pp. 305–18. Berkeley, CA: University of California Press.

Owen, E. C. G., Wells, R. S. and Hofmann, S. (2002) Ranging and association patterns of paired and unpaired adult male Atlantic bottlenose dolphins, *Tursiops truncatus*, in Sarasota, Florida, provide no evidence for alternative male strategies. *Canadian J. Zool.*, **80**, 2072–89.

Pereira, M. E. and Fairbanks, L. A., eds (2002) *Juvenile Primates: Life History, Development, and Behavior*. Chicago: University of Chicago Press.

Pfeifer, S. (1985) Sex differences in social play of scimitar-horned oryx calves (*Oryx dammah*). *Z. Tierpsychol.*, **69**, 281–92.

Rose, N. A. (1992) *The social dynamics of male killer whales*, Orcinus orca, *in Johnstone Strait, British Columbia*. Ph.D. thesis, University of California, Santa Cruz.

Schroeder, J. P. (1990) Breeding bottlenose dolphins in captivity. In *The Bottlenose Dolphin*, eds. S. Leatherwood and R. R. Reeves, pp. 435–46. San Diego: Academic Press.

Scott, E., Mann, J., Watson-Capps, J. J., Sargeant, B. L. and Connor, R. C. (2005) Aggression in bottlenose dolphins: evidence for sexual coercion, male–male competition and female tolerance through analysis of tooth-rake marks and behaviour. *Behaviour*, **142**, 21–44.

Signoret, J. P., Adkins-Regan, E. and Orgeur, P. (1989). Bisexuality in the prepubertal male pig. *Behav. Process.*, **18**, 133–40.

Smolker, R. A., Richards, A. F., Connor, R. C. and Pepper, J. (1992). Sex differences in patterns of association among Indian Ocean bottlenose dolphins. *Behaviour*, **123**, 38–69.

Smolker, R. A., Mann, J. and Smuts, B. B. (1993). The use of signature whistles during separations and reunions among wild bottlenose dolphin mothers and calves. *Behav. Ecol. Sociobiol.*, **33**, 393–402.

Smolker, R. A., Richards, A. F., Connor, R. C. Mann, J. and Berggren, P. (1997) Sponge-carrying by Indian Ocean bottlenose dolphins: possible tool-use by a delphinid. *Ethology*, **103**, 454–65.

Smuts, B. B. and Watanabe, J. M. (1990) Social relationships and ritualized greetings in adult male baboons (*Papio cynocephalus anubis*). *Int. J. Primatol.*, **11**, 147–72.

Terry, R. P. (1984) Intergeneric behavior between *Sotalia fluviatilis guianensis* and *Tursiops truncatus* in captivity. *Z. Säugetierk.*, **49**, 290–9.

Vasey, P. L. (1995) Homosexual behavior in primates: a review of evidence and theory. *Int. J. Primatol.*, **16**, 173–204.

Wells, R. S., Scott, M. D. and Irvine, A. B. (1987) The social structure of free-ranging bottlenose dolphins. In *Current Mammalogy*, Vol. I., ed. H. Genoways, pp. 247–305. New York: Plenum Press.

Williams, T. M., Friedl, W. A. and Haun, J. E. (1993) The physiology of bottle-nosed dolphins (*Tursiops truncatus*) – heart-rate, metabolic-rate and plasma lactate concentration during exercise. *J. Experiment. Biol.*, **179**, 31–46.

Zahavi, A. (1977) The testing of a bond. *Anim. Behav.*, **25**, 246–7.

5

Going with the herd: same-sex interaction and competition in American bison

HILDE VERVAECKE AND CATHERINE RODEN

Introduction

American bison bulls have been described as homosexually active animals whereas female–female interactions are considered to be rarer (review in Bagemihl, 1999). Bison are enormous grazers, with males that can be up to 2 m tall. Both males and females sport horns. They were formerly found throughout central North America, but wild bison do now only survive in protected areas. Older bulls generally live alone or in male groups. Females form their own herds with calves, yearlings and younger males. During the two months of the rutting season, female groups aggregate and males join these larger herds.

We addressed the topic of same-sex sexual interactions in bison while studying herds on a large-surface farm in the Belgian Ardennes where more than one hundred adult bison graze year round. Our observations revealed that the existing picture of homosexual behaviour in bison is far from complete. We therefore present our data in conjunction with previous descriptions of and explanations for same-sex mounting in bovids.

Males

Same-sex mounting has been regularly observed in American bison (*Bison bison*) males and can be as frequent as heterosexual mounting (Reinhardt, 1985a). Such homosexual behaviour has been interpreted as a display of dominance between bulls (McHugh, 1958), as play behaviour (McHugh, 1958; Lumia, 1972) or as a consequence of exclusion from breeding (Lott, 1983). Some age-classes of males may ever only mount members of their own sex, because they do not have access to females (review in Bagemihl, 1999).

Same-sex mounting also occurs in related bovines. Male–male mounting in zebu (*Bos indicus*) calves can be more frequent than female–female and male–female interactions (Reinhardt, 1983). Male calves of musk ox (*Ovibos moschatus*) mounted other males as often as females (Reinhardt and Flood, 1983), whereas sub-adult, sexually mature male musk ox directed sexual behaviour primarily – but not exclusively – towards females (Reinhardt, 1985b). In Scottish highland cattle homosexual mounting between male calves and adult bulls has also been observed (Reinhardt *et al.*, 1986).

Male–male mounting can be so frequent in domestic bovids (*Bos taurus*) that it is considered to be a problem for the cattle industry. The 'buller–steer syndrome' is characterized by the persistent following and repeated mounting of a steer (referred to as 'buller') by one or a group of steers (known as 'rider'). Due to excessive riding, the buller steer becomes exhausted, often shows loss of hair, swelling and trauma on the rump and tail head. In extreme cases, bullers can suffer broken bones or may even die from injuries (Stookey, 1997). The buller–steer syndrome may be partially related to the establishment of social hierarchies among unfamiliar animals. Indeed, it increases after entry into the feedlot or after regrouping. Riders are believed to mount repeatedly to overcome any challenge from the buller steer (Klemm *et al.*, 1983). Riders rarely display the sequence of 'normal' sexual behaviour, which includes flehmen, penile erections and intromission, whereas bullers are often low ranking, have an unfamiliar odour as well as higher oestrogen and testosterone levels. A historic increase in incidences of buller–steer sydrome in the USA has been related to feed containing oestrogenic substances, anabolic agents (diethylstilbestrol, DES), which promote growth, anabolic hormone implants and an increase in the number of animals per pen (Stookey, 1997; Edwards, 1995; Blackshaw *et al.*, 1997; Irvin *et al.*, 1979; Taylor *et al.*, 1997).

Females

Female–female mounting has never been reported for free-ranging bison (Lott, 1981). However, female wood-bison (*Bison bison athabascae*) kept in a zoo did mount each other (Matsuda *et al.*, 1996). This all-female group was monitored for hormone cyclicity (urinary PdG, faecal P4). 'Standing to be mounted' was considered to express behavioural oestrous and did indeed coincide with the hormonally detected oestrous. Similarly, two oestrous female *Bison bonasus* (wisent or European bison) kept without a mate attempted to jump on each other, and the same was observed twice in female wisent in a breeding reserve (Krasinski and Raczynski, 1967: 435). Four European bison engaged in mounting or were mounted by another female the day before calving. This occurred in the presence of an adult bull (Daleszczyk and Krasinski, 2001).

Female zebu (*Bos indicus*) may allow riding by subordinate females (Upham, 1995), and it occurs also among pregnant cows (Reinhardt, 1983). That mounting was uncommon between oestrous and non-oestrous cows may be due to the restricted contact with other females by oestrous cows, who remained in association with males. In oestrous water buffalo (*Bubalus bubalis*), homosexual behaviour was low (Ocampo et al, 1989) or absent (Perera *et al.*, 1977). Female–female mounting was never observed in the gaur (*Bos gaurus*) (Godfrey *et al.*, 1991).

This relative rarity contrasts with the frequent female–female riding amongst *Bos taurus*, where it is used as a sign to detect heat (for example, Pranee *et al.*, 1996). Peri-oestrous cows cluster in 'sexually active groups' who display mounting and chin-on-back behaviour (Williamson *et al.*, 1972a). They walk around to find other cows to mount, they stand to be mounted, do not let milk down fully, may vocalize frequently and have clear viscous vaginal discharge. This conspicuous syndrome is thought to have evolved as a signal to bulls. In Holstein cows the standing oestrous lasted 7.5 ± 2.4 hour and the total duration of female–female mounting rarely exceeded 12 hours unless two or more animals were in or near oestrous at the same time (Hurnik *et al.*, 1975).

In dairy cows (Williamson *et al.*, 1972b) and zebu (Medrano *et al.*, 1996), when one animal was stimulated to express oestrous using progestagen treatment, she had to wait for another cow to come into heat to fully exhibit behavioural oestrous. The intensity and duration of oestrous was diminished if only one cow was in heat at a given time (Hurnik *et al.*, 1975). Dairy cows also show individual preferences for certain cows as sexual partners (Cecim and Houster, 1988).

Ultimate explanations suggest that homosexual behaviour may have evolved to improve survival or, indirectly, reproductive abilities. For male bison, we will test our data against hypotheses which view same-sex behaviour as an expression of dominance, as a tool in the formation of mutual bonds or as training for heterosexual interactions. For females, we will focus in particular on the ideas that same-sex sexual behaviour stimulates heterosexual activity, that it might impede a competitor's reproduction or that it facilitates oestrous synchrony.

Methods

Species portrait

Wild bison males spend the winter mostly alone or in a bachelor herd, where, after working out their dominance positions, they peacefully graze in the same herd. Female herds contain on average nine adults (>2 years). Prior to the rut, which peaks in the first half of August, bison will merge into herds of about 40 cows with about four adult bulls per ten females (McHugh, 1958). However, female groups can vary between 10 and 40 members, and aggregations during the rut may contain 100–200 individuals (Lott, 1981).

During the rut, bulls attempt to monopolize oestrous females through tending bonds which typically last for a few hours up to a few days. Males try to remain continuously parallel to one particular cow and walk, graze and stand next to her. They may briefly leave her to patrol other females, check urine and vulvas. After a breeding season, a male may have lost 100 kg. Bulls frequently chase lower-ranking males away. Oestrous females are conspicuous in a herd by their attempts to escape from lower-ranking males or by the presence of tending males. They also move around more than other herd members (Lott, 2002). Most females calve in April and May. Weaning occurs over 8–20 months later (Green, 1987).

Study herds

Bison were studied from 1999 to 2002. The animals are held in semi-natural conditions on commercial farms in southern Belgium (Samrée and Bastogne). All bison were imported from the USA at one year of age. The animals graze year round on meadows averaging 25 ha with a total surface of 180 ha (Samrée) and 60 ha (Bastogne) in a rotation system. They are fed additional hay from September until April. All are individually recognizable through ear tags and phenotypic differences.

In winter, males are gathered in a bachelor group. In summer, they are divided into subgroups which are assigned to one of four different female herds: grand troupeau (GT), petit troupeau (PT), Francais (FR) and Recogne (RE), with a ratio of one male to 10–15 females. Females of GT and PT were of very similar age (GT: all born in 1992 except for two born in 1991; PT: all born in 1991) unlike those of FR (seven born in 1992, two in 1995, six in 1996, 27 in 1998). Each year, calves are weaned in December at about seven months of age and removed from the herd. In total, 30 breeding males older than one year were part of the study (1999–2002).

Thus, natural group dynamics and sex ratios are at least partially replicated, since breeding males join the herd in the pre-rut and leave mid autumn. The female herd generally moves in a large cluster. Even in the absence of predators, the bison graze only a few lengths apart and females with young rest within a body length of a neighbour. Male–female contacts occur occasionally and are accompanied by wallowing displays at the fences and by male bellowing during rut. At peak rut, a lone female may occasionally leave the herd to seek proximity and visual contact with a neighbouring herd.

In early August, females are synchronously in oestrous and up to 38% ovulates on the same day (Vervaecke *et al.*, 2002). They are seasonally polyoestrous, showing one or more oestri prior to conception (Vervaecke *et al.*, 2003). Male reproductive success is highly skewed with a single male fathering up to 80% of the calves of a herd (Roden *et al.*, 2002, 2003).

Data collection

Observations (Table 5.1) were carried out from a 4 × 4 vehicle within the herd. The animals were habituated to its presence. C. Roden focused on bulls and H. Vervaecke on cows. Observations were randomly distributed throughout the day, totalling more than 2000 hours.

For various types of interactions, all occurrences (Altmann, 1974) were scored, as well as identity of actor and receiver.

With respect to *reproductive behaviour*, we noted daily which bull tended a female (parallel position, following for more than 30 min, mounting attempts, licking, chin-on-back). A female was considered to be near oestrous in the two days before a tending day. Approximate conception dates were calculated from birth dates in the subsequent year minus the average gestation period of 268 days (Vervaecke *et al.*, 2003). The last day of tending or copulation in the week surrounding the calculated conception date was taken as the cut-off point, from which date on the cow was considered to be pregnant.

For *rank-order analysis* we used only dominant behaviours (threat, chase, contact aggression), which were followed by a submissive behaviour (walk away, jump away, flee), as well as displacements in which the approached individual yielded. For bulls, dominance was determined by scoring agonistic behaviours in the all-male groups during non-rut. During rut, too few males ($n = 2$, 3 or 4) lived in mixed-sex herds to measure linearity of ranks. For this period we determined dominance status of the bulls from submissive behaviours (yield to, keep distance, get up upon approach), which were always unidirectional.

Homosexual behaviour as defined in our study included components which are usually also shown by adults during heterosexual courtship behaviour but not in other contexts. Male same-sex behaviour (Figure 5.1) included (a) *chin-on-back* (actor puts chin on back or shoulder of receiver); (b) *attempt to mount* (the chest is thrown on to the behind of recipient in an attempt to mount but the actor does not leave the ground with both front legs); (c) *mount* (chest is thrown on to the back of recipient and the actor leaves the ground with both front legs); (d) *parallel* (parallel standing, walking or grazing, maintaining close proximity <1m, standing head to tail); (e) *persistent follow* (follow another individual from <1m over at least 5m, while the group is not in linear progression). Female same-sex

Table 5.1 *Observation schedule: bison herd identity: GT = grand troupeau, PT = petit troupeau, FR = francais, RE = Recogne*

Sex	Season	Date	Herd	Females (n)	Males (n)	Hours
Males	Non-rut	21 Nov–03 Dec 1999	All male (Go-Je-Gr-Ro-Bo-Ci-Or-Su)	–	8	48
	Non-rut	12 Feb–21 Feb 2000	All male (Go-Je-Gr-Ro-Bo-Ci-Or-Su)	–	8	70
	Non-rut	05 Jun–07 Jun 2000	All male (Ro-Bo-Ci-Or-Su-Lo-Ka-Pi)	–	8	26
	Non-rut	27 Nov–03 Dec 2000	All male (Bs-Gr-Ro-Bo-Or-Su)	–	6	39
	Non-rut	18 May–11 Jun 2001	All male (Bs-Gr-Ro-Bo-Or-Ha-Ax-Ju)	–	8	108
	Rut	8 Jun–8 Sep 2000	Mixed-sex: GT / PT / FR / RE	52 / 34 / 45 / 54	4 / 2 / 4 / 4	578
	Rut	12 Jun–6 Sep 2001	Mixed-sex: GT / PT / FR / RE	42 / 30 / 42 / 49	3 / 3 / 2 / 3	394
	Rut	9 Jul–5 Sep 2002	Mixed-sex: GT / PT / FR / RE	41 / 33 / 42 / 50	3 / 3 / 3 / 3	255
Females	Rut	26 May–14 Sep 2001	Mixed-sex GT	42	3 (9, 4, 2 yr)	325 (a)
	Rut	13 Jun–27 Oct 2001	Mixed-sex GT	32	4 (9, 4, 2 yr)	255 (b)
	Rut	9 Jul–13 Sep 2002	Mixed-sex GT	41	3 (10, 2, 1 yr)	172
	Rut	9 Jul–13 Sep 2002	Mixed sex PT	33	3 (2, 2, 1 yr)	46
	Rut	9 Jul–13 Sep 2002	Mixed sex FR	42	3 (6, 2, 1 yr)	40

Notes: (a) *Ad libitum* scoring of all occurrences of agonistic interactions.

(b) *Ad libitum* scoring of all occurrences of same-sex interactions.

Figure 5.1. American bison on a large-surface farm in the Belgian Ardennes: an incident of male–male mounting (photo: Hilde Veraecke).

behaviour included all of the above plus (f) *panting* (a panting sound produced while tossing the head, similar to that observed during heterosexual courtship); (g) *licking of genitals*; (h) *restrict* (standing perpendicular in front of an animal or moving in front ['cutting'] to prevent another individual from moving).

Data analyses

Ranks were calculated using MatMan, a program for the analysis of sociometric matrices (De Vries *et al.*, 1993). To determine if rank influenced which individual acted as actor and which as recipient of homosexual behaviour, the dyadic distributions were compared with an expected proportion of 50% by means of a non-parametric binomial test.

We used the rowwise matrix correlation method (Hemelrijk, 1990; De Vries, 1993) to analyse whether individuals modulated the frequency of homosexual acts as a function of their own or the recipients' dominance rank or age and to reveal reciprocity of same-sex interactions. To determine the relative importance of age and rank, a partial correlation analysis kept the effect of a specific factor (for example, age) constant (which is thus 'controlled for') while calculating the correlation between two other factors (for example, homosexual acts and rank).

Random permutations ($n = 20.000$) were carried out to obtain estimates of the
p-values. Values <0.05 were considered statistically significant and values from
0.05 to 0.1 indicated a trend (Siegel and Castellan, 1988).

If female reproductive status (pregnant, oestrous, non-oestrous) influenced,
homosexual activity was estimated by calculating expected frequencies. Since
all females entered the study period as non-pregnant and since all are expected
to conceive, an average of 50% of females should be pregnant on any given
observation day. The other 50% are either in oestrous or non-oestrous. Within
the oestrous cycle (maximum 21 days), four days were defind as oestrous (two
tending days, two near-oestrous days). Thus we obtained expected proportions
of 50% (pregnant), 9.5% (oestrous) and 40.5% (non-oestrous).

Inter-birth intervals (days between births in year 2001 and year 2002) of
females who interacted homosexually in 2001 were compared for the GT herd
by means of a t-test. For all herds, a Mann–Whitney U test analysed whether
timing of births in the preceding season (2001) was related to the actor role of
same-sex interactions in 2002. The reverse, that is if engaging as actor in homosexual interactions in 2001 influenced births in 2002, was only analysed for the
GT-herd.

We used the following dates as cut-off points for the timing of births in
2001 ('previous season'): herd GT, early 28 April–15 May, mid 16 May–2 June, late
3 June–19 June; herd PT, early 12 April–5 May, mid 6 May–29 May, late 30 May–
21 June; herd FR, early 29 April–20 May, mid 21 May–11 June, late 12 June–2 July.
Dates for births in 2002 (subsequent season) were as follows for herd GT: early
23 April–8 May, mid 9 May–23 May, late 24 May–9 June.

Results

Same-sex interactions between males

We recorded 313 homosexual interactions in 63 dyads (24 actors, 22
recipients), which included 142 attempts to mount and mounts plus 169 chin
rests. On one occasion full mounting with anal penetration was recorded. Only
once was it also observed that two males grazed parallel for just some minutes
and that one followed another male persistently.

Homosexual behaviour occurred almost exclusively outside the breeding season (Mann–Whitney U test, 2-tailed, rut vs. non-rut, $p < 0.05$; Figure 5.2). During
rut, most males did also interact homosexually (Figure 5.3), but the frequency of
same-sex interactions was much rarer. In the rut of the year 2000, 11 out of 14
males interacted (mean 4, median 2, range 0–14 mounts per actor). In 2001, three
out of 11 males interacted (mean 3, median 0, range 0–12 mounts per actor). In

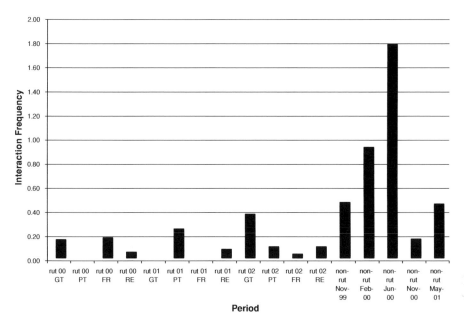

Figure 5.2. Average frequency of bull homosexual behaviour during the rut and non-rut.

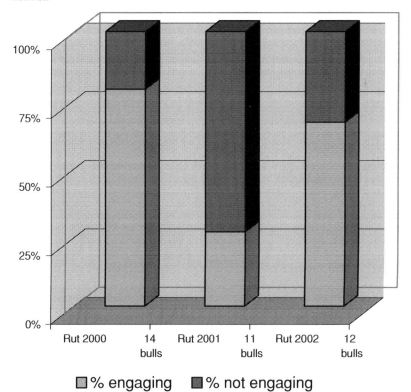

Figure 5.3. Proportion of bulls that engaged in homosexual behaviour.

2002, eight out of 12 males interacted (mean 3, median 1, range 0–16 mounts per actor). Most of these homosexual interactions were seen at the beginning of the breeding season, some at the end and virtually none during the peak of the rut.

Rank orders were found to be significantly linear among bulls in all-male groups (h' range = 0.96–1; $p < 0.001$; DC range = 0.88–1). Amongst the males that were put together with the female herds during rut (mixed-sexed groups), the direction of displays of dominance versus submission was likewise unidirectional.

In three of the five bachelor groups, younger bulls performed more homosexual interactions than older bulls (Kr matrix correlation, $p < 0.05$). Subdominants initiated and received more interactions than dominant bulls (Kr matrix correlation, $p < 0.005$) but this was a side effect of the correlations with the factor age (Kr partial matrix correlation). This indicates that age had a more important effect on homosexual behaviour than dominance rank. Homosexual interactions were found to be reciprocal in all but one bachelor group (Kr matrix correlation, $p < 0.005$) even after partialling out the factor age ($p < 0.005$, except for one bachelor group).

Same-sex interactions between females

We recorded 395 homosexual interactions for 123 dyads (GT 2001: 0.82 events/h; GT 2002: 0.63/h; PT 2002: 0.08/h; FR 2002: 1.80/h; Figure 5.4). In 16 cases in 2001 and 15 cases in 2002, the parallel position and following lasted for more than 30 min, accompanied by one or more attempts to mount, mounting, licking or chin-on-back. This pattern was therefore similar to what is defined as a tending bond for males.

In five cases a female was tended simultaneously by two others. Once a female was persistently followed by three and once by four females. The latter group also contained two males, which led to a cluster of seven sexually active individuals attempting to manoeuvre into a parallel position with one particular bison. We observed 31 instances (12 in 2001, 19 in 2002) where a female was being tended simultaneously by a male on one side and a female on the other.

The herd with the most homosexual interactions between females was the 2001 GT herd. It was made up of a majority of younger females of mixed ages that were kept with bulls up to ten years old. The PT herd, made up of older females of equal ages and males of maximally two years old, performed few homosexual acts. Individuals that interacted frequently during one year did not necessarily do so during the subsequent year, even if a herd had a similar bull age composition.

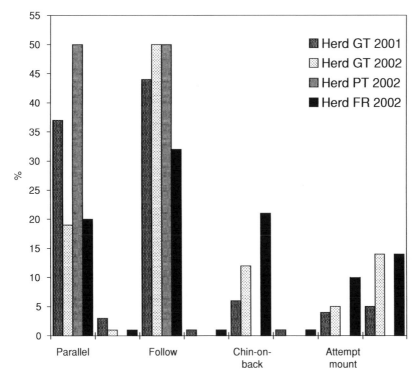

Figure 5.4. Distribution of female homosexual interactions.

As for males, rank orders were linear in female herds (2001 GT: $h' = 0.51$, $p < 0.0001$, DC $= 0.94$; 2002 GT: $h' = 0.53$, DC $= 0.93$; FR: $h' = 0.21$, DC $= 0.92$, $p < 0.0001$; PT $h' = 0.23$, DC $= 0.92$, $p < 0.0004$). In most of the 123 dyads, the actor was dominant over the recipient (binomial test, 2001: 2-tailed $p = 0.000$; 2002: $p = 0.021$). The behaviour was typically evenly distributed over the different group members. However, in herd GT of 2001, two out of ten actors were involved in 41% and 35% of all interactions.

Same-sex interactions of oestrous, non-oestrous and pregnant individuals were not evenly distributed (chi-square $= 34.371$, df $= 2$, $p < 0.0001$). Oestrous females were significantly more often than expected actors as well as recipients of homosexual behaviour. Non-oestrous females were significantly more often recipients, but were actors significantly less often than expected (Table 5.2). All but three cases of a non-oestrous female mounting another cow were directed towards individuals who were reintroduced after they were briefly removed from the herd for medical treatment of their calf.

Birth intervals tended to be longer for actors (average 343 days, SD $= 18.4$, SEM $= 8.2$) than receivers (average 358 days, SD $= 18.4$, SEM $= 4.6$; T-test:

Table 5.2 *Distribution of same-sex interactions (%) in relation to reproductive state of actors and recipients of the different dyads, significance = p < 0.05 (chi-square test on absolute frequencies of the sums)*

		Recipient				Significance
		Non-oestrous	Oestrous	Pregnant	Sum	
Actor	Non-oestrous	6.0	0.5	0.0	6.5	less*
	Oestrous	4.5	26.0	8.0	38.5	more*
	Pregnant	6.0	25.0	24.0	55.0	as expected
	Sum	16.5	51.5	32.0	100.0	
	Significance	more*	more*	less		

Note: Expected values: non-oestrous 40.5%, oestrous 9.5%, pregnant 50.0%.

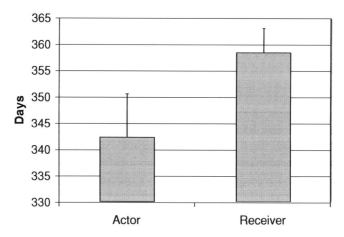

Figure 5.5. Average interbirth interval (days, SEM) of homosexual actors and receivers.

df = 19, two-tailed p = 0.133, one-tailed p = 0.066; data for GT-herd only; Figure 5.5).

In no herd did the timing of birth in the preceding season (2001) differ between individuals who engaged as actors or recipients in the next season (Mann–Whitney U test = 856,500 Wilcoxon W = 2081,500 Z = −1,156, two-tailed p = 0.248). Also, the timing of birth in the subsequent season (2002) was not different between females who acted as actors and recipients in the preceding season (Mann–Whitney U test = 67, Wilcoxon W = 95, Z = −0.589, two-tailed p = 0.556; GT herd only).

Finally, the number of dyads engaging in homosexual behaviour did neither correlate with the number of females being tended by males in general (rs = 0.132) nor with the number tended by the alpha male (rs = 0.016; data for GT herd in 2002).

Discussion

Our observations reveal that bison do frequently engage in homosexual behaviour. We will, in the following, point out similarities and differences with previous reports about the distribution across ages and sexes and test some of the functional explanations for same-sex behaviour which have been suggested.

Male–male interactions: general features

Bagemihl (1999), in his exhaustive review of animal same-sex activities, stated that homosexual behaviour is so prevalent among American bison bulls – especially during the rut, when it may be seen several times a day – that homosexual mounting is more common than heterosexual mounting. However, we observed most male–male mounting outside of the breeding season and it never exceeded the frequency of heterosexual behaviour.

Lott (1983) reported homosexual interactions only for young males excluded from breeding. Bagemihl (1999) reports they are most frequent in three-year-old bulls, decline thereafter and are virtually absent in five-year olds. In semi-wild herds, more than half of all mounting (55%) may be same-sex, and for some ages all mounting may be homosexual (reviewed in Bagemihl, 1999). We also found more homosexual behaviour in young bulls. However, all bulls also engaged in heterosexual behaviour such as tending and mounting, although some did not reproduce (Roden et al., 2002, 2003). Our study set-up may differ from the wild where bull to cow ratios are much higher (1:1 instead of 1:15) and where younger males might be excluded more often from access to females by the presence of multiple males.

Further, none of our study bison displayed the typical 'buller syndrome'. In cattle, an animal is considered a buller if it had received at least seven mounts and ten chin rests, followed by seven riders throughout the duration of a 4 minute period (Clavelle, 2002). Homosexual behaviour was never that prevalent nor persistent in our study animals.

Sexual behaviour of male mammals is, on the proximate level, largely stimulated by testosterone (Davidson, 1977; Davidson et al., 1978; Nelson, 2000). In male cattle, early castration markedly decreases the serum T level and reduces sexual behaviour (Baker and Gonyou, 1986; Tennessen et al., 1985). In contrast, steers can be sexually stimulated by injections of testosterone (Dykeman *et al.*, 1982). In cattle, intact males mount more often than castrates (Hinch *et al.*, 1982), while bulls respond to anti-testosterone treatment with decreased mounting (Andreae *et al.*, 1981). However, aggressive and sexual behaviour in bulls – including the onset and frequency of male–male mounts – is additionally influenced by Gonadotropin Releasing Hormone (GnRH), which is secreted in a pulsatile manner by the hypothalamus (Evans *et al.*, 1995).

Several potential functions of same-sex behaviour in male bison will be investigated in the following.

Male–male interactions: expression of dominance?

Same-sex sexual behaviour amongst males is often considered to be linked to the expression of social hierarchies. However, our study did not reveal a correlation of male–male mounting with social rank, confirming results for a small confined herd with eight males (Reinhardt, 1985a). This contrasts with a finding for sub-adult, sexually mature male musk ox, where male–male courtship was almost exclusively performed by the dominant partner (Reinhardt, 1985b). But then, this was not true for male musk ox calves where mounting was considered to represent play (Reinhardt and Flood, 1983). In young zebu males, mounting and dominance were negatively correlated, but higher ranking partners mounted subordinates more often than vice versa (Reinhardt, 1983). Here, mounting is also regarded as play, since the behaviour seemed to lack serious motivation ('wrong' sex), was discontinuous, could easily be replaced by other activities and the roles were interchangeable (Reinhardt *et al.*, 1978). In Scottish highland cattle, mounting did not reflect dominance either (Reinhardt *et al.*, 1986). Thus, most of these results are in strong contrast to what has been reported for cattle (Klemm *et al.*, 1983; Taylor *et al.*, 1997).

Male–male interactions: social bonding?

It is conceivable that male homosexual behaviour serves to form a mutual bond. We would then expect to find reciprocity and expect homosexual behaviour to be more pronounced when more males are present in the group.

Homosexual behaviour in our study bison was indeed mostly reciprocal and occurred outside the breeding season in bachelor herds. For musk ox, male–male courtship has been suggested to neutralize aggression and promote low-risk co-existence of males (Reinhardt, 1985b). In zebu calves, homosexual interactions were considered to lead to the formation of friendships that lasted more than half a year (Reinhardt *et al.*, 1978). Such aggregations could conceivably reduce predation on younger bulls, whereas older males can risk becoming solitary, due to their enormous size and strength (McHugh, 1972; Lott, 2002). Older males should thus have less need to bond, which is consistent with our finding that young bison bulls performed more homosexual interactions.

Male–male interactions: acquisition of experience?

As a functional explanation for male–male homosexual interactions, it has been suggested for both musk ox calves (Reinhardt and Flood, 1983) and male bison (Reinhardt, 1987) that this behaviour trains proper coordination and orientation of heterosexual mating. Similarly, Lumia (1972) argued that bison which perform a 'useful' type of play are more likely to survive, as are its offspring. We did indeed find a prevalence of male–male interactions in younger bulls, which were often accompanied by playful butting. Such mounting could conceivably function as a motor training for copulation, although penetration, ejaculation and persistent following are missing.

In summary, our data are compatible with the hypothesis that bison bulls use homosexual interactions for experience and training. When males are in an all-male group, the homosexual behaviour could also express social bonding. However, our results did not support the dominance expression hypothesis.

Female–female interactions: general features

Homosexual interactions between study females were more frequent than expected on the basis of previous reports about bison. Again, this might be due to our study set-up with its rather skewed bull:cow ratio of 1:15, which might have encouraged female–female interactions more than the ratio of 1:1 often found in the wild.

Still, the participation of individual cows fluctuated from one year to the next and depended largely on the female reproductive state. Females that tended other females were all tended by bulls in the same season and all the females were of proven fertility. The two females which interacted homosexually most pertinently displayed male-like behaviours, including the typical male panting sound, while performing chin-on-back, and one of them bellowed in an inter-group confrontation, which is common in male contest. However, such females were not 'pseudohermaphrodites', which have been described as being inter-mediate in size between bulls and cows, with male-like horns, female external genitals and a uterus combined with testes (Lott *et al.*, 1993, in: Bagemihl, 1999).

Hormones seem to be proximate triggers for female homosexual behaviour in bovids, too. In cattle, cows mount more when treated with testosterone (Kesler *et al.*, 1981). Active mounting has also been related to oestrogen secreted by the ovaries of females in heat (Nalbandov, 1958). Oestrogen seems to also elicit homosexual mounting and mountee behaviour in wisent that are about to give birth (Daleszczyk and Krasinski, 2001). This is due to increased oestrogen secretion of the placenta which renders the uterus susceptible for oxytocin, which in turn controls uterine contractions (Catchpole, 1972).

In the following, we will test our data against some proposed functional hypotheses for females.

Female–female interactions: improving chances to copulate with a male?

Same-sex interactions such as persistent following, parallel postures or mounting attract attention from other herd members. Males in particular become attentive when a female is being mounted. Similarly, dominant males who watch a female being mounted by a subordinate bull are known to quickly mount the female themselves. The occasional formation of clusters of females who interact homosexually suggests that the behaviour is 'contagious'. The over-representation of oestrous females also supports the hypothesis that same-sex interactions may stimulate sluggish males and speed up the inter-sexual tending bond. Homosexual behaviour of oestrous females should also be more common on days when more males are engaged in tending activities, since competition for male attention would then be higher. This, however, was not the case.

Same-sex interactions could in a different way be related to copulations, in that it trains female actors how to mount males – a behaviour seen prior to copulation, which, along with pushing the male during the tending bond, seems to stimulate the bull (Lott, 1981). Oestrous recipients would be ideal trainees since they would be more willing to stand parallel or to allow another female to mount. However, we do not know if homosexually interacting females do mount males during the tending bond – not all females do so – and also not if this behaviour has reproductive benefits, for example a faster conception.

Female–female interactions: related to dominance?

Same-sex interactions were clearly not related to 'conflict management' since they did not occur in the context of tension or competition over resources, except perhaps for a few events after briefly removed females were reintroduced into the herd. An increase in chin-on-back and mounting has been previously described for first contacts amongst domestic bovines (Bouissou, 1974). This could conceivably reduce stress or serve to evaluate physical condition and thus dominance status.

A connection of homosexual interactions with the expression of social hierarchies can also be construed from the fact that actors in same-sex sexual interactions were mostly dominant over the recipients. This is in line with suggestions that homosexual behaviour reaffirms the dominance hierarchy. Subordinate females tried at times to escape from same-sex actors but would ultimately not resist. This can be easily understood since resistance would carry the risk of injury from the dominant female's horns.

Nevertheless, such observations cannot easily explain why reproductive status was an important predictor of same-sex interactions.

Female–female interactions: impeding a competitor's reproduction?

Females frequently sniffed each other's urine and vulva and showed flehmen behaviour, probably gathering information about the reproductive status of others. This pattern is in line with the hypothesis that actors in female same-sex interactions may aim to decrease a competitor's reproductive success.

Various potential mechanisms could foster this function. In its simplest form, female actors hamper tending males to contact oestrous females, in that their tending disturbs the attempts of bulls to do the same. Potentially, female mounters could also reduce the mountee's receptivity by providing alternative sexual stimulation. In addition, homosexual interactions may stress the recipient, which would negatively influence conception or gestation. A recipient could also be prevented from reproducing with the dominant male, giving an edge to actors whose sons were fathered by the top male. Actors would potentially benefit from reduced resource competition, particularly during the subsequent winter, if a female recipient remains barren in a given year. A similar effect would be achieved if a female recipient reproduces later in the season, given the lower competitive value of late-born calves in their age cohort. With respect to all these scenarios, actors should preferentially target oestrous or pregnant females.

Our observations showed that the presence of an additional tending animal clearly disturbed the bull who had to manoeuvre constantly in attempts to remain in parallel stance with an oestrous female. Often the female actor placed herself between the male and the oestrous female. Reactions towards homosexual acts, however, were neutral or at best mildly negative. In cases of double tending – when an oestrous female was tended by both a male and a female – the bull would repeatedly displace the female actor, sometimes directing mild head threats and loud vocal threats. Nevertheless, bulls were certainly much more tolerant towards female actors than they were towards interfering males. Moreover, much of the female homosexual behaviour occurred away from the male.

The competition hypothesis predicts resistance of the female recipient against same-sex actors. Indeed, females tended by other females would sometimes try to escape, a behaviour also seen in the initial stages of heterosexual tending or when a low-ranking bull attempted to monopolize a female. Female recipients repeatedly galloped away, while the tending females tried to restrict their movements. These pursuits continued until both were panting with protruding tongues. That female actors were normally dominant is in line with the hypothesis, since tending can only be enforced by those who are physically capable of controlling the actions of others. We referred to the behaviour of same-sex actors as 'stalking', reflecting our impression that tending was imposed by the actor.

Costs for actors seem to be minimal. Calves may be at risk if their mothers frequently approach other females since the latter displace and aggress calves irrespective of their mothers' rank. This may explain why same-sex behaviour of two top-ranking females was very frequent in the one year in which they were barren and then dropped considerably in the subsequent year when both had calves.

The hypothesis does also predict a reproductive disadvantage to the recipient of same-sex behaviour. However, evidence for this is weak. Although birth intervals of recipients compared with actors were lengthened by an average of 16 days, birth dates were not delayed further into the season.

Female–female interactions: facilitating oestrous synchrony?

Homosexual interactions could serve to synchronize oestrous cycles via pheromonal signals. To have like-aged calves in a herd could be advantageous because it would allow a more coordinated movement, for example when a predator attacks, or it would increase the dilution effect because same-aged calves are also potential prey. That females sniffed each other's urine and vulva, showed flehmen behaviour and that homosexually interacting females were rarely pregnant is in line with this potential function.

However, several other results do not support the hypothesis. Firstly, same-sex dyads were not closer in subsequent birth dates than random dyads of the same herd. Secondly, females that had given birth early during the preceding season should be the first to resume oestrous and be preferred recipients in order to induce cyclicity. However, no significant difference existed between actors and recipients in whether they had had early, mid or late calves in the preceding season. Thirdly, both oestrous and non-oestrous females should actively initiate homosexual interactions but the latter were the least active. Finally, earlier births were neither recorded for recipients during the previous season nor for actors during the subsequent season.

In summary, although not all females engaged in homosexual behaviour, these interactions seem part of the normal female bison ethogram. We found some evidence for hypotheses that relate the behaviour to the expression of social dominance and reproductive competition, but benefits of the behaviour appear to be very subtle.

Acknowledgements

We thank Jean-Francois d'Hoffschmidt and the cooperation 'Bison d'Ardenne' where the study animals are held. We thank Geert Van den Broeck for logistic support and daily collection of birth dates. Hilde Vervaecke is supported by a grant of the Fund for Scientific Research, Flanders (FWO). Catherine Roden is supported by a Dehousse research grant of the Centre for Research and Conservation (CRC). We thank the Flemish Government for structural support to the CRC of the Royal Zoological Society of Antwerp (RZSA).

References

Altmann, J. (1974) Observational study of behavior: sampling methods, *Behaviour*, **49**, 227–67.

Andreae, U., Kuphaldt, D., Schütte, G. and Smidt, D. (1981) Sozial- und Sexualverhalten von Mastbullen in Rahmen von Versuchen zur geschlechtlichen Ruhigstellung. *Landbauforsch. Völkenrode*, **31**, 23–9.

Bagemihl, B. (1999) *Biological Exuberance: Animal Homosexuality and Natural Diversity*. New York: St. Martin's.

Baker, A. M. and Gonyou, H. W. (1986) Effects of Zeranol implantation and late castration an sexual, agonistic and handling behavior in male feedlot cattle. *J. Anim. Sci.*, **62**, 1224–32.

Blackshaw, J. K., Blackshaw, A. W. and McGlone, J. J. (1997) Buller steer syndrome: review. *Appl. Anim. Behav. Sci.*, **54**, 97–108.

Bouissou, M. F. (1974) Etablissements des relations de dominance-soumission chez les bovins domestiques. *Ann. Biol. Anim. Bioch. Biophys.*, **14**, 383–410.

Catchpole, H. R. (1972) Hormonal mechanisms during pregnancy and parturition. In *Reproduction in domestic animals*, eds. H. H. Cole and P. T. Cupps, pp. 415–40. New York and London: Academic Press.

Cecim, M. D. S. and Houster, C. L. (1988) Social preferences affect mounting activity in dairy heifers. *J. Anim. Sci.*, **66**, 231–8.

Clavelle, J. L. (2002) A description of mount behaviour during the Buller Steer Syndrome in a western Canadian feedlot. *Abstracts, 6th ISAE North American Regional Meeting*, University of Laval, Quebec Canada, 20–1 June.

Daleszczyk, K. and Krasinski, Z. A. (2001) Parturition behaviour of European bison *Bison bonasus* living in reserves. *Folia Zoolog.*, **50**, 75–8.

Davidson, J. M. (1977) Neuro-hormonal bases of male sexual behavior. In *Reproductive Physiology II (International Review of Physiology*, Vol. 13), ed. R. O. Greep, pp. 225–54. Baltimore, Maryland: University Park Press.

Davidson, J. M., Gray, G. D. and Smith, E. R. (1978) Animal models in the endocrinology of reproductive behavior. In *Animal Models for Research on Contraception and Fertility*, ed. N. Alexander, pp. 61–81. Hagerstown, Maryland: Harper & Row.

De Vries, H. (1993) The rowwise correlation between two proximity matrices and the partial rowwise correlation. *Psychometrika*, **58**, 53–69.

De Vries, H., Netto, W. J. and Hanegraaf, P. L. H. (1993) MATMAN: a program for the analysis of sociometric matrices and behavioral transmission matrices. *Behaviour*, **125**, 157–75.

Dykeman, D. A., Katz, L. S. and Foote, R. H. (1982) Behavioral characteristics of beef steers administered estradiol, testosterone and dihydrotestosterone. *J. Anim. Sci.*, **55**, 1303.

Edwards, T. A. (1995) Buller Syndrome, what's behind this abnormal sexual behavior? *Large Animal Vet.*, **50**, 6.

Evans, A. C. O., Davies, F. J., Nasser, L. F., Bowman, P. and Rawlings, N. C. (1995), Differences in early patterns of gonadotrophin secretion between early and late maturing bulls, and changes in semen characteristics at puberty. *Theriogenol.*, **43**, 45–51.

Godfrey, L. W., Lunstra, D. D., French J. A., Schwartz, J., Armstrong, D. L. and Simmons, L. (1991) Oestrous synchronization in the gaur (*Bos gaurus*): behavior and fertility to artificial insemination after prostaglandin treatment. *Zoo Biol.*, **10**, 35–41.

Green, W. C. H. (1987). Mother–daughter interactions in American bison (*Bison bison*): factors associated with individual variation. Ph.D. thesis, City University of New York.

Hemelrijk, C. K. (1990) A matrix partial correlation test used in investigations of reciprocity and other social interaction patterns at group level. *J. Theor. Biol.*, **143**, 405–20.

Hinch, G. N., Lynch, J. J. and Thwaites, C. J. (1982) Patterns and frequency of social behaviour in young grazing bulls and steers. *Appl. Anim. Ethol.*, **9**, 15–30.

Hurnik, J. F., King, G. J. and Robertson, H. A. (1975) Estrus and related behaviour in postpartum Holstein cows. *Appl. Anim. Ethol.*, **2**, 55–68.

Irvin, M. R., Melendy, D. R., Amoss, M. S. and Hutcheson, D. P. (1979) Roles of predisposing factors and gonadal hormones in the buller syndrome of feedlot steers, *J. Am. Vet. Med. Assoc.*, **174**, 367–70.

Kano, T. (1992) *The Last Ape: Pygmy Chimpanzee Behavior and Ecology*. Stanford, CA: Stanford University Press.

Kesler, D. J., Troxel, T. R., Vincent, D. L., Scheffrahn, N. S. and Noble, R. C. (1981) Detection of estrus with cows administered testosterone via injections and/or silastic implants. *Theriogenol.*, **15**, 327–34.

Klemm, W. R. C., Sherry, C. J., Schake, L. M. and Sis, R. F. (1983) Homosexual behaviour in feedlot steers: an aggression hypothesis. *Appl. Anim. Ethol.*, **11**, 187–95.

Komers, P. E., Messier, F. and Gates, C. C. (1992) Search or relax: the case of bachelor wood bison. *Behav. Ecol. Sociobiol.*, **31**, 195–203.

Krasinski, Z. and Raczynski, J. (1967) The reproduction biology of European bison living in reserves and in freedom. *Acta Theriolog.*, **12**, 407–44.

Lott, D. F. (1974) Sexual and aggressive behavior of adult male American bison (*Bison bison*). In *Behavior in Ungulates and Its Relation to Management*, Vol. 1, eds. V. Geist and F. Walther, pp. 382–94. Morges, Switzerland: International Union for Conservation of Nature and Natural Resources.

Lott, D. F. (1981) Sexual behavior and intersexual strategies in American bison (*Bison bison*). *Z. Tierpsychol.*, **56**, 115–27.

Lott, D. F. (1983) The buller syndrome in American bison bulls. *Appl. Anim. Ethol.*, **11**, 183–6.

Lott, D. F. (2002) *American Bison. A Natural History*. Berkeley, Los Angeles: University of California Press.

Lott, D. F., Benirschke, K., McDonald, J. N., Stormont, C. and Nett, T. (1993) Physical and behavioral findings in a pseudohermaphrodite American bison. *J. Wildlife Diseases*, **29**, 360–3.

Lumia, A. R. (1972) The relationship between dominance and play behavior in the American buffalo, *Bison bison*. *Z. Tierpsychol.*, **30**, 416–19.

Matsuda, D. M., Bellem, A. C., Gartley, C. J., Madison, V., King, W. A., Liptrap, R. M. and Goodrowe, K. L. (1996) Endocrine and behavioral events of oestrous cyclicity and synchronization in wood bison (*Bison bison athabascae*). *Theriogenol.*, **45**, 1429–41.

McHugh, T. (1958) Social behavior of the American buffalo (*Bison bison bison*). *Zoologica*, **43**, 1–40.

McHugh, T. (1972) *The Time of the Buffalo*. New York: Knopf.

Medrano, E. A., Hernandez, O., Lamothe, C. and Galina, C. S. (1996) Evidence of asynchrony in the onset of oestrous signs in Zebu cattle following Synchromate B treatment. *Res. Vet. Sci.*, **60**, 51–4.

Nalbandov, A. V. (1958) *Reproductive Physiology*. San Francisco: W. H. Freeman.

Nelson, R. J. (2000) *An Introduction to Behavioral Endocrinology*. Sunderland, MA: Sinauer Associates Inc.

Ocampo, M. B., Ocampo, L. C., Rayos, A. A. and Kanagawa, H. (1989) Present status of embryo transfer in water buffalo (a review). *Japan. J. Vet. Res.*, **37**, 167–80.

Perera, B. M. A. O., Pathiraja, N., Kumaratilake, W. L. J. S., Abeyratne, A. S. and Buvanendran, V. (1977) Synchronisation of oestrus and fertility in buffaloes using a prostaglandin analogue. *Vet. Rec.*, **101**, 520–1.

Pranee, R., King, G. J., Subrod, S. and Pongpiachan, P. (1996) Estrus behaviour of Holstein cows during cooler and hotter tropical seasons. *Anim. Reprod. Sci.*, **45**, 47–58.

Reinhardt, V. (1983) Flehmen, mounting and copulation among members of a semi-wild cattle herd. *Anim. Behav.*, **31**, 641–50.

Reinhardt, V. (1985a) Social behavior in a confined bison herd. *Behaviour*, **92**, 209–26.

Reinhardt, V. (1985b) Courtship behavior among musk-ox males kept in confinement. *Zoo Biol.*, **4**, 295–300.

Reinhardt, V. (1987) The social behaviour of North American bison. *Int. Zoo News*, **34**, 3–8.

Reinhardt, V. and Flood, P. F. (1983) Behavioural assessment in musk ox calves. *Behaviour*, **87**, 1–21.

Reinhardt, V., Mutiso, F. M. and Reinhardt, A. (1978) Social behaviour and social relationships between female and male prepubertal bovine calves (*Bos indicus*). *Appl. Anim. Ethol.*, **4**, 43–54.

Reinhardt, C., Reinhardt, A. and Reinhardt, V. (1986) Social behaviour and reproductive performance in semi-wild Scottish highland cattle. *Appl. Anim. Behav. Sci.*, **15**, 125–36.

Roden, C., Vervaecke, H. and Van Elsacker, L. (2002) Genetic paternity and reliability of observed mating behaviour in American bison bulls (*Bison bison*). In *Advances in Ethology, 37 (Supplements to Ethology)*, eds. M. Dehnhard and H. Hofer, p. 69. Berlin, Vienna: Blackwell.

Roden, C., Vervaecke, H., Mommens, G. and Van Elsacker, L. (2003) Reproductive success of bison bulls (*Bison bison bison*) in semi-natural conditions. *Anim. Reproduct. Sci.*, **79**, 33–43.

Siegel, S. and Castellan, N. J. (1988) *Non-parametric Statistics for the Behavioral Sciences.* 2nd edn. New York: McGraw-Hill.

Stookey, J. M. (1997) *Buller Steer Syndrome: Alberta Feedlot Management Guide.* Barrhead/AB: Alberta Agriculture, Food and Rural Development.

Taylor, L. F., Booker, C. W., Jim, G. K. and Guichon, P. T. (1997) Epidemiological investigation of the buller steer syndrome (riding behaviour) in a Western Canadian feedlot. *Aust. Vet. J.*, **75**, 45–51.

Tennessen, T., Price, M. A., and Berg R. T. (1985) The social interactions of young bulls and steers after re-grouping. *Appl. Anim. Behav. Sci.*, **14**, 37–47.

Upham, L. (1995) Aspects of reproduction in female *bos indicus* cattle: a review. Nebraska Veterinary and Biomedical Sciences Newsletter. http://nvdls.unl.edu/dec95.htm.

Vervaecke, H., Roden, C., Schwarzenberger, F. and Van Elsacker, L. (2002) Oestrus synchrony in female American bison (*Bison bison*): how tight is the race? *Abstracts*, 9th Benelux Congress of Zoology, Antwerp, Belgium, 8–9 November.

Vervaecke, H., Roden, C., Schwarzenberger, F., Palme, R. and Van Elsacker, L. (2003) Duration of gestation in American bison (*Bison bison*): behavioural and hormonal data. *Abstracts*, 5th Annual Research Symposium. Marwell Zoo, 7–8 July.

Williamson, N. B., Morris, R. S., Blood, D. C. and Cannon, C. M. (1972) A study of oestrous behaviour and oestrus detection methods in a large commercial dairy herd: II. Oestrous signs and behaviour patterns. *Veterin. Rec.*, **91**, 58–72.

6

Exciting ungulates: male–male mounting in fallow, white-tailed and red deer

LUDEK BARTOŠ AND JANA HOLEČKOVÁ

Introduction

Intrasexual mounting in ungulates

Same-sex mounting is quite common among animals (Weinrich, 1980; Dagg, 1984; Vasey, 1995; Bagemihl, 1999). It is particularly well documented for hoofed mammals where it has been observed in at least 12 deer species (Table 6.1) and many other ungulates (see review in Bagemihl, 1999). Same-sex mounting occurs in both sexes and at all ages, and in the wild as well as in captivity. The behaviour is, however, generally more common in the young than in adults, often occurring during play. Also, captive animals tend to exhibit the behaviour more often than those in the wild (Dagg, 1984).

While male–male mounting is a common same-sex behaviour in cervids and ungulates, female–female mounting, though uncommon, is also widespread (Dagg, 1984). For farm animals in particular, intra-female mounting has been understood to be a normal element of behaviour associated with oestrous.

Nevertheless, the topic of same-sex sexual interactions amongst cervids is still poorly explored and understood. We will therefore review findings from earlier studies of fallow deer (Holečková, 1999; Holečková et al., 2000), and present new research on white-tailed and red deer. We focus on male–male mounting, as female–female and female–male mounts were very rarely observed in our studies. They were restricted to the period of rut, and thus, like in cattle, seem to be associated with oestrous. Observations of female–female mounting outside the rut in red deer (Hall, 1983a) could not be replicated by us.

Table 6.1 *Intra-sex mounting in cervids*

Species	Latin name	Conditions (a)	Sex (F = female, M = male)	Young individuals either sex	Source
Fallow deer	*Dama dama*	C S	M		Holečková et al. (2000), Holečková (1999)
Pere David's Deer	*Elaphurus davidianus*	C	F M	yes	Schaller and Hamer (1978), Wemmer et al. (1983)
Muntjac	*Muntiacus reevesi*	C	M		Barrette (1977)
Caribou	*Rangifer tarandus*	C W	F M	yes	Bergerud (1974), Bagemihl (1999)
Moose	*Alces alces*	C W	F M	yes	Bagemihl (1999)
Red deer	*Cervus elaphus*	C S W	F M	yes	Bützler (1974), Darling (1937), Hall (1983a), Lincoln et al. (1970), Bagemihl (1999), Bartoš (1985, unpubl.)
Mule deer	*Odocoileus hemionus*	C W	F M	yes	Geist (1981), Halford et al. (1987)
White-tailed deer	*Odocoileus virginianus*	C W	F M	yes	Hirth (1977), Bagemihl (1999), Rue (1989), Hesselton and Hesselton (1982), Bartoš and Holečková (2004, this volume)
Sika deer	*Cervus nippon*	C	M		Chapman et al. (1984)
Elk	*Cervus elaphus roosevelti*	C W	M		Harper et al. (1967), Rue (1989)
Axis deer	*Axis axis*	W	M		Schaller (1967)
Southern pudu	*Pudu pudu*	C	M		Pluhek 2002 (personal communication)

Note: (a) C – captivity, S – semi-wild, W – wild.

The question of function

While same-sex mounting has been described in some detail for deer, it is difficult to link it to a single underlying cause. The behaviour might indeed have multiple functions, both sexual and social, as we have postulated for red deer calves mounting their mothers (Vaňková and Bartoš, 2002).

Male–male mounting is often considered to express intra-male competition. Among feedlot steers, for example, it is thought to indicate dominance and aggression (Klemm *et al.*, 1983; Franc *et al.*, 1988; Bartoš *et al.*, 1988). Mounting has also been linked to the dominance hierarchy in Grant's gazelle (Walther, 1965) and white-tailed deer and elk (Rue, 1989). In fact, Walther (1984, p. 293) has strongly argued that 'mounting is not a sexual behaviour beyond any possible doubt in hoofed mammals, and that mounting a subordinate animal of the same sex may be more related to an act of aggression (jumping at the other and throwing the weight of the body on him) and comes close to a demonstration of dominance in agonistic interactions'.

However, not all researchers agree with this position. Resko *et al.* (1996) suggest that same-sex mounting in rams is in fact a component of sexual activity alone and does not serve any further purpose. It is also not always the dominant individual that mounts the subordinate. Captive red deer stags, for example, tend to mount animals dominant to themselves (Hall, 1983a). Walther (1978) reported intra-male mounting among Thomson's gazelles during migrations – seemingly unrelated to dominance disputes. In blackbucks this behaviour is very common and usually involves play fighting. The mounter is not necessarily of a higher rank than the mountee (Dubost and Feer, 1981).

In cattle, mounting increases after social disruption (Dalton *et al.*, 1967) such as formation of new herds (Tennessen *et al.*, 1985) or when bulls were regrouped at abattoirs (Bartoš *et al.*, 1988; Franc *et al.*, 1988). Vice versa, mounting decreased in socially stabilized groups (Bartoš *et al.*, 1988, 1993). In domestic bulls and some other mammals, including captive red deer (Hall, 1983b; Smith and Dobson, 1990) and fallow deer (Holečková, 1999; Holečková *et al.*, 2000), intra-male mounting arises during disturbances caused by humans (Dalton *et al.*, 1967; Dagg, 1984).

Testicular hormones influence same-sex mounting at the proximate level. Cattle females mount more when treated with testosterone (Kiser *et al.*, 1977; Bouissou, 1978; Laaser *et al.*, 1981; Kesler *et al.*, 1981). Similarly, intact males mount other males more often than castrates (Hinch *et al.*, 1982; Tennessen *et al.*, 1985), while bulls respond to anti-testosterone treatment with decreased mounting (Andreae *et al.*, 1981). Moreover, castrated bucks of fallow deer treated with testosterone were more likely to mount, though there was no apparent dependence of mounting activity on testosterone levels (Holečková, 1999; Holečková *et al.*, 2000).

Juvenile intra-sex mounting has been interpreted as play (Altmann, 1963; Geist, 1981; Schilder, 1988). In semi-domestic cattle, male calves engaged more frequently in same-sex mounting than females (Reinhardt, 1983), suggesting that the behaviour serves as 'training' for future successful copulations (see also Vaňková and Bartoš, 2002). Similarly, intra-sex mounting in free-ranging zebra young has been linked to establishing or confirming dominance hierarchies (Schilder and Boer, 1987).

Thus, no consensus exists on the function of mounting in cervids in general and that of male–male mounting in particular. This chapter will examine male–male mounting in fallow, white-tailed and red deer, and specifically its relationship with dominance. We will also consider alternative explanations, including hormone levels, age, environmental disturbance and the general level of 'excitability' of a given species.

Methods

Basic social organization of deer

Fallow, white-tailed and red deer have similar social organization. During most of the year, they live in sex-segregated groups (Bützler, 1974; Chapman and Chapman, 1975; Clutton-Brock *et al.*, 1988; Marchinton and Hirth, 1984). Females aggregate with their offspring in matrilineal groups. Males live either in bachelor herds, often of rather unstable membership, or on their own (Bützler, 1974; Marchinton and Hirth, 1984; Bartoš and Perner, 1985). White-tailed deer groups are usually smaller than those of fallow and red deer (Hawkins and Klimstra, 1970; Brown Jr, 1974; Chapman and Chapman, 1975; Clutton-Brock *et al.*, 1988).

All three species are seasonal in their reproduction, with an autumn rut lasting from a few weeks to several months. Fallow and red deer usually form harems with a single male, though different populations may adopt different mating systems (Apollonio, 1998; Carranza, 1995). White-tailed deer bucks usually form short-lived, consecutive tending bonds with individual does (Marchinton and Hirth, 1984).

Cycles of male antler growth reflect underlying hormonal levels. For example, antler growth in fallow deer depends on seasonal levels of androgene. A plasma androgene concentration above a minimal threshold is a necessary prerequisite for normal antler re-growth (Bartoš *et al.*, 2000). This occurs in the spring, immediately after a period of the lowest seasonal levels, when antlers are cast. High androgen levels stop antler growth and induce antler mineralization followed by velvet shedding (Bubenik, 1990). Like fallow deer, red deer also cast their antlers in the spring, followed by immediate antler re-growth. White-tail males follow a different pattern and cast their antlers as seasonal testosterone levels drop, with re-growth following only after an interval (Sempéré and Boissin, 1982).

Study groups

White-tailed deer (Odocoileus virginianus)

Groups of white-tailed deer bucks were studied in an enclosure at the University of Georgia's Whitehall Deer Research Facility in Athens, Georgia. Group 1, observed in 1999, consisted of 16 deer. About two months before the expected date of antler casting, we confined 12 bucks (1.5–7.5 years) in a 0.6 ha enclosure until they shed velvet the following year. From 10 February to 3 June, we conducted almost daily observations which began with the provision of food and finished when all animals left the feeding site or laid down. We assessed dominance relationships via approaches, displacements, attacks, threat gestures, sparring and fighting (as defined by Hirth, 1977) and recorded mounting. In 2002, we observed two separated groups at the same facility. Group 2 contained both young and adult bucks ($n = 18$), while group 3 consisted predominantly of yearlings and two-year olds ($n = 12$).

In our analysis of dominance we avoided dependence among repeated observations made on the same bucks by applying a log-linear repeated measurement model with weighted least squares (PROC CADMOD, SAS). Differences between the groups were calculated using contrasts of maximum likelihood estimates (Stokes et al., 2000). When mounting incidences were low, we applied a Poisson regression (PROC GENMOD in SAS V9.0), with mount counts plus attacks as a measure of exposure. A log transformed count of attacks was treated as the offset. Because some of the bucks mounted others more times, the initiator of the activity was taken as a repeated subject in the model. The following factors were available: categorical variables 'season' (1999, 2002), 'group' (groups 1, 2 and 3), 'month' (February–August) and numeric variables 'age' (up to ten years) and 'body weight'. A full model containing all explanatory variables and first-order interaction terms was initially fitted. Each term was then dropped from the full model unless doing so had a significant effect on model fit.

Fallow deer (Dama dama)

Records of intra-male mounting were obtained during three seasons in the deer park in the Czech Republic and covered all months except September.

Concurrently, we took blood samples from two captive intact fallow deer bucks to compare the intra-male mounting frequency with seasonal changes of serum testosterone (Holečková *et al.*, 2000).

Red deer (Cervus elaphus)

We observed a herd of white coloured red deer – which has a rare colouration – for more than 17 years (Bartoš, 1990). We recorded male–male mounting on a single day only, when a subordinate individual mounted a higher-ranking mountee twice (Bartoš, unpublished).

However, we learned that male–male mounting could be elicited during a specific husbandry manoeuvre. In 1998, on a red deer farm at Bohumilice, Southern Bohemia, a mixed-sex herd of red deer with stags in velvet was being pushed through a corridor into a handling facility, where they would be divided into breeding groups for the coming rut. When the deer became confined by walls of solid timber, they appeared anxious and many mounts occurred. During the next time of handling at the beginning of September we video-recorded the behaviour of a mixed group of continuously changing composition. The number of animals in the group varied from 22 to 31, with some moving into the facility and some remaining outside. All animals came from the same paddock and had known each other for a long time.

Results

Qualitative description of male–male mounts

Intra-male mounting follows the same form in all three species and is similar to heterosexual mounting. Occasionally, mounts were preceded by one male nuzzling the other's rear end. Mounts were attempted mostly from behind, as in heterosexual intercourse. Sometimes, in what appeared to be lack of experience, a male tried to mount another male from the side or even over the head. Fallow and white-tailed deer mountees would, in such situations, attack the mounter's belly, though this did not deter further attempts. Single mounts, not exceeding 10 seconds, were the most common for all species. Captive fallow and white tailed deer bucks, however, also mounted two to six times in a row.

We were able to record occassional erections of the mounter in small captive facilities, but not under semi-wild conditions. Although some males, particularly captive fallow deer, tended to prefer certain individuals for repeated mounting, no real pair-bonding was observed (Holečková, 1999). On occasion, while one male mounted another, a third male approached the pair and mounted the mountee, too (Figure 6.1).

Figure 6.1. A fallow deer buck mounts another one, while being mounted by a third (photo: Ludik Barto).

For white-tailed deer, it was typical that subordinates approached a dominant with protruding tongue and lowered the body's back as if lifting the front part up. However, they would not go through with the mount and instead displayed submissive postures and looked away. It was twice recorded that mountees attacked the mounter vigorously. Also, a subordinate mountee kicked a quiet bystander. In one incident, yearlings engaged in mutual grooming of head, ears and neck, while standing in front of each other with necks crossed. A third yearling mounted one of the grooming animals who moved away. The mounter then groomed the remaining yearling. Another case was triggered when a hot-air balloon flying above alerted the whole group. A three-year-old dominant buck then attempted to mount a two-year-old mate. The mountee moved away, but the mounter kicked and mounted him for 10 seconds, while displaying an erect penis.

The following quantitative data will in particular explore the inter-relationship of mounting and dominance relations between the bucks.

White-tailed deer

Male–male mounting occurred during all study months, including periods when the deer had hard antlers, when they cast antlers and during re-growth (Figure 6.2, lower part).

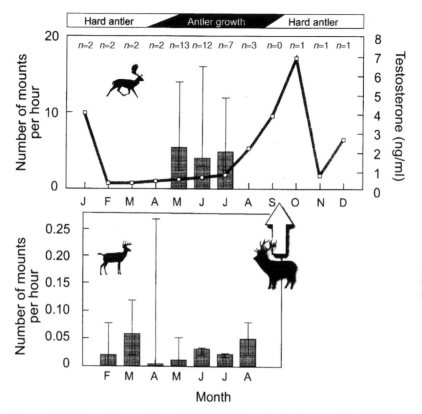

Figure 6.2. Male–male mounting of white-tailed deer bucks (lower part) and fallow deer (upper part; adapted from Holečková *et al.*, 2000), measured as frequency per hour of observations. Bars express median with range between lower and upper quartile. Seasonal changes in mean testosterone are shown for fallow deer. $n =$ number of observations sessions. Arrow indicates when handling of red deer was video recorded.

Mountees often responded to mounts by moving away (31% of cases in group 1; 43% in group 2; 75% in group 3; Figure 6.3). The log-linear model revealed an interaction between dominance and group (chi-square (2) = 6.46, $p = 0.039$, Figure 6.4), indicating that dominance and group composition cannot be analysed separately. The contrast test results indicated that proportion of dominant mounters in group 1 was different from both group 2 (chi-square (1) = 9.50, $p = 0.002$) and group 3 (chi-square (1) = 7.52, $p = 0.006$), while groups 2 and 3 did not differ significantly. Thus, dominance could at best partially explain mounting direction.

Moreover, compared with other activities, the incidence of mounting was very low (1.0% in group 1, 1.1% in group 2, 2.2% in group 3). The final step of the iterated model based on a Poisson regression contained only the effect of 'age' (chi-square (1) = 10.28, $p = 0.0013$). Odds ratio for age was equal to 0.4497. Thus,

Figure 6.3. Male–male mounting in white-tailed deer. The mountee moves away in response to being mounted (photo: Ludik Barto).

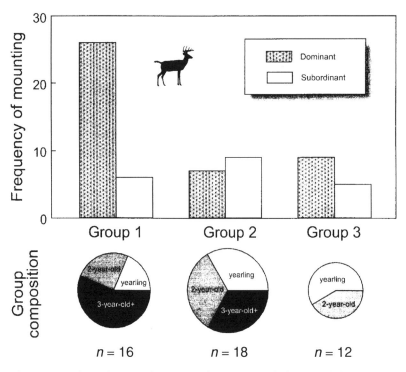

Figure 6.4. Male–male mounting among three groups of white-tailed deer. Upper part indicates frequency of mount initiation in relation to dominant or subordinate rank. Lower part details group composition.

those with the highest probability to mount others were the youngest bucks (yearlings). With each year of increasing age, probability to mount decreased to 45.0% of the previous age, reaching 0.03% of probability of the yearlings at the age of ten.

Fallow deer

We recorded 314 male–male mounts during 30 observation sessions totalling 120 hours. The mountee did not normally respond to being mounted. Mounting was restricted to May, June and July (Figure 6.2, upper part) and coincided with an increase of testosterone levels following the seasonal minimum, when antlers are cast. Thus, mounting did clearly not depend on high testosterone levels per se. In 78.0% of all cases, a smaller buck mounted a larger one. Therefore, dominance was again no good predictor for the direction of a mount.

Red deer

Data on red deer were, as mentioned above, collected, while a farmed herd was pushed through a corridor into a handling facility for division into breeding groups (Figure 6.2, arrow). On average, the corridor held eight stags, seven hinds, and 12 calves at a time. The entire handling episode lasted 2 hours 14 minutes during which 70 mounts were recorded. Hinds and calves did not mount any other animal. Stags were more likely to mount hinds ($n = 59$) than other stags ($n = 12$; comparison between mount frequencies and presence of animals minus one according to sex; Fisher exact probability test, $p = 0.008$). Mounters were larger than mountees on seven occassions, equally sized on three and smaller on two. Only two of the seven smaller mountees and two of the three equally sized stags responded by moving away. One of the equally sized stags attacked the mounter.

It seemed unselective whether hinds or stags were mounted, and the sequence of events appeared to be unguided. For instance, a stag mounted a yearling male who moved away, whereupon the stag mounted a hind twice. The mounted hind moved away, whereupon the stag approached the same yearling male as before and mounted him once more. A stag who had been recorded mounting another stag during this event was later used as a sire and produced progeny.

It was possible to discern that the mounting stag had penile erection for 18 of the mounts (25.7%). In this sample, mounters always had an erection while mounting a hind and ejaculated on three occassions, while only three out of four mounts of males were accompanied by an erection (Figure 6.5). However, this difference is not significant (Fisher exact probability test, $p = 0.211$). In addition, we noticed some erections in stags which stood close to mounters.

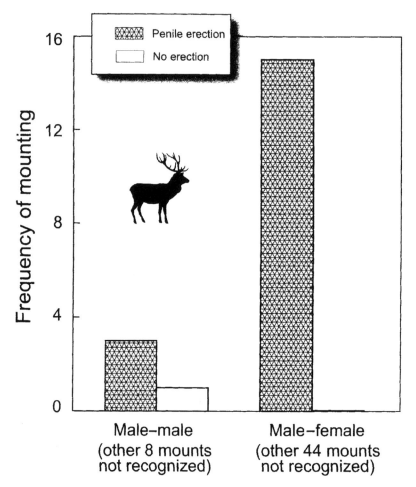

Figure 6.5. Penile erection of red deer stags during mounts of other males compared with mounts of hinds.

Discussion

Male–male mounting was commonly observed amongst three deer species but its frequency varied widely, depending on circumstances. We will now address the predictive power of various proximate and ultimate explanations for the behaviour.

Dominance

Our data set is not consistent with the often expressed view that actors in male–male mounting are dominant and recipients subordinate. In white-tailed deer, a strong preponderance of dominant mounters was only found in one of three study groups. Responses of mountees also differed strikingly. The strongest resistance against being mounted was recorded in a group which consisted only of immature bucks, whose social relationships were least established. Otherwise, we did not usually notice opposition against being mounted. Thus, under certain social conditions, intra-male mounting may serve social cohesion, similar to what has been suggested for, for example, rhesus monkeys (Reinhardt *et al.*, 1986). Nevertheless, in contrast, we did not find evidence of bonding between individuals involved.

Season and hormones

Intra-male mounting in fallow deer was clearly seasonal and restricted to the first three months of antler regrowth. High testosterone levels per se did not correlate with mounting. Nevertheless, antler re-growth is associated with an initial increase in testosterone levels, after minimum levels in February when the antlers are cast. Thus, the bucks seem to respond to the onset of higher testosterone levels (Holečková *et al.*, 2000).

In contrast, white-tailed deer intra-male mounting occurred across all study months, including periods when the deer had hard antlers, when they cast antlers and during re-growth. Published photos of male–male mounting of bucks with hard antlers later in the season (Rue, 1989, p. 313) and on snow (Bagemihl, 1999, p. 379) suggest that the behaviour may occur year-round.

Disturbance and excitement

It is known that disturbance and tension can be linked to intra-male mounting (Dagg, 1984). Vice versa, in fallow deer who lived almost undisturbed in a park, only smaller bucks mounted larger ones and mounting did not elicit resistance (Holečková *et al.*, 2000). However, incidences increased sharply in a captive setting, and all bucks mounted others, regardless of size (Holečková, 1999). Red deer also reacted with frequent mounting to temporary confinement with many others. Thus, despite differences in baseline levels of male–male mounting (almost none in red deer, but seasonal in fallow deer), both species responded similarly to what was probably a stressful situation.

Close proximity and excitement in general seems to easily trigger male–male mounting (Hall, 1983a; Smith and Dobson, 1990), which then tends to become 'contagious'. For example, in Roosevelt elk, a spike horn bull 'attempted to mount the older animal. The mature bull ran for several yards and resumed feeding. A few minutes later, the older bull attempted to mount the spike horn. On the same date, several cows were observed attempting to mount other cows and a calf made a similar attempt on a yearling cow' (Harper *et al.*, 1967, p. 37). Similarly, in Newfoundland moose 'a larger adult bull approached and the young adult left the yearling to face the larger animal. As these two butted and pushed, the yearling ran up to the rear of the larger bull, sniffed and stomped his front feet. Then he reared up as if attempting to mount' (Dodds, 1958, p. 415).

General differences in excitability may explain the varying behavioural pattern in cervids including our three study species. Red deer seem to be the least excitable, whereas white-tails are easy to panic and fallow deer lie in between (Bartoš, unpublished). Correspondingly, under undisturbed conditions, virtually no male–male mounting was displayed by red deer, whereas fallow deer showed it when seasonal elevation of testosterone levels might have likewise increased excitability. The easily excitable white-tailed deer, on the other hand, exhibit mounting across the whole year.

Excitability also decreases with increasing age and body size. This may explain why it was invariably smaller bucks who mounted larger ones in fallow deer. A correlation with age was also very obvious in white-tailed deer, where male–male mounting decreased from high levels in young bucks until it was practically zero in adults.

Penile erection and male–male mounting

Erection has been regarded as indicative of sexual arousal and thus evidence of a sexual motivation underlying same-sex mounting (for example, Weinrich, 1980; Bagemihl, 1999). Both Old World and New World deer, however, may show erections in situations which are removed from a direct sexual context, for example, when males dig and root into the ground with antlers. Such 'ground-rutting' (Heymer, 1977) is, for example, observed in red deer (Bartoš unpublished), North American elk (Geist, 1991) and sambar (Geist, 1998). Red deer in serious combat also frequently have an erect penis (Figure 6.6), whereas sparring, which is a playful pushing with antlers, is not associated with erections. Thus, penile erections during male–male mounting per se cannot be taken as real evidence of sexual motivation and might be attributed to excitement associated with serious fighting (Heymer, 1977).

Further important evidence suggests different arousal levels in same-sex versus opposite sex encounters: All male–female mounts for which we had complete observations were associated with penile erections, whereas a small proportion of the male–male mountings were not.

Male–male mounting as side effect

In many ungulates including deer, only few males will reproduce each year under natural conditions (for example, Clutton-Brock *et al.*, 1988;

Figure 6.6. Serious combat fight of red deer stags is often accompanied by erections (photo: Ludik Barto).

Marchinton and Miller, 1994; Moore *et al.*, 1995) due to intense male competition, which is risky and may lead to injuries or even death (Clutton-Brock *et al.*, 1988; McElligott *et al.*, 1998; Mattiangeli *et al.*, 1999). Evolution might therefore favour a rather opportunistic approach to mating. This inclination may also result in a male mounting another male instead of a female. Costs of this behaviour will be low for an individual who had no other opportunities anyway and counter selection will therefore not take place.

In conclusion, male–male mounting does not seem to have any special function in cervids. In particular, it cannot be clearly linked to roles of a dominant actor and subordinate recipient. Instead, intra-male mounting is likely to be a side effect of excitement and not necessarily sexual. Moreover, that it is typical for young animals and significantly decreases with age suggests that the animal needs experience to cope with changing hormonal states or that the behaviour is a form of learning. Thus, depending on circumstances, various proximate factors and ultimate functions may be involved.

Acknowledgements

Obtaining data presented here has been supported by the following grants: US-Czech Science Technology Program (95036); Fulbright Foundation (LB); Grant Agency of the Czech Republic (505/95/0291; 523/99/0984); Ministry of Agriculture of the Czech Republic (MZe 002701402). General support for LB during repeated visits in conjunctionn with the white-tailed deer study is highly acknowledged, provided by the Daniel B. Warnell School of Forest Resources, University of Georgia, Athens, GA, USA, and from Karl V. Miller in particular. We would also like to thank Vratislav Ksada for analysing the video records of red deer behaviour. Finally, thanks are due to V. Sommer for his careful editing of our manuscript.

References

Altmann, M. (1963) Naturalistic studies of maternal care in moose and elk. In *Maternal Behavior in Mammals*, ed. H. D. Rheingold, pp. 233–53. New York, London: John Wiley.

Andreae, U., Kuphaldt, D., Schütte, G. and Smidt, D. (1981) Sozial- und Sexualverhalten von Mastbullen in Rahmen von Versuchen zur geschlechtlichen Ruhigstellung. *Landbauforsch. Völkenrode*, **31**, 23–9.

Apollonio, M. (1998) Relationships between mating system, spatial behaviour, and genetic variation in ungulates, with special reference to European cervids. *Acta Theriol.*, 155–62.

Bagemihl, B. (1999) *Biological Exuberance: Animal Homosexuality and Natural Diversity.* New York: St. Martin's.

Barrette, C. (1977) Social behavior of captive mutjacs *Muntiacus reevesi* (Ogilby 1839). *Z. Tierpsychol.*, **43**, 188–213.

Bartoš, L. (1990) Social status and antler development in red deer. In *Horns, Pronghorns and Antlers: Evolution, Morphology and Social Significance,* eds. G. A. Bubenik and A. B. Bubenik, pp. 442–59. New York: Springer.

Bartoš, L. and Perner, V. (1985) Integrity of a red deer stag social group during velvet period, association of individuals, and timing of antler cleaning. *Behaviour*, **95**, 314–23.

Bartoš, L., Franc, C., Albiston, G. and Beber, K. (1988) Prevention of dark cutting (DFD) beef in penned bulls at the abattoir. *Meat Sci.*, **22**, 213–20.

Bartoš, L., Franc, Č., Rehák, D. and Štípková, M. (1993) A practical method to prevent dark-cutting (DFD) in beef. *Meat Sci.* **34**, 275–82.

Bartoš, L., Schams, D., Kierdorf, U., Fischer, K., Bubenik, G. A., Šiler, J., Losos, S., Tománek, M. and Laštovková, J. (2000) Cyproterone acetate reduced antler growth in surgically castrated fallow deer. *J. Endocrinol.*, **164**, 87–95.

Bergerud, A. T. (1974) Rutting behaviour of Newfoundland caribou. In *The Behaviour of Ungulates and its Relation to Management,* eds. V. Geist and F. Walther, pp. 395–435. IUCN Publ. New Ser. 24, Vol. 1.

Bouissou, M. F. (1978) Effect of injections of testosterone propionate on dominance relationships in a group of cows. *Hormon. Behav.*, **11**, 388–400.

Brown Jr, B. A. (1974) Social organization in male groups of white tailed deer. In *The Behaviour of Ungulates and its Relation to Management,* eds. V. Geist and F. Walther, pp. 436–46. IUCN Publ. New Ser. 24, Vol. 1.

Bubenik, G. A. (1990) Neuroendocrine regulation of the antler cycle. In *Horns, Pronghorns, and Antlers: Evolution, Morphology, Physiology, and Social Significance*, eds. G. A. Bubenik and A. B. Bubenik, pp. 265–97. New York: Springer.

Bützler, W. (1974) Kampf- und Paarungsverhalten, soziale Rangordnung und Aktivitätsperiodik beim Rothirsch. Supplement, *Z. Tierpsychol.* 16. Hamburg, Berlin: Paul Parey.

Carranza, J. (1995) Female attraction by males versus sites in territorial rutting red deer. *Anim. Behav.*, **50**, 445–53.

Chapman, D. I. and Chapman, N. G. (1975) *Fallow Deer: Their History, Distribution and Biology*. Lavenham, Suffolk: Terence Dalton.

Chapman, D. I., Chapman, N. G., Horwood, M. T. and Masters, E. H. (1984) Observations on hypogonadism in a perruque sika deer (*Cervus nippon*). *J. Zool., Lond.*, **204**, 579–84.

Clutton-Brock, T. H. (1982) *Red Deer: Behavior and Ecology of Two Sexes*. Edinburgh: Edinburgh University Press.

Clutton-Brock, T. H., Albon, S. D. and Guinness, F. E. (1988) Reproductive success in male and female red deer. In *Reproductive Success*, ed. T. H. Clutton-Brock, pp. 325–43. Chicago and London: University of Chicago Press.

Dagg, A. I. (1984) Homosexual behaviour and female–male mounting in mammals – a first survey. *Mammal Rev.*, **14**, 155–85.

Dalton, D. C., Pearson, M. E. and Sheard, M. (1967) The behaviour of dairy bulls kept in groups. *Anim. Prod.*, **9**, 1–5.

Darling, F. (1937) *A Herd of Red Deer*. London: Oxford University Press.

Dodds, D. G. (1958) Observations of pre-rutting behavior in Newfoundland moose. *J. Mammal.*, **39**, 412–16.

Dubost, G. and Feer, F. (1981) The behaviour of male *Antilope cervicapra* L., its development according to age and social rank. *Behaviour*, **76**, 62–127.

Franc, Č., Bartoš, L., Hanyš, Z. and Tomeš, Z. (1988) Preslaughter social activity of young bulls relating to the occurrence of dark-cutting beef. *Anim. Prod.*, **46**, 153–61.

Geist, V. (1981) Behaviour: adaptive strategies in mule deer. In *Mule and Black-tailed Deer of North America*, ed. O. C. Wallmo, pp. 157–223. Lincoln, NE: University of Nebraska Press.

Geist, V. (1991) *Elk Country*. Minnetonka, MN: NorthWord Press.

Geist, V. (1998) *Deer of the World: Their Evolution, Behavior, and Ecology*. Mechanicsburg, PA: Stackpole Book.

Halford, D. K., Arthur III, W. J. and Alldredge, A. W. (1987) Observations of captive Rocky Mountain mule deer behavior. *Great Basin Naturalist*, **47**, 105–9.

Hall, M. J. (1983a) Social organisation in an enclosed group of red deer (*Cervus elaphus* L.) on Rhum: 1. The dominance hierarchy of females and their offspring. *Z. Tierpsychol.*, **61**, 250–62.

Hall, M. J. (1983b) Social organisation in an enclosed group of red deer (*Cervus elaphus* L.) on Rhum: 2. Social grooming, mounting behaviour, spatial organisation and their relationships to dominance rank. *Z. Tierpsychol.*, **61**, 273–92.

Harper, J. A., Harn, J. H., Bentley, W. W. and Yocum, C. F. (1967) The status and ecology of Roosevelt elk in California. *Wildlife Monogr.*, **16**, 1–49.

Hawkins, R. E. and Klimstra, W. D. (1970) A preliminary study of the social organization of white-tailed deer. *J. Wildlife Managem.*, **34**, 407–19.

Hesselton, W. T. and Hesselton, R. M. (1982) White-tailed deer. In *Wild Mammals of North America: Biology, Management, and Economics*, eds. J. A. Chapman and G. A. Feldhamer, pp. 878–901. Baltimore and London: Johns Hopkins University Press.

Heymer, A. (1977) *Ethologisches Wörterbuch, Ethological Dictionary, Vocabulaire Ethologique.* Berlin, Hamburg: Paul Parey.

Hinch, G. N., Lynch, J. J. and Thwaites, C. J. (1982) Patterns and frequency of social behaviour in young grazing bulls and steers. *Appl. Anim. Ethol.,* **9**, 15–30.

Hirth, D. H. (1977) Social behavior of white-tailed deer in relation to habitat. *Wildlife Monogr.,* **53**, 1–55.

Holečková, J. (1999) Vzájemnépokládánísamců daňka skvrnitého (*Dama dama.* [Inter-male mounting in fallow deer (*Dama dama*)]. M.Sc. thesis. Department of Zoology, Faculty of Science, Charles University, Praha.

Holečková, J., Bartoš, L. and Tománek, M. (2000) Inter-male mounting in fallow deer (*Dama dama*) – its seasonal pattern and social meaning. *Folia Zool.,* **49**, 175–81.

Kesler, D. J., Troxel, T. R., Vincent, D. L., Scheffrahn, N. S. and Noble, R. C. (1981) Detection of estrus with cows administered testosterone via injections and/or silastic implants. *Theriogenol.,* **15**, 327–34.

Kiser, T. E., Britt, J. H. and Ritchie, H. D. (1977) Testosterone treatment of cows for use in detection of estrus. *J. Anim. Sci.,* **44**, 1030–35.

Klemm, W. R. C., Sherry, C. J., Schake, L. M. and Sis, R. F. (1983) Homosexual behaviour in feedlot steers: an aggression hypothesis. *Appl. Anim. Ethol.,* **11**, 187–95.

Laaser, G. H., Kiracofe, G. H., Heekin, H. D. and Ward, H. S. (1981) Effect of age and type of testosterone treatment on cows used for heat detection. *J. Anim. Sci.,* **47**, 35.

Lincoln, G. A., Youngson, R. W. and Short, R. V. (1970) The social and sexual behaviour of the red deer stag. *J. Reprod. Fert.,* Suppl. **11**, 71–103.

Marchinton, R. L. and Hirth, D. H. (1984) Behavior. In *White-tailed Deer: Ecology and Management,* ed. L. K. Halls, pp. 129–68. Harrisburg, PA: Stackpole Books.

Marchinton, R. L. and Miller, K. V. (1994) The rut. In *Deer,* eds. D. Gerlach, S. Atwater and J. Schnell, pp. 109–21. Harrisburg, PA: Stackpole Books.

Mattiangeli, V., Mattiello, S. and Verga, M. (1999) The fighting technique of male fallow deer (*Dama dama*): an analysis of agonistic interactions during the rut. *J. Zool., Lond.,* **249**, 339–46.

McElligott, A. G., Mattiangeli, V., Mattiello, S., Verga, M., Reynolds, C. A. and Hayden, T. J. (1998) Fighting tactics of fallow bucks (*Dama dama,* Cervidae): reducing the risks of serious conflict. *Ethology,* **104**, 789–803.

Moore, N. P., Kelly, P. F., Cahill, J. P. and Hayden, T. J. (1995) Mating strategies and mating success of fallow (*Dama dama*) bucks in a non-lekking population. *Behav. Ecol. Sociobiol.,* **36**, 91–100.

Reinhardt, V. (1983) Flehmen, mounting and copulation among members of a semi-wild cattle herd. *Anim. Behav.,* **31**, 641–50.

Reinhardt, V., Reinhardt, A., Bercovitch, F. B. and Goy, R. W. (1986) Does intermale mounting function as a dominance demonstration in rhesus monkeys? *Folia Primatol.,* **47**, 55–60.

Resko, J. A., Perkins, A., Roselli, C. E., Fitzgerald, J. A., Choate, J. V. A. and Stormshak, F. (1996) Eadocrine correlates of partner preference behavior in rams. *Biol. Reprod,* **55**, 120–6.

Rue III, L. L. (1989) *The Deer of North America*. Danbury, CN.: Outdoor Life Books.

Schaller, G. B. (1967) *The Deer and the Tiger: A Study of Wildlife in India*. Chicago: University of Chicago Press.

Schaller, G. and Hamer, A. (1978) Rutting behavior of Pere Davids deer, *Elaphurus davidianus*. *Zool. Garten N. F. Jena*, **48**, 1–15.

Schilder, M. B. H. and Boer, L. P. (1987) Ethological investigation on a herd of plains zebra in a safari park: time-budgets, reproduction and food competition. *Appl. Anim. Behav. Sci.*, **18**, 45–56.

Schilder, M. B. H. (1988) Dominance relationships between adult plains zebra stallions in semi-captivity. *Behaviour*, **104**, 300–19.

Sempéré, A. J. and Boissin, J. (1982) Neuroendocrine and endocrine control of the antler cycle in roe deer. In *Antler Development in Cervidae*, ed. R. D. Brown, pp. 109–22. Kingsville, TX: Caesar Kleberg Wild. Res. Inst.

Smith, R. F. and Dobson, H. (1990) Effect of preslaughter experience on behaviour, plasma cortisol and muscle pH in farmed red deer. *Vet. Rec.*, **126**, 155–8.

Stokes, M. E., Davis, C. S. and Koch, G. G. (2000) *Categorical Data Analysis Using the SAS System*. 2nd edn, Cary, NC: SAS Institute Inc.

Tennessen, T., Price, M. A., and Berg, R. T. (1985) The social interactions of young bulls and steers after re-grouping. *Appl. Anim. Behav. Sci.*, **14**, 37–47.

Vaňková, D. and Bartoš, L. (2002) The function of mounting behaviour in farmed red deer calves. *Ethology*, **108**, 473–82.

Vasey, P. L. (1995) Homosexual behavior in primates: a review of evidence and theory. *Int. J. Primatol.*, **16**, 173–204.

Walther, F. (1965) Verhaltensstudien an der Grantgazelle (*Gazella granti* Brooke 1872) im Ngorongoro-Krater. *Z. Tierpsychol.*, **22**, 167–208.

Walther, F. R. (1978) Forms of aggression in Thomsons Gazelle their situational motivation and their relative frequency in different sex, age, and social classes. *Z. Tierpsychol.*, **47**, 113–72.

Walther, F. R. (1984) *Communication and Expression in Hoofed Mammals*. Bloomington: Indiana University Press.

Weinrich, J. D. (1980) Homosexual behavior in animals: a new review of observations from the wild, and their relationship to human sexuality. In *Medical Sexology: The Third International Congress*, eds. R. Forleo and W. Pasini, pp. 288–95. Littleton, MA: PSG Publishing.

Wemmer, C., Collins, L. R., Beck, B. B. and Rettberg, B. (1983) The ethogram. In *The Biology and Management of an Extinct Species, Pere David's Deer*, eds. B. B. Beck and C. Wemmer, pp. 91–124. Park Ridge, NJ: Nozes.

7

Frustrated felines: male-male mounting in feral cats

AKIHIRO YAMANE

Introduction

The domestic cat (*Felis catus*) is one of the most familiar animals to humans and many scientists have investigated its ecology and social behaviour. Nevertheless, studies and information on homosexuality in these cats are very limited. This lack of attention can be accounted for, in large part, by the infrequent incidence of the homosexual behaviours observed in felids. In addition, there may be a tendency for investigators to consider such infrequent behaviours as merely abnormal events with no particular evolutionary significance.

Michael (1961) conducted detailed observations on both male and female homosexual mounting in domestic cats under laboratory conditions. He found relative body size did not predict the role that males took during same-sex mounting interactions, but territorial ownership did. He showed that a male cat introduced into an experimental pen was mounted by another male that had previously been introduced into the pen and allowed to establish a territory. The introduced male did not attempt to escape, but, rather, accepted the mounting and pelvic thrusting of the territorial male. Homosexual behaviour was also observed between oestrous females, but it occurred only rarely (Michael, 1961). Kerby and Macdonald (1988) found that, under free-ranging conditions, non-breeder males occasionally engaged in same-sex mounting with males and this appeared to be part of a larger pattern of non-conceptive mounting that also occurs with with young females and kittens. Among wild felids, homosexuality has been observed, albeit infrequently, in male lions (*Panthera leo*) and male and female cheetahs (*Acinonyx jabatus*) (Schaller, 1972; Eaton, 1974; Chavan, 1981)

In this chapter, I first describe the instances of male–male mounting that have been observed in my study population of feral cats. Second, I examine five adaptive or maladaptive hypotheses that might account for these homosexual interactions. The hypotheses that I will examine are: (a) mistaken identity, (b) dominance-assertion, (c) sexual play for practice, (d) sexual inhibition of a competitor and (e) outlet for sexual frustration.

To begin with, the 'mistaken identity hypothesis' proposes that male–male mounting occurs when males mistake other males for females. This seemingly maladaptive behaviour is commonly used to explain same-sex mounting in sexually monomorphic avian species (for example Birkhead *et al.*, 1985; Hatchwell, 1988). Second, the 'dominance-assertion hypothesis' claims that dominant males mount subordinates to reaffirm their position in the dominance hierarchy and thereby reduce aggression (Wickler, 1967). This explanation has been widely invoked to explain same-sex mounting in species that form social hierarchies (review in Bagemihl, 1999). Third, the 'sexual play for practice hypothesis' holds that same-sex mounting is a form of play, which functions as a type of rehearsal for heterosexual copulation. Homosexual behaviour between immature primates is frequently observed during play, and, in several species, this play may serve as practice for future adult heterosexual interactions (Dagg, 1984; Vasey, 1995). Fourth, the 'sexual inhibition of a competitor hypothesis' claims that male–male mounting is a tactic to inhibit competitors from achieving a heterosexual mating (Tyler, 1984). Finally, the 'outlet for sexual frustration hypothesis' suggests that sexual frustration leads to homosexual behaviour when females refuse a male's sexual advances (reviewed by Dagg, 1984).

In the discussion, I will examine the validity of each of these explanations for the occurrence of male homosexual mounting in the feral cat population under investigation.

Materials and methods

Species

The feral cat is a domestic cat that is essentially not tamed by humans and not attached to a particular household. The feral cats are not bred by humans and are allowed complete freedom of movement. The mating system of free-ranging feral cats can be characterized as promiscuous (Natoli and De Vito, 1991), or scramble competition polygyny (Liberg, 1983; Yamane *et al.*, 1996; see also Davies, 1991) with no paternal care (Natoli and De Vito, 1991). Males and females are sexually dimorphic in body mass (Liberg, 1981).

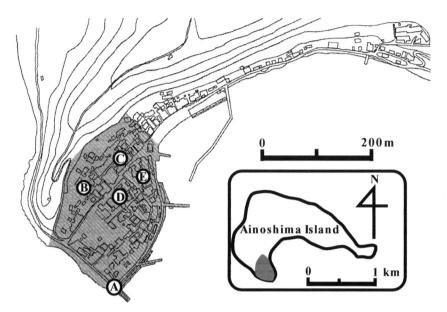

Figure 7.1. Map of feral cat study area on Ainoshima Island (125 ha), with main study site in the southwest part (shaded area; 6.0 ha). A-E = locations of garbage sites.

Study population

This study was carried out on Ainoshima Island (33°45′ N, 130°21′ E), located about 7 km off the coast of Shingu in Fukuoka, Japan (Figure 7.1). The island is 125 ha in size and about 90% of it is covered with vegetation. Approximately 500 people live on this island, and the majority of the inhabitants make a living by fishing. All cats within the main study site (6.0 ha) have been identified individually since 1989. In January 1993 there were 74 sexually mature cats (48 males, 26 females). Five garbage sites (A, B, C, D, E) provided the cat population with abundant food resources (Yamane *et al.*, 1994, 1997). Fifty-nine cats (79.7%) in the population foraged for food at the garbage sites regularly. Since each cat utilized only one of the five garbage sites for feeding and resting, it was concluded that the cat population on this island formed feeding groups (Izawa *et al.*, 1982; Yamane *et al.*, 1994, 1996, 1997). By the age of five, most males had dispersed from their natal groups and were living in other groups (Yamane, 1994). Group-living females remained in their natal groups, where they bore kittens (Yamane *et al.*, 1997).

Terminology

Courtship occurred when males followed and aggregated around an oestrous female. Males that courted females were quick to follow them if they attempted to move away.

Figure 7.2. Homosexual mounting between male feral cats. Just after an oestrous female ran away from courting males, one male suddenly bit and mounted a nearby immature male.

Heterosexual copulations (involving intromission) were distinct from heterosexual mountings because, when intromission occurred, the female emitted a sharp cry, turned towards the male, threatened him and then usually rolled on the ground vigourously (Natoli and De Vito, 1991). *Copulation rates* were calculated by dividing the number of copulations a male performed by the total amount of courting time in which the male engaged. Because copulations occurred infrequently, copulation rates were not calculated for males that were observed to court females for less than five hours. During heterosexual mounts and copulations, the male would give 'curr' calls (a courtship vocalization) and bite the female's neck. Curr calls ('crrrrrrrr') were distinct from 'purring' ('grrr grrr grrr') and could be differentiated by ear in terms of their sound cycle.

Male–male mounting is defined as the behaviour that occurs when one male bites the neck of another male and then mounts him in a ventral–dorsal manner (Figure 7.2). Pelvic thrusting sometimes occurred in association with male–male mounting.

For the purposes of data analysis, I classified males into three groups: males that mount males (MM males), males that are mounted by males (RM males) and males that show no homosexual behaviour (NM males).

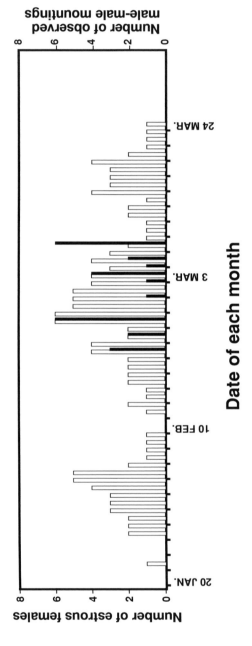

Figure 7.3. Distribution of female oestrous and observed male–male mountings (20 January–31 March 1993). Open bars = oestrous females/day; solid bars = male–male mountings.

Data collection and analysis

Focal animal observations of oestrous females were conducted during the 1993 mating season from 20 January to 31 March (Figure 7.3). Each morning during the 1993 mating season, I monitored the oestrous condition of all the females in the study area using a 'route censusing' technique. This involved walking a route that I devised, through the study area. The route passed through back alleys, gardens and fields frequented by the cats. It typically took one or two hours to complete this route. By walking the study site route once or twice each morning during the oestrous period, I was able to observe all the animals in the study population and determine which female(s) were in oestrous. During the 1993 mating season, I spent approximately 400 hours conducting route censuses. Once oestrous female(s) were identified, I collected focal animal data on those females for the remainder of the day. Oestrous females were usually followed for a minimum of one to two hours per day. Observation times varied depending on the number of other females in oestrous on the same day. Some females could not be followed because they were not habituated. During focal animal observations, I recorded heterosexual copulation rates, the female's distance from the males that were courting her, agonistic encounters between the males in the females' vicinity and male homosexual behaviour. If no oestrous female was found, I continued the route censusing until sunset so as to check the location and condition of the study subjects. A total of 160 hours of focal animal data were on oestrous females (for detailed methods, see Yamane *et al.*, 1994, 1996).

Cat behaviour outside the 1993 mating season was monitored by 'route censusing'. I usually performed route censuses three to four times per day, which allowed me to check all the cats in the study area. During route censusing, I checked presence, location, physical condition, births, deaths and the behaviour of the animals in my study population. I conducted route censuses for approximately five hours per day for a total 284 days (minimum three days per month) between 1989 and 1993 ($n = 1420$ hours).

Receptive females were usually courted by several males simultaneously (Figure 7.4; Yamane *et al.*, 1996, Yamane, 1998). Male dominance ranks were based on the distance between an oestrous female and the courting males. The measurement unit was oestrous female body length (approximately 50 cm, since the mean body length of a sexually mature female was 49.3 ± 0.6 cm for $n = 11$ cats measured in 1990). Proximity measurements were taken at one-minute intervals. A mean courtship distance value was calculated for each male based on proximity measurements collected during the oestrous season. This value was considered as an index of the male's dominance rank during courtship. The

Figure 7.4. A group of males courting an oestrous female. A black arrow points towards the oestrous female. A white arrow points towards the dominant male. The other individuals in the photo are more subordinate males. Note how the dominant male is oriented towards, and in close spatial proximity to, the oestrous female. Note also how the oestrous female is surrounded by subordinate males.

shorter the distance between an oestrous female and a courting male, the higher the male's dominance rank (see Yamane *et al.*, 1996). Male heterosexual mating success was estimated by copulation rate throughout the oestrous season (see Yamane *et al.*, 1996; Yamane, 1998).

Male cats were weighed using a basket containing dried fish as bait, which was attached to a scale and then placed in the study area. The cat's body mass was measured when it jumped into the basket to eat the bait. The body mass of about 96% (46/48) of the male cats within the study area was measured in this manner.

To examine differences between MM, RM and NM males in terms of heterosexual mating ability, I calculated the mean body mass, courtship distance and copulation rates for each type of male.

Statistics were calculated using the program, StatView J-4.5. All tests are two-tailed.

Results

General description of homosexual behaviours

From 1989 to 1993, male homosexual behaviours, including neck biting, mounting and pelvic thrusting, were never observed outside the oestrous season (1420 observation hours). During the 1993 mating season, 26 cases of male–male mounting behaviour involving 14 male pairs were recorded. This represents approximately one male–male mount for every 21.5 hours of observation time. Typically, a MM male would neck bite a RM male just before and during a male–male mount. Half of the homosexual mounts were accompanied by pelvic thrusting. Intromission and ejaculation were never observed. Overall, male–male mounting did not appear to differ much in form from heterosexual mounts. Unlike heterosexual mounts, however, 'curr' calls were not exchanged between males during homosexual mounts. All of the MM ($n = 13$), RM ($n = 5$) and NM ($n = 30$) males observed formed mutually exclusive groups. In other words, MM males were never mounted by other males and RM males never mounted other males.

In the study area, female cats came into oestrous in early spring (Figure 7.3; Yamane et al., 1996; Yamane, 1998). The mean duration of females' oestrous cycles $\pm$ SD was 5.25 $\pm$ 3.63 days ($n = 30$). During the 1993 mating season, there were three peaks in the number of oestrous females. Occurrences of male homosexual mounting seemed to be concentrated around the largest of these three peaks in oestrous female availability (Figure 7.3).

Oestrous females were typically followed by more than one male simultaneously (mean $\pm$ SD = 3.34 $\pm$ 2.29, range = 1–20). All homosexual mounts were observed in the presence of an oestrous female, or just after a female left a male that was courting her. Of the 26 male–male mounts observed, 76.9% ($n = 20$) were performed by courting MM males that had suddenly switched their attention from the female to a nearby RM male. In most such cases (55%) MM males had been courting oestrous females for more than an hour. In some of these cases, the courting MM males had fallen asleep because they had been following the oestrous female for hours. Upon waking, the MM males realized the oestrous female was gone and immediately mounted a RM male. The additional 23.1% ($n = 6$) of male–male mounts were executed by courting males who had lost sight of the oestrous female as she ran away from them.

None of the female study subjects was observed to engage in homosexual behaviour during the study.

Comparative analysis of characters among males

Mean body mass varied significantly between different classes of males (mean $\pm$ SE, MM = 3.90 $\pm$ 0.16 kg, RM = 3.02 $\pm$ 0.16 kg, NM = 3.60 $\pm$ 0.10 kg; Kruskal–Wallis test, $H = 7.58$, $p < 0.05$; Figure 7.5). RM males were significantly lighter than MM males (Mann–Whitney U test, $n_{1,2} = 5, 11$, $U = 16$, $p < 0.001$) and NM males (Mann–Whitney U test, $n_{1,2} = 5, 30$, $U = 16$, $p < 0.01$). There was no significant difference in body mass between MM and NM males (Mann–Whitney U test, $n_{1,2} = 11, 30$, $U = 115.5$, $p = 0.147$).

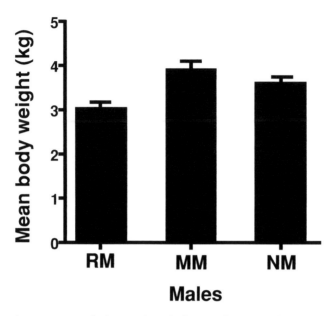

Figure 7.5. Mean body mass (± SE) of MM males, RM males, and NM males.

The average body mass of sexually mature males and females ± SE was 3.6 ± 0.1 kg ($n = 46$) and 2.8 ± 0.1 kg ($n = 19$), respectively (Yamane *et al.*, 1996). Overall, males' body mass was 1.28 times greater than that of females. This difference was significant (Mann–Whitney U test, $n_{1,2} = 46, 18, U = 84, p < 0.0001$). However, the body mass of RM males and females did not differ significantly (Mann-Whitney U test, $n_{1,2} = 5, 18, U = 29.5, p = 0.257$).

The age distributions of MM males, RM males and NM males are shown in Figure 7.6. Overall, 33% of older males (≥5 years) mounted other males, whereas only 11% of younger males did so. Conversely, none of the older males (≥5 years) was mounted by other males, but 28% of the younger males were mountees. RM males were significantly younger than MM males (Mann–Whitney U test, $n_{1,2} = 5, 11, U = 6.5, p < 0.05$) and NM males (Mann–Whitney U test, $n_{1,2} = 5, 31, U = 24.5, p < 0.05$). MM and NM males did not differ significantly in age (Mann–Whitney U test, $n_{1,2} = 11, 31, U = 129.5, p < 0.245$).

Out of the 26 cases of male–male mounting, only 26.9% ($n = 7$) occurred between members of the same social group, the other 73.1% occurred between males of different social groups. Among extra-group male–male mountings, all the MM males were visitors to the group.

All RM males were sexually inactive. RM males occasionally followed oestrous females, but the manner in which they did so differed from courting males. Specifically, they were slow in reacting to movement by oestrous females.

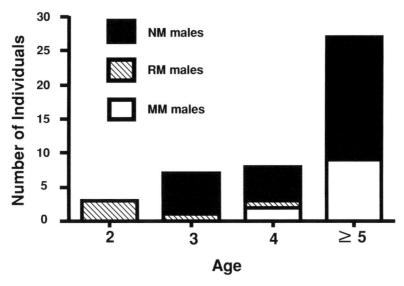

Figure 7.6. Age distribution of MM males RM males and NM males.

Moreover, RM males never directed neck bites towards oestrous females, nor did they mount them. In contrast, MM males performed neck bites and mounted not only males, but also sexually receptive females.

There was a significant negative correlation between male body weight and courtship distance (Spearman's rank correlation, $n = 40$, $r_s = -0.37$, $p < 0.05$). Mean $\pm$ SE courtship distance throughout the oestrous season was 1.41 ± 0.12 m ($n = 30$) for NM males and 1.24 ± 0.13 m ($n = 12$) for MM males (Figure 7.7a). As such, the dominance rank of MM males throughout the oestrous season was higher than that of NM males, although the difference between them was not significant (Mann–Whitney U test, $U = 164.5$, $Z = -0.43$, $p = 0.67$).

There was a significant positive correlation between body weight and the number of copulations a male performed (Spearman's rank correlation, $n = 46$, $r_s = 0.42$, $p < 0.005$). Likewise, there was a significant positive correlation between male body weight and copulation rate (Spearman's rank correlation, $n = 20$, $r = 0.48$, $p < 0.05$). MM males (mean $\pm$ SE $= 0.16 \pm 0.07$, $n = 9$) and NM males (mean $\pm$ SE $= 0.19 \pm 0.07$, $n = 10$) did not differ significantly in their copulation rates throughout the mating season (Mann–Whitney U test, $I = 41.0$, $Z = -0.33$, $p = 0.73$; Figure 7.7b)

Discussion

Homosexual behaviour appears to be rare in domestic and wild felids. Accordingly, most of the available information on this topic has been anecdotal,

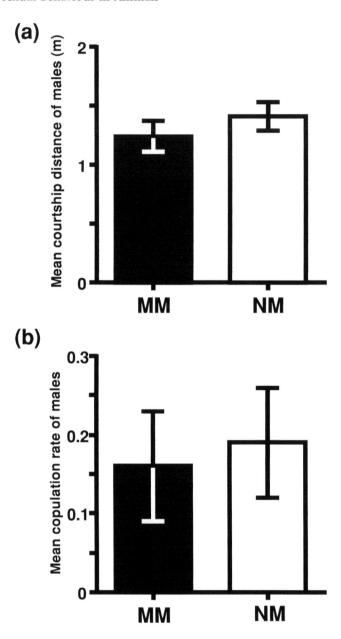

Figure 7.7. (a) Mean courtship distance ($\pm$ SE m) of MM males and NM males. Note that the shorter the distance, the higher the dominance rank during courtship. (b) Mean copulation rate ($\pm$ SE) of MM males and NM males.

to date. This study represents the first attempt, to empirically investigate the functional basis of homosexual behaviour among a free-ranging felid, the feral cat. During the study, male–male mounting was observed, but not female–female mounting. Male–male mounting only occurred during the mating season and in the presence of an oestrous female. Compared with male mountees, male mounters tended to be older, heavier, dominant individuals that were experienced in terms of heterosexual courtship and copulation. Male mounters and mountees tended to be from different groups.

On the basis of these observations a number of functional explanations that might account for same-sex mounting in feral cats can be rejected.

The *mistaken identity hypothesis* holds that male–male mountings occur because the mounting male mistakes other males for females. This hypothesis is commonly invoked to account for same-sex mounting in sexually monomorphic species, particularly birds (Birkhead *et al.*, 1985). Unlike monomorphic birds, feral cats show sexual dimorphism in size. Nevertheless, RM males are similar in size to sexually mature females. It is, therefore, possible that MM males might have confused RM males for females. Moreover, some of the male–male mounts were performed just after a receptive female had suddenly disappeared from the area. As such, it is possible that the mounting male mistook the male mountee for a recently departed oestrous female. In addition, most instances of male–male mountings occurred when MM males conducted extra-group visits. It is conceivable that an extra-group MM male, unfamiliar with the members of another group, might have difficulty distinguishing between oestrous females and younger, smaller RM males. This being said, it seems unlikely that older, sexually experienced males would mistake males for females so frequently. Indeed, the behaviour of MM males suggests that they were fully able to differentiate between male and female mountees. Mounting males never 'curr' called to male mountees, but they did do so to female mountees. Likewise, mounting males did not court other males by following them, they only behaved this way towards oestrous females. As such, the 'mistaken identity hypothesis' appears to be an implausible explanation for homosexual mounting between males in this population of feral cats.

The *sexual play for practice hypothesis* holds that same-sex mounting is a form of play, which functions as a type of rehearsal for heterosexual copulation. If this hypothesis is correct, the mounters should be immature individuals that are inexperienced with regard to heterosexual copulation. Data from this study do not support this prediction, since the male cats that mounted tended to be older (>4 years of age), sexually experienced individuals, whereas younger (<4 years), sexually inexperienced males were *never* observed to mount. Evidence that older males tend to mount other males suggests that the 'sexual play for practice hypothesis' cannot account for the expression of male–male mounting of the feral cat.

The *dominance assertion hypothesis* holds that dominant individuals mount subordinates to reaffirm their position in the dominance hierarchy (Wickler, 1967). Accordingly, dominant and subordinate individuals should adopt specific roles during mounting interactions. My observations indicate that dominant males invariably mounted subordinate males. Instances of reverse mountings, in which a subordinate male mounted a dominant individual, were never observed. On the face of it, these findings suggest that male–male mounting among feral cats may function as a form of dominance assertion. However, if males seek to express their dominance *vis-à-vis* other same-sex individuals through mounting interactions, then they should do so at any time throughout the year. In other words, we would predict that same-sex mounting would not be restricted to a specific temporal period. Nevertheless, the data do not support this prediction. Same-sex mounting among males was never observed outside of the early spring when female cats in the study population came into oestrous. It is possible, of course, that a dominance assertion function for male–male mounting exists, but is limited to situations characterized by competition over receptive females during the oestrous season. Outside the mating season other expressions of dominance might be employed, such as territorial marking (Ishida and Shimizu, 1998). Even if one accepts this alternative line of reasoning, we would expect that during the mating season high-ranking males would assert their dominance against sexually active males who constitute potential competitors for copulations with females. Nonetheless, all the RM males were sexually inactive males that did not appear to be serious rivals to MM males. Taken together, these results suggest that dominance assertion may play a role in the expression of homosexual mounting between male feral cats during the mating season, but it provides only a partial explanation, at best, as to why these interactions occur.

The *sexual inhibition of a competitor hypothesis* claims that male–male mounting is a tactic that mounters use to inhibit competitors from achieving heterosexual matings (Tyler, 1984). It is possible that during such interactions mounting males reduce same-sex mountees' interest in opposite-sex mates by providing alternative sexual stimulation. As a result, a mountee's probability of copulating is reduced to the mounter's reproductive advantage. For example, Wagner (1996) investigated male–male mounting in razor-bills (*Alca torda*) and found that those males that were mounted by other males were less successful in mating heterosexually. The majority of male–male mounts observed during this study (76.9%) occurred when a MM male quickly switched his attention from a female he was courting to a nearby RM male whom the MM male then mounted. While courting an oestrous female, the MM males stay in close proximity, guarding her from other males in the vicinity. RM males that approached the oestrous female too closely may have been perceived by the MM male as sexual competitors. Under such circumstances, MM males may have been able to reduce the RM males ability to access, and ultimately copulate with, oestrous females simply by mounting them and, as such, physically constraining them. Although this scenario is possible, it does not explain why courting males could not simply threaten RM males away. Why was it necessary to mount them? Perhaps MM males were attempting to reduce the RM males' sexual interest in the oestrous female by providing alternative sexual stimulation. It is not clear, however, if mounting a male rival would, indeed, provide the male mountee with sexual stimulation, let alone reduce their desire to mount, and copulate with, an oestrous female. Finally, it is questionable as to whether the younger, lighter RM males who remained at a substantial distance from oestrous females and were slow to react to the movement of oestrous females, posed any real threat as sexual rivals to MM males. Consequently, the value of this hypothesis in accounting for male–male mounting in feral cats remains equivocal at present.

The *outlet for sexual frustration hypothesis* suggests that sexual frustration leads to homosexual behaviour when females refuse a male's sexual advances (reviewed by Dagg, 1984). The rebuffed male then mounts a same-sex individual. Several lines of evidence lend support to this hypothesis as an explanatory framework for understanding some of the male–male mounting interactions observed during this study. First, homosexual behaviour was never observed outside the oestrous season as would be predicted on the basis of this hypothesis. Second, 23.1% of the male–male mounts occurred when a courting MM male had lost sight of the oestrous female as she ran away. The courting MM male then turned his attention towards a RM male and mounted him. Third, the majority (55%) of instances in which a MM male switched his attention from courting an oestrous female to mounting a RM male occurred after the MM male had been courting the female for over one hour. In such instances, it is possible that the MM male was sexually frustrated because he wanted to copulate with the female, but she would not let him. These observations fit very well with the predictions of an explanatory framework that invokes sexual frustration.

In general, a male's threshold for mating attempts is relatively low compared with females. This lower threshold for sexual arousal increases the male's chances of fertilizing as many females as possible, but this, in turn, can lead to relatively indiscriminate male mating behaviour (Alcock, 1993). Thus, homosexual mounting in feral cats may merely be a by-product of males' low threshold for sexual arousal, a mating strategy that is, on balance, adaptive. When viewed from this perspective, it is, therefore, not surprising that a male that has had a heterosexual copulation thwarted might respond by indiscriminately engaging in a homosexual mount.

A number of other studies suggest a link between male homosexual mounting and sexual frustration. Rose *et al.* (1991) demonstrated, for example, that sub-adult male northern elephant seals (*Mirounga angustirostris*) that were unable to access female mates routinely mounted same-sex immature substitutes. In felids, Kerby and Macdonald (1988) reported that non-breeding, subordinate male domestic cats tended to mount other males. Eaton (1974) observed one inter-action among semi-wild cheetahs during which four males approached a het-erosexual mating pair. The male that was mating threatened the approaching same-sex conspecifics, at which point one of the threatened males immediately mounted another.

In conclusion, my observations suggest that male–male mounting in feral cats cannot be explained in terms of mistaken identity or as a form of sexual play which functions as practice for heterosexual copulation. Certainly, for a subset (23.1%) of the male–male mounts observed, the sexual frustration hypothesis seems to be the most appropriate explanation because MM males mounted RM males after the former lost sight of an oestrous female they had been courting, as she ran away from them. For an additional 42.3% of the mounts the 'sexual frustration hypothesis' seems like a highly plausible explanation because MM males mounted RM males only after they had been courting an oestrous female for a protracted period of time (>1 hour).

The remainder of the male–male mounts observed are more difficult to inter-pret. There is only weak support for the hypotheses that male–male mounting in feral cats functions as a type of dominance demonstration and as a mech-anism to inhibit the sexual behaviour of a male competitor. It is possible that as a consequence of these same-sex interactions the participant's relative domi-nance ranks are reinforced and male rivals are impeded from accessing oestrous females. However, evidence indicating that MM males initiate male–male mount-ing for the express purpose of reinforcing their dominance or blocking a sexual rival's access to an oestrous female, is ambiguous at present. Some mixture of the 'dominance assertion', 'sexual inhibition of a competitor' and 'sexual frus-tration' hypotheses might account for the majority of mounts observed (76.9%). If so, this raises the possibility that different male–male mounting interactions in feral cats serve different functions and, the same male–male mounts may serve multiple functions. Finally, if sexual frustration accounts for why some MM males mount same-sex individuals, then we must conclude that sexual motivation plays some role in the expression of these same-sex interactions, at least from the perspective of the male mounter. In instances where the 'sexual frustration hypothesis', alone, accounts for the occurrence of male–male mounting, it would appear that such interactions are best characterized as functionless by-products of the males' adaptive tendency towards having a low threshold for sexual arousal. In these sorts of cases, functional explanations cannot be invoked to account for the existence of such behaviour.

More research is needed in order to fully understand the evolutionary basis of male homosexual behaviour in group-living feral cats.

References

Alcock, J. (1993) *Animal Behavior*, 5th edn. Sunderland, MA: Sinauer Associates.

Bagemihl, B. (1999) *Biological Exuberance: Animal Homosexuality and Natural Diversity*. New York: St Martins Press.

Birkhead, T. R., Johnson, S. D. and Nettleship, D. N. (1985) Extra-pair matings and mate-guarding in the common murre *Uria aalge. Anim. Behav.*, **33**, 608–19.

Chavan, S. A. (1981) Observations of homosexual behaviour in Asiatic lion *Panthera leo persica. J. Bombay Nat. Hist. Soc.*, **78**, 363–64.

Dagg, A. I. (1984) Homosexual behaviour and female–male mounting in mammals – a first survey. *Mammal Rev.*, **14**, 155–85.

Davies, N. B. (1991) Mating system, In *Behavioural Ecology*, 3rd edn, eds. J. R. Krebs and N. B. Davies, pp. 268–94. Oxford: Blackwell.

Eaton, R. L. (1974) *The Cheetah*. New York: Van Nostrand Reinhold.

Hatchwell, B. J. (1988) Interspecific variation in extra-pair copulation and mate defense in Common Guillemots (*Uria aalge*). *Behaviour*, **107**, 157–85.

Ishida, Y. and Shimizu, M. (1998) Influence of social rank on defending behaviors in feral cat. *J. Ethol.*, **16**, 15–21.

Izawa, M., Doi, T. and Ono, Y. (1982) Grouping patterns of feral cats (*Felis catus*) living on a small island in Japan. *Japan J. Ecol.*, **32**, 373–82.

Kerby, G. and Macdonald, D. W. (1988) Cat society and the consequence of colony size. In *The Domestic Cat*, eds. T. C. Turner and P. Bateson, pp. 67–81. Cambridge: Cambridge University Press.

Liberg, O. (1981) Predation and social behaviour in a population of domestic cats: an evolutionary perspective. Ph.D. thesis, University of Lund.

Liberg, O. (1983) Courtship behaviour and sexual selection in the domestic cats. *Applied Anim. Ethol.*, **10**, 117–32.

Michael, R. P. (1961) Observation upon the sexual behaviour of the domestic cat (*Felis catus* L.) under laboratory conditions. *Behaviour*, **18**, 1–24.

Natoli, E. and De Vito, E. (1991) Agonistic behaviour, dominance rank and copulatory success in a large multi-male feral cats, *Felis catus* L., colony in central Rome. *Anim. Behav.*, **42**, 227–41.

Rose, N. A., Deutsch, C. J. and Le Boeuf, B. J. (1991) Sexual behaviour of the male northern elephant seals III: the mounting of weaned pups. *Behaviour*, **119**, 171–92.

Schaller, G. B. (1972) *The Serengeti Lion*. Chicago: University of Chicago Press.

Tyler, P. A. (1984) Homosexual behaviour in animals. In *The Psychology of Sexual Diversity*, ed. K. Howells, pp. 42–62. Oxford: Basil Blackwell.

Vasey, P. L. (1995) Homosexual behavior in primates: a review of evidence and theory. *Int. J. Primatol.*, **16**, 173–204.

Wagner, R. H. (1996) Male–male mountings by a sexually monomorphic bird: mistaken identity or fighting tactics? *J. Avian Biol.*, **27**, 209–14.

Wickler, W. (1967) Socio-sexual signals and their intra-specific imitation among primates. In *Primate Ethology*, ed. D. Morris, pp. 69–147. London: Weidenfeld & Nicolson.

Yamane, A. (1994) Mating system and male reproductive tactics of the feral cat (*Felis catus*). Ph.D. thesis, Kyushu University.

Yamane, A. (1998) Male reproductive tactics and reproductive success of the group-living feral cat (*Felis catus*). *Behav. Proc.*, **43**, 239–49.

Yamane, A., Ono, Y. and Doi, T. (1994) Home range size and spacing pattern of a feral cat population on a small island. *J. Mammal. Soc. Japan*, **9**, 9–20.

Yamane, A., Doi, T. and Ono, Y. (1996) Mating behaviors, courtship rank and mating success of male feral cat (*Felis catus*). *J. Ethol.*, **14**, 35–44.

Yamane, A., Emoto, J. and Ota, N. (1997) Factors affecting feeding order and social tolerance to kittens in the group-living feral cat (*Felis catus*). *Appl. Anim. Behav. Sci.*, **52**, 119–27.

PART III PRIMATES

8

The pursuit of pleasure: an evolutionary history of female homosexual behaviour in Japanese macaques

PAUL L. VASEY

Introduction

Courtship displays, mounting and genital contact/stimulation between same-sex individuals have been observed in a variety of macaque species including *Macaca silenus, M. tonkeana, M. maura, M. nigra, M. nemestrina, M. radiata, M. arctoides, M. fascicularis, M. mulatta* and *M. fuscata* (reviewed in Nadler, 1990; Vasey, 1995; Bagemihl, 1999). These same-sex interactions are widely referred to as 'homosexual behaviour' in the macaque literature, although some researchers (for example, Wallen and Parsons, 1997; Dixson, 1998) argue that this label is misleading because homosexuality is a uniquely human phenomenon. However, the term 'homosexual behaviour' refers to discrete acts or interactions and does not denote sexual preference (that is, the sex with which one prefers to engage in sexual activity when given a choice), sexual orientation (that is, an overall pattern of sexual attraction/arousal over some unit of time as measured by multiple parameters such as behaviour, fantasy, feelings, attraction, genital blood flow etc.), sexual orientation identity (that is, the sexual orientation that an individual perceives themselves to be) or categories of sexual beings (that is, a homosexual, a heterosexual, a bisexual; for non-Western examples see Nanda, 2000). For the remainder of this chapter, I refer to same-sex courtship displays, mounting and/or genital contact/stimulation in animals as 'homosexual behaviour'.

Detailed studies on same-sex mounting and courtship in macaques are only available for rhesus (Fairbanks *et al.*, 1977; Huynen, 1997; Kapsalis and Johnson, Chapter 9) and Japanese macaques and, at present, more is known about female–female interactions than male–male interactions in both these species. A comparison of the rhesus and Japanese macaque literature quickly highlights an important point: homosexual behaviour is not a uniform phenomenon. In female rhesus macaques, multiple lines of research appears to be converging around the same conclusion, namely, female homosexual behaviour in this species is an adaptive mechanism for alliance formation (Fairbanks *et al.*, 1977; Huynen, 1997; this volume, Chapter 9). In Japanese macaques, however, research indicates that females do not engage in homosexual behaviour to mediate alliance formation (Vasey, 1996; see below). This lack of behavioural uniformity makes generalizations difficult and, arguably, inappropriate. Indeed, it is probably more accurate to speak in terms of macaque homosexual *behaviours* that vary both within species and between species in terms of their form, frequency, motivation, cause, development, function and evolutionary history.

Female Japanese macaques (*Macaca fuscata*), in certain captive and free-ranging populations, routinely engage in same-sex sexual behaviour in adulthood (for example, Arashiyama-East [Texas]: Fedigan and Gouzoules, 1978; Wolfe, 1984; Arashiyama [Japan]: Takahata, 1982; Wolfe, 1984; Université de Montréal colony: Vasey, 2002a). McDonald-Pavelka (1993) mentioned the case of a single Japanese macaque female that engaged in same-sex sexual behaviour exclusively over the course of her lifespan. This case is, however, exceptional. In all other studies, female subjects have been reported to alternate between heterosexual behaviour and homosexual behaviour usually over the course of the same mating season. As such, female Japanese macaques that engage in same-sex sexual behaviour should not be thought of as exclusively homosexual.

Female–female sexual behaviour in Japanese macaques takes the form of mounting. Single mounts between females are rarely observed. Instead, multiple or series mounting between consort partners is the norm. Throughout these interactions, partners mount each other in a bi-directional manner and routinely employ various types of mounting postures including double foot-clasp mounts (Figure 8.1), sitting mounts (Figure 8.2), standing mounts and reclining mounts (Vasey *et al.*, 1998). During mounts, mounters often perform pelvic thrusts against mountees who, in turn, commonly reach back to grasp the mounter and gaze intently into her eyes.

In addition to actual mounting behaviour, females solicit each other to mount, and to be mounted, using vocalizations and a variety of postural and facial gestures. Sexual solicitations include pushing, hitting, grabbing, slapping the ground, head bobbing, screaming, lip quivering, body spasms and intense gazing. A female can specifically request to be mounted by presenting her inclined back or her hindquarters to a potential mounter. Conversely, a female can specifically request to mount another female by placing her hands on the other's hindquarters (Vasey *et al.*, 1998). Female–female mounting and

Figure 8.1. Female–female double foot-clasp mount in Japanese macaques (photo: Stefani Kovacovsky).

Figure 8.2. Female–female sitting mount in Japanese macaques. (photo: Stefani Kovacovsky).

solicitations in Japanese macaques take place during temporary associations termed 'consortships'. These relationships are sexually exclusive in that, during a consortship, females only engage in sexual behaviour with their consort partner. Consortships between females can be brief, lasting less than one hour, or they can continue for over a week (Wolfe, 1984). While engaging in same-sex consortship activity, females exhibit a reddening of the face and perineum that is associated with increased receptivity and proceptivity.

Mounting and solicitations are clearly the defining feature of same-sex consortships. However, between bouts of sexual activity, partners exhibit highly synchronized activity, huddling and sleeping together, as well as grooming, following and defending each other for prolonged periods of time (Gouzoules and Goy, 1983; Wolfe, 1984; Vasey, 1996). Partners may also forage side-by-side during same-sex consortships, although this form of co-feeding is not a prominent feature of these relationships.

Same-sex consortship activity between female Japanese macaques takes place during the fall/winter mating season (primarily October–February). Outside this period, female–female consortships are never observed. Prior to conception, females direct and receive mounts to and from other females most frequently during the periovulatory period (that is, day of ovulation $\pm$ 3 days), somewhat less frequently during the follicular phase and not at all during the luteal phase (O'Neill et $al.$, 2004). During the follicular phase, pregnanediol (PdG) drops to baseline levels; during the periovulatory phase, PdG begins to rise slightly and estrone (E1) levels reach their peak; and during the luteal phase, PdG levels peak for an extended duration, while E1 levels return to baseline. Following conception there appears to be a second peak in same-sex consortship activity between the sixth and tenth week of gestation, coinciding with a marked decrease in pregnanediol levels at that time (Gouzoules and Goy, 1983; O'Neill et $al.$, 2004).

The frequency of female–female sexual behaviour varies both within and between Japanese macaque populations, raising the possibility that these behaviours represent group-specific social 'traditions' (Stephenson, 1973). In some populations, the behaviour has never been reported and appears to be non-existent. In other populations, some females engage in same-sex sexual behaviour, while others do not, and the proportion of females engaging in the behaviour varies from year to year. In the Arashiyama West (Texas) population, for example, the proportion of females engaged in same-sex consortship activity has been reported to vary from 61% (Fedigan and Gouzoules, 1978), to 78% (Wolfe, 1984), to 51% (Gouzoules and Goy, 1983). Similarly, in the Arashiyama (Japan) troop of Japanese macaques the proportion of females engaged in same-sex consortship activity has been reported to vary from 23% (Wolfe, 1984) to

47% (Takahata, 1982). It remains to be determined whether all females in such populations engage in same-sex consortship activity at some point in their lives.

In still other populations, every sexually mature female engages in same-sex consortship activity and they do so frequently. For example, every sexually mature female in a captive, mixed-sex, group at the Université de Montréal engaged in same-sex sexual behaviour (Vasey, 2002a). During the 1993–94 mating season, 14 of these females were sexually active and the majority of consortships formed ($n = 95$) were female–female ones (55%), not male–female ones (45%). During these same-sex consortships, females ($n = 14$) solicited their partners for sex, on average, 28 times per hour and they mounted each other, on average, 31 times per hour ($n = 129$ h).

Observational (Wolfe, 1984) and experimental (Vasey and Gauthier, 2000) research suggest that female homosexual behaviour in Japanese macaques increases in the context of female-biased operational sex ratios (that is, the number of sexually active males to receptive females). This pattern of female sexual activity is typically explained in proximate terms as owing to a lack of male mates. However, some males are invariably available *and* motivated to mate with females under these demographic conditions (Wolfe, 1984; Vasey and Gauthier, 2000). Many females, nevertheless, rebuff solicitations by such males in favour of same-sex sexual partners. As such, heterosexual deprivation is not an adequate explanation to account for the increase in female homosexual behaviour under these demographic conditions. Fedigan and Gouzoules (1978, pp. 494) suggested that female–female consortship activity 'appeared to be part of a larger pattern of female sexual initiative' and, as such, 'females were not forced to choose other females due to a lack of males'. Wolfe (1984) has hypothesized that the expression of female homosexual behaviour in the context of female-skewed operational sex ratios reflects a quest on the part of the females for sexual novelty. Vasey and Gauthier (2000) have argued that higher levels of female homosexual activity observed under these demographic conditions is not caused by a lack of males, *per se*, but, rather, results from a paucity of preferred male mates coupled with an abundance of preferred female sexual partners. In addition, they note that male sexual competitors are scarce under these demographic conditions and, as such, females are more able to access and maintain preferred, same-sex sexual partners. This particular perspective takes into account the fact that female Japanese macaques that engage in homosexual behaviour are bisexual in orientation, not preferentially heterosexual, as is commonly assumed and they must often compete inter-sexually for access to female sexual partners (Vasey, 1998, 2002a; Vasey and Gauthier, 2000).

The goal of this chapter is to review my research on female homosexual behaviour in Japanese macaques. I begin by examining whether female homosexual behaviour in Japanese macaques evolved to facilitate adaptive socio-sexual goals. Within ethological circles, socio-sexual behaviours are commonly defined as behaviours that are sexual in terms of their outward form, but enacted to facilitate some type of adaptive social goal or breeding strategy (Wickler, 1967). Specifically, I examine whether female Japanese macaques engage in homosexual behaviour to (a) acquire reproductive partners, (b) form alliances, (c) communicate about dominance, (d) acquire alloparental care or (e) reconcile following conflicts. In addition, I present new data that suggest that female Japanese macaques do not use homosexual behaviour to reduce social tension. Following this, I review the literature to ascertain whether female homosexual behaviour in Japanese macaques is a sexual behaviour, as opposed to a strictly socio-sexual one. Finally, I attempt to lay the foundation for future investigations into the evolution of female homosexual behaviour in Japanese macaques by speculating about the evolutionary history of female mounting in this species.

Methods

Subjects and study site

The study group totalled 37 individuals (18 adult females, ages 3.5--23 years; five adult males, ages 4.5–9.5 years; seven immature females and seven immature males) housed at the Université de Montréal's Laboratory of Behavioural Primatology (established 1992) near St-Hyacinthe, Québec, Canada. The group was comprised of three unrelated matrilines. The founding members of the study group originated from the Arashiyama West colony of Japanese macaques that were translocated from Arashiyama, Japan, to southern Texas in 1972 (Fedigan, 1991). The study group's age, sex and matrilineal composition were typical for this species as observed under free-ranging conditions (Yamagiwa and Hill, 1998).

The monkeys lived in five indoor rooms (25 × 15.5 × 3.6 m) and two outdoor enclosures (17 × 16 × 4 m), which were furnished with swinging and climbing devices. Animals were fed daily with a mixture of grains, monkey chow, fruit and vegetables that was distributed in deep wood-chip litter covering the indoor floors. Water was available at will.

Socio-ecology

Japanese macaques are cercopithecine monkeys that are members of the *fascicularis* clade, whose other members include long-tailed (*M. fascicularis*), rhesus (*M. mulatta*) and Formosan (*M. cyclopsis*) macaques. They are found only on the islands of Japan, where they range from 31 to 41° latitude. Their typical habitat is mountainous broad-leafed or montane forest. They are quadrupedal, semi-terrestrial and omnivorous, feeding on nuts, fruit, buds, leaves, bark and, occasionally, insects. Japanese macaques are one of the very few primate species that lives in temperate regions and is able to tolerate cold winters and deep snow. Temperatures in their most northerly distribution fall to –5° C with 2–3 meters of snowfall (Izawa and Nishida, 1963). Japanese macaques live in multi-male, multi-female troops in which females are philopatric and male typically disperse at puberty. These troops are structured around sets of matrilineal kin groups.

Data collection and analysis

Homosexual consortships occurred when two females engaged in series mounting (three or more mounts within a ten-minute period). These consortships were deemed to have terminated if the two female partners were not in proximity (separated by a distance of more than one meter) and exhibited no mounting for ten minutes.

Hindquarter presentations were performed by potential mountees, which stood quadrupedally with their arms and legs flexed and their perineum oriented towards the potential mounter. *Back presentations* were performed by potential mountees, which sat with their forearms slightly bent and their backs inclined and oriented towards the potential mounter. *Hands-on-hindquarter solicitations* were performed by potential mounters, which grasped or placed both hands on the hindquarters of the potential mountee.

Displacement occurred when an individual provoked the immediate departure of another by moving within one arm's length from the other individual while looking at them. *Aggressive behaviours* included open-mouth stare threats, smacking the ground, grunt vocalizations, lunging, chasing, hitting, grabbing, fur pulling and biting that occurred either singly or in various combinations.

The *pre-conflict period* was defined as the 10-min period prior to an aggressive interaction between female consort partners. This 10-min period was divided into the *pre-conflict period-A* and the *pre-conflict period-B*. *Pre-conflict period-A* was defined as the 5-min period prior to an aggressive interaction between female consort partners. *Pre-conflict period-B* was defined as the 5-min period preceding pre-conflict *period-A*. The *post-conflict period* was defined as the 10-min period following an aggressive interaction between female consort partners.

Dominance relationships were determined by the direction of a combination of submissive behaviour patterns (displacements, fear grimaces, rapid flights, piercing screams, defensive hunching). Submission had to be uni-directional between two individuals for the recipient of submissive signals to be considered dominant.

Interventions referred to the intrusion by an individual (supporter) into an ongoing conflict between two opponents, during which the intruding animal supported one of the opponents (recipient of support) against the other (target). *Opportunities for interventions* included all incidences of dyadic aggression during which a third female could have intervened in support of the one of the opponents. Rates of intervention were determined by dividing the number of actual interventions by the total number of opportunities for interventions.

Inter-sexual competition for female sexual partners occurred when a sexually motivated male and female (*competitors*) simultaneously sought access to the same anovulatory female sexual partner (*focus of competition*). Competitors were considered to be sexually motivated *vis-à-vis* the focus of competition if they solicited her for sex (a) while competing intersexually for her or (b) at some time during the same day. Inter-sexual competition for female sexual partners was manifested during intrusion and counter challenges. *Intrusions* occurred when a male competitor attempted to acquire exclusive access to an infertile or postconception female engaged in a homosexual consortship by targeting that female as the focus of competition and her partner as his competitor. *Counter-challenges* occurred following an intrusion when a female competitor attempted to maintain exclusive access to her consort partner, the focus of competition, while preventing the male competitor from doing the same. *Choice of a sexual partner* occurred following inter-sexual competition interactions when the female focus of competition chose to mount with either the male competitor or the female competitor.

Approaches involved one individual moving within one meter of another. Only dyadic approaches between consort partners were analysed where one consort partner approached the other who was alone, while the nearest other individual was more than one meter away.

Females from different matrilines were considered non-kin, as were cousins and aunt–niece dyads. Evidence indicates that female Japanese macaques treat each other as non-kin when their level of relatedness is less than $r = 0.25$ (Chapais *et al.*, 1997).

All observations were recorded on paper in 30-min blocks. Observations took place from dawn until dusk (approximately 7:00–16:00 h). I collected 129 hours of focal animal data on female homosexual consortships between 17 November 1993 and 13 February 1994, excluding 25–29 December 1993. I collected focal data for both consort partners simultaneously, because they were invariably interacting together. When two homosexual consortships occurred simultaneously, I observed the one for which I had the least data.

All statistical tests are two-tailed. Z' indicates a Z score corrected for ties. Sample sizes and compositions varied depending on whether subjects performed the particular behaviours during the particular periods under analysis.

Results

Female homosexual activity

During the 1993–94 mating season, two adult females did not engage in sexual activity. All of the others ($n = 16$) were observed to perform sexual solicitations and 15 of these engaged in series mounting during consortships. Focal animal data were obtained for every female that engaged in homosexual consortships, except for one elderly female (age 23) whose consortships were brief and infrequent. Focal data were collected on 21 different homosexual consortships. Focal females ranged in age from 4.5 to 23.

Alliance formation

If female Japanese macaques engage in homosexual behaviour to form alliances, than the following predictions should hold true:

Prediction 1: There should be a close temporal relationship between homosexual consortship formation and alliance formation between the consort partners.

This prediction was supported. Consort partners intervened in support of each other on 167 occasions. When engaged in homosexual consortships, females intervened for each other during 17% of the opportunities to do so ($n = 1012$), compared with only 0.002% before the consortship ($n = 1290$) and 0.004% after the consortship period ($n = 1138$). Thus, when given the opportunity to do so, dominant and subordinate consort partners intervene for each other significantly more often during same-sex consortships, than either before (dominant partners, $p < 0.05$; subordinate partners, $p < 0.01$) or after (dominant partners, $p < 0.05$; subordinate partners, $p < 0.01$) these relationships.

Prediction 2: During consortships, consort partners should intervene more for each other than for other individuals including kin.

This prediction was supported. During same-sex consortships, females ($n = 9$) intervened significantly more for their sexual partners (mean $= 1.33$ interventions/h) than for other sexually mature, non-kin females (mean $= 0.09$ interventions/h) (dominant partner, $p < 0.05$; subordinate partner, $p < 0.01$).

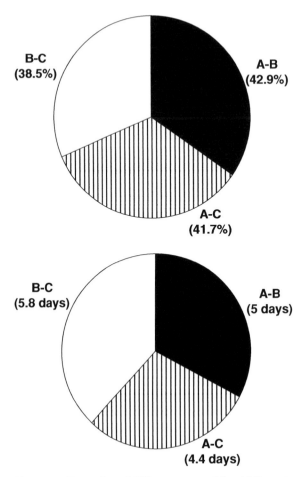

Figure 8.3. Formation of different consortships. (a) Percentages; (b) duration (days).

Prediction 3: Higher ranking non-kin females should be preferred as consort partners because they represent more powerful allies.

This prediction was not supported. Thus, non-kin individuals from the two highest ranking matrilines (A & B) did not form consortships together significantly more often than they did with individuals from the lowest ranking matriline (that is, A–C, B–C; $x^2 = 0.066$, $df = 2$, $p > 0.95$; Figure 8.3a).

Prediction 4: Females should spend more time consorting with higher ranking same-sex partners because they represent more powerful allies.

This prediction was not supported. There was no significant difference in the number of days over which each of the three different types of consortships took place (Kruskal–Wallis test: $H' = 0.1$, $df = 2$, $n = 28$, $p = 0.93$; Figure 8.3b).

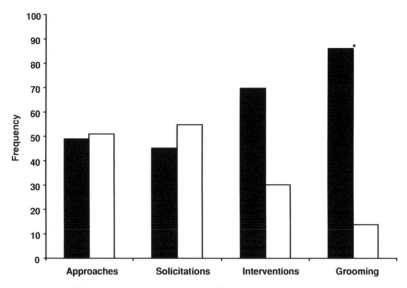

Figure 8.4. Affiliation between dominant and subordinate consort partners. Black columns: dominant partner; white columns: subordinate partner. *p = 0.001.

Prediction 5: Subordinate partners should be more responsible for maintaining homosexual consortships because they stand to gain more alliance-related benefits from these associations.

This prediction was not supported. A total of 2386 approaches were observed. Approaches occurred bi-directionally within consort dyads (Wilcoxon test, n = 21 consortships, Z' = 0.82, p = 0.41). A total of 3649 sexual solicitations were observed. Sexual solicitations also occurred bi-directionally within consort dyads (Wilcoxon test, n = 21 consortships, Z' = 0.3, p = 0.77). When given the opportunity to do so, consort partners intervened (n = 167 interventions) at similar rates to support each other during conflicts (Wilcoxon tests, n = 16 consortships, Z = 1.4, p = 0.16). Grooming interactions were recorded during 1328 instantaneous scan samples. Dominant females groomed their subordinate partners significantly more often than the reverse (Wilcoxon test, n = 18 consortships, Z' = 3.7, p < 0.01; Figure 8.4).

Dominance demonstration

If female Japanese macaques engage in homosexual behaviour to communicate about their relative dominance ranks, then the following predictions should hold true:

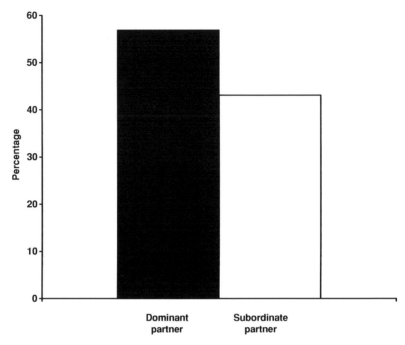

Figure 8.5. Patterns of mounting between dominant and subordinate consort partners.

Prediction 1: Dominant consort partners should mount subordinates more often than the reverse.

This prediction was not supported. A total of 3945 female–female mounts were observed. Dominant and subordinate consort partners mounted each other in a bi-directional manner (Wilcoxon test, $n = 21$, $Z' = 0.2$, $p = 0.85$; Figure 8.5). This pattern of bi-directional mounting held regardless of the type of mount analyzed (Table 8.1).

Prediction 2: Dominant consort partners should perform 'hands-on-hindquarters' solicitations significantly more often than their subordinate counterparts because these gestures function as requests to mount and they prompt the receivers to prepare to be mounted.

This prediction was not supported. A total of 720 hands-on-hindquarters solicitations were observed. Subordinate consort partners performed hand-on-hindquarters solicitations as often as their dominant counterparts (Wilcoxon test, $n = 20$, $Z' = 1.6$, $p = 0.1$).

Table 8.1 *Comparison of dominant and subordinate consort partners'*
behaviour during mounting interactions using Wilcoxon tests for
consortship dyads

Mount type	Mounts	Consortships	Z'/T	P
Double foot-clasp mounts, thrusting	1578	20	1.8*	0.07
Double foot-clasp mounts, no thrusting	1043	20	1.7*	0.09
Sitting mounts, thrusting	749	12	53	0.15
Sitting mounts, no thrusting	297	14	55.5	0.88
Standing mounts, thrusting	40	10	39	0.28
Standing mounts, no thrusting	84	12	44	0.73
Reclining mounts, thrusting	30	6	11.5	0.92
Reclining mounts, no thrusting	124	9	28	0.57

Note: * indicates Z value corrected for ties. All other values are *T* values.

Prediction 3: Subordinate consort partners should perform hindquarter
presentations and back presentations significantly more often than their
dominant counterparts because these gestures function as requests to be
mounted.

This prediction was not supported. A total of 1225 hindquarter presentations and
463 back presentations were observed. Dominant consort partners performed
hindquarter and back presentations as often as their subordinate counterparts
(Wilcoxon test, hindquarter presentations: $n = 20$, $Z' = 1.6$, $p = 0.1$; back
presentations, $n = 17$, $Z' = 0.2$, $p = 0.81$).

Acquisition of alloparental care

If female Japanese macaques engage in homosexual behaviour to acquire
parental care, then the following predictions should hold true:

Prediction 1: During homosexual consortships, females should behave in an
affiliative manner with their partner's immature offspring.

This prediction was not supported. Females ($n = 17$) were within arm's reach
of their same-sex consort partners' immature offspring during 9.7% of the scan
samples ($n = 6190$ mean $= 6.3\%$, range $= 1.5$–21%). Despite this proximity, they
virtually never groomed these immature individuals (0.0005% of scan samples).
During focal observations, 16 females had the opportunity to intervene in support of their same-sex consort partners' immature offspring when the latter were

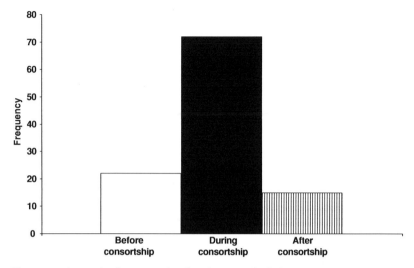

Figure 8.6. Aggression by consorting females towards their partners immature offspring.

involved in conflicts ($n = 134$ opportunities, mean $= 8.4$, range 1–25). None of the females supported their partners' immature offspring in this manner.

> *Prediction 2: During homosexual consortships, females should not behave in an aggressive manner with their partner's immature offspring.*

This prediction was not supported. Females intervened against their consort partners' immature offspring on 10.5% of the occasions that they had the opportunity to do so ($n = 134$ opportunities, mean $= 0.88$, range $= 0$–6). Females intervened against their consort partners' immature offspring significantly more than they supported them (Wilcoxon test, $n = 6$, $T = 21$, $p < 0.05$). In addition, females aggressed their consort partners' immature offspring significantly more often during homosexual consortships ($n = 73$ aggressive interactions) than either before ($n = 22$ aggressive interactions; Wilcoxon test, $n = 10$, $T = 47$, $p < 0.05$) or after them ($n = 15$ aggressive interactions; Wilcoxon test, $n = 13$, $T = 88.5$, $p = 0.001$; Figure 8.6).

Acquisition of opposite-sex mates

If female Japanese macaques engage in homosexual behaviour to acquire opposite-sex mates, then the following prediction should hold true:

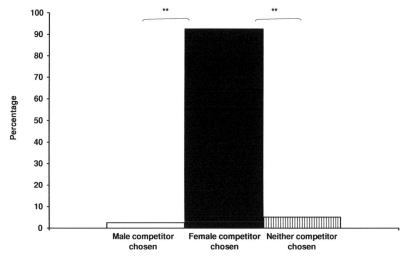

Figure 8.7. Patterns of partner choice by foci of competition following inter-sexual competition episodes ($n = 120$). $**p = 0.002$.

Prediction: Females should cease homosexual activity when a sexually motivated male approaches them for sex.

This prediction was not supported. Following episodes of inter-sexual competition, females that were the foci of competition could choose between two sexually motivated competitors, one male and the other female, who where simultaneously available. In such instances ($n = 120$), females that were the foci of competition chose to continue mounting with the female competitor (that is, their same-sex consort partner) significantly more often then they chose to begin mounting with the male competitor (Wilcoxon test, $n = 14$, $T = 105$, $p = 0.002$) or to cease mounting altogether (Wilcoxon test, $n = 13$, $T = 91$, $p = 0.002$; Figure 8.7).

Reconciliation

If female Japanese macaques engage in homosexual behaviour to reconcile conflicts, then the following prediction should hold true:

Prediction: Homosexual behaviour should occur more often in the period immediately following a conflict between the consort partners.

This prediction was not supported. The average rate at which dominant female consort partners performed sexual solicitations did not differ significantly ($n = 7$, $T = 5$, $p = 0.13$) between the pre-conflict (mean $= 2.6$) and post-conflict

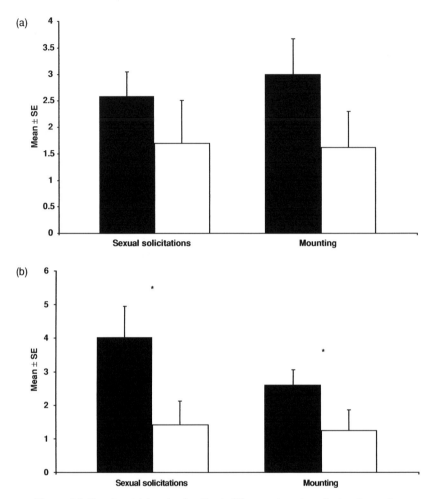

Figure 8.8. Dominant (a) and subordinate (b) consort partners' rate of sexual solicitations and mounting ($M \pm$ SE) during pre- and post-conflict periods. Black columns: pre-conflict period; white columns: post-conflict period. *$p < 0.05$.

(mean $=$ 1.7) periods. Likewise, the average rate at which dominant female consort partners mounted their subordinate counterparts did not differ significantly ($n = 7, T = 3.5, p = 0.08$) between the pre-conflict (mean $=$ 3.01) and post-conflict periods (mean $=$ 1.62; Figure 8.8a).

In contrast, the average rate at which subordinate female consort partners performed sexual solicitations was significantly higher ($n = 7, T = 1, p = 0.03$) during the pre-conflict period (mean $=$ 4.04) compared with the post-conflict period (mean $=$ 1.42). Similarly, the average rate at which subordinate female consort partners mounted their dominant counterparts was significantly higher

Table 8.2 *Rate of sexual solicitations and mounts performed by dominant consort partners during the pre-conflict periods*

	Solicitations				Mounts			
	Pre-conflict period-A		Pre-conflict period-B		Pre-conflict period-A		Pre-conflict period-B	
Consort pair*	Interactions	Rate	Interactions	Rate	Interactions	Rate	Interactions	Rate
A-B5	1	2	1	0	1	0	1	1
A-B2	6	2.17	6	2.17	5	1.8	5	2.4
A3-B3	1	1	1	1	1	2	1	1
A3-B2	1	1	1	4	1	1	1	0
A3-B21	3	0.75	3	0.33	1	2	1	0
A31-B3	1	0	1	3	0	0	0	0
A31-A2	7	1.86	7	2	7	2.57	7	3.57
A2-B9	14	1.14	14	2.29	7	2.57	7	4
B7-C	0	0	0	0	3	1.67	3	2.67

Note: * Listed in descending rank order.

($n = 7, T = 0, p = 0.03$) during the pre-conflict period (mean $= 2.63$) compared with the post-conflict period (mean $= 1.25$; Figure 8.8b).

The regulation of social tension

If female Japanese macaques engage in homosexual behaviour to reduce social tension, then the following prediction should hold true:

Prediction: Homosexual behaviour should increase in frequency at the impending approach of an aggressive interaction between the consort partners.

This prediction was not supported. The frequency with which dominant female consort partners performed sexual solicitations did not differ significantly (Wilcoxon test, $Z' = 0.95, n = 8, p = 0.34$) between pre-conflict period-A (mean $= 1.2$) or pre-conflict period-B (mean $= 1.8$). Likewise, the frequency with which dominant female consort partners mounted their subordinate counterparts did not differ significantly (Wilcoxon test, $Z = 0.287, n = 8, p = 0.774$) between pre-conflict period-A (mean $= 1.7$) or pre-conflict period-B (mean $= 1.8$; Table 8.2).

Similarly, the frequency with which subordinate female consort partners performed sexual solicitations did not differ significantly (Wilcoxon test, $Z = 1.27, n = 7, p = 0.2$) between pre-conflict period-A (mean $= 1.34$) or pre-conflict period-B (mean $= 2.14$). Likewise, the frequency with which subordinate female

Table 8.3 *Rate of sexual solicitations and mounts performed by subordinate consort partners during the pre-conflict periods*

	Solicitations				Mounts			
	Pre-conflict period-A		Pre-conflict period-B		Pre-conflict period-A		Pre-conflict period-B	
Consort pair*	Interactions	Rate	Interactions	Rate	Interactions	Rate	Interactions	Rate
A-B2	5	2	5	1.8	3	3.67	3	2.33
A3-B3	1	0	1	1	1	1	1	2
A3-B2	0	0	0	0	1	0	1	2
A3-B21	4	1	4	1.75	3	1	3	2
A31-B3	1	2	1	1	1	2	1	1
A31-A2	7	1.71	7	2.57	3	0.33	3	1
A2-B9	18	1.67	18	1.89	11	2.09	11	2.09
B7-C	2	1	2	5	3	0.67	3	2.67

Note: *Listed in descending rank order.

consort partners mounted their dominant partners did not differ significantly (Wilcoxon test, $Z' = 1.02$, $n = 8$, $p = 0.31$) between pre-conflict period-A (mean $= 1.35$) or pre-conflict period-B (mean $= 1.89$; Table 8.3).

Discussion

Is homosexual behaviour between female Japanese macaques a socio-sexual behaviour?

Alliance formation

A number of primate species appear to use same-sex sexual behaviour as a mechanism for alliance formation (for example, Fairbanks *et al.*, 1977; Idani, 1991; Parish, 1994; Smuts and Watanabe, 1990). Among female Japanese macaques, same-sex consortships and alliance formation are temporally related. Dominant and subordinate consort partners intervene for each other significantly more often during homosexual consortships, compared with before or after the consortship period. During consortships, females intervened significantly more for their sexual partners than for other sexually mature, non-kin females. This indicates that the tendency for consorting females to intervene more for each other did not reflect a generalized motivation to intervene for any sexually mature, non-kin females. Despite this causal relationship, patterns of partner choice and affiliation between consort partners suggest that females were not engaging in same-sex consortships for the express purpose of alliance formation (Vasey, 1996).

If the primary reason for engaging in same-sex consortships is to mediate alliance formation, then two key behavioural patterns should be expected to occur. First, higher-ranking females should be preferred as consort partners because they represent more powerful allies and, as such, are able to challenge targets that they dominate but that their subordinate consort partners do not. By providing support against targets that rank above their subordinate partners, but below themselves, dominant consort partners can help their partners increase in dominance temporarily, at little risk to themselves (Chapais, 1992; Vasey, 1996). Females should, thus, prefer to consort with dominant partners and competition for such partners should ensue. The most successful competitors should also be the most dominant ones. Given the well-demonstrated constraint that incest avoidance places against the formation of consortships involving close kin in Japanese macaques (Chapais and Mignault, 1991; Wolfe, 1984; Chapais et al., 1997), this should have resulted in a concentration of consortships between members of the two highest ranking matrilines in the study group (that is, A–B consortships versus A–C or B–C consortships where A matriline > B females > C females). Non-kin individuals of higher rank (A–B) did not form consortships together significantly more often than other types of consortships (that is, A–C, B–C). Moreover, there was no significant difference in the number of days over which each of the three different types of consortships took place.

Second, in light of the dominance-related benefits they derive from same-sex consortships, subordinate partners should be more responsible for showing the affiliative behaviours that maintain these associations. This prediction was not supported, however. During same-sex consortships, the overwhelming majority of affiliation (for example, approaches, sexual solicitations, interventions) flowed in a remarkably bi-directional manner with no statistically significant differences between dominant and subordinate consort partners (Chapais and Mignault, 1991; Vasey, 1996; Vasey et al., 1998).

The sole difference in affiliative behaviour involved grooming, which the dominant consort partners performed significantly more often than their subordinate counterparts. Thus, dominant consort partners were actually *more* responsible for maintaining same-sex consortships than their subordinate partners. This may have been because dominant females need to overcome subordinate females initial reticence to form consortships with them, given that, outside of such relationships, subordinate females would have rarely, if ever, interacted affiliatively with their dominant consort partners.

In sum, although same-sex consortships and alliance formation are temporally related, patterns of partner choice and affiliation between consort partners suggest that females were not engaging in these relationships for the express purpose of alliance formation. As such, associations with useful allies during same-sex consortships, let alone powerful ones, would be variable at best. This suggests that same-sex sexual behaviour was not designed by natural selection as a socio-sexual adaptation for alliance formation in female Japanese macaques.

Dominance demonstration

Arguably, the most often-cited explanation for why animals engage in same-sex mounting is because they are communicating about aspects of their dominance relationships (Wickler, 1967; Chadwick-Jones, 1989). According to this perspective, mounting is an expression of dominance, while allowing oneself to be mounted expresses subordinate status *vis-à-vis* the mounter. As such, the roles individuals adopt during same-sex mounting should be differentiated in line with their rank.

Several lines of evidence indicate that female Japanese macaques do not use same-sex sexual behaviour to communicate about their dominance relationships, because the roles dominant and subordinate consort partners adopt during these interactions are not differentiated on the basis of their relative dominance status (for further details see Vasey *et al.*, 1998). Neither consort partner was significantly more or less responsible for mounting or being mounted. Instead, dominant and subordinate partners tended to take turns mounting and being mounted, in a bi-directional manner.

Likewise, dominant consort partners did not perform hands-on-hindquarters solicitations significantly more often than their subordinate counterparts, despite the fact that these gestures function as requests to mount and they prompt the receivers to prepare to be mounted. Moreover, subordinate consort partners did not perform hindquarter or back presentations significantly more often than their dominant counterparts, despite the fact that these gestures function as requests to be mounted. Taken together, the available evidence indicates that same-sex mounting in female Japanese macaques is not a socio-sexual adaptation to communicate about dominance relationships.

Acquisition of alloparental care

Research on various bird species, particularly gulls and terns, suggests that females will form pair bonds with other females to obtain alloparental care for their dependant offspring (for reviews see: Bagemihl, 1999; Diamond, 1989). Several lines of evidence suggest that female Japanese macaques do not engage in same-sex consortships to obtain alloparental care for their immature offspring (for further details see Vasey, 1998).

Grooming and alliance support are the most overt expressions of non-sexual affiliation in Japanese macaques but these activities do not characterize interactions between females and their consort partners' immature offspring. Females were frequently within arm's reach of their consort partners but despite this proximity, they virtually never groomed these individuals.

During same-sex consortships, females had various opportunities to intervene in support of their same-sex consort partners' immature offspring when the latter were involved in conflicts. Nevertheless, none of the females supported their partners' immature offspring in this manner. It is striking, however, that during 10.5% of these conflicts, females intervened against their partners' immature offspring. In addition, females aggressed their partner's immature offspring significantly more often during same-sex consortships than either before, or after, the consortship period.

In sum, there is no evidence that female Japanese macaques provided alloparental care to the immature offspring of their same-sex consort partners and this was not for lack of opportunities to do so. In fact, the relationship between females and their same-sex consort partners' immature offspring could best be described as agonistic, not alloparental, in nature. This points to the conclusion that female–female sexual behaviour in Japanese macaques is not a socio-sexual adaptation for the acquisition of alloparental care.

Acquisition of male mates

Parker and Pearson (1976) proposed that female–female 'sexual' behaviour functions to increase the reproductive success of the mounting female. By mimicking the copulatory pattern of rival males, they argued that the mounting female can attract dominant male sexual partners and increase her chances of insemination. If female Japanese macaques engage in same-sex consortships to attract male reproductive partners, then three key behavioural patterns should be expected to occur. First, same-sex mounting between females should only occur in the presence of males. Second, same-sex mounting between females should only occur when the mounting female is fertile. Third, same-sex mounting should cease after a male solicits one of the female partners for sex.

Several lines of evidence indicate that female Japanese macaques do not use same-sex mounting to attract male mates. First, consorting females often attempt to spatially and visually separate themselves from other group members, including males (Gouzoules and Goy, 1983; Vasey, 1996, 1998). Second, same-sex sexual behaviour between female Japanese macaques frequently occurs following conception when the issue of insemination is moot (Fedigan and Gouzoules, 1978;

Gouzoules and Goy, 1983; Wolfe, 1984; Vasey, 1998). Third, females engaged in same-sex consortships typically tend to ignore or even threaten males that solicit them for sex (Wolfe, 1984; Vasey, 1998). These threats sometimes escalate into direct attacks on the males and attempts on the part of consorting females to drive soliciting males away (Vasey, 1998). Taken together, the available evidence indicates that same-sex mounting in female Japanese macaques is not a socio-sexual adaptation to attract male reproductive partners.

Reconciliation

Research on a number of species, particularly bonobos, indicates that homosexual behaviour may function to mediate reconciliations following the outbreak of aggression (Kano, 1980; de Waal, 1987; Furuichi, 1989). The 'reconciliation hypothesis' predicts that the frequency with which homosexual behaviours are manifested should be very context specific. Namely, although homosexual behaviour may occur outside of a post-conflict period, it should be exhibited more frequently following an aggressive incident, as compared with a matched control period preceding the conflict (Veenema, 2000). For example, genital–genital rubbing between female bonobos appears to serve a reconciliatory function within a post-conflict context, but females also engage in these types of interactions outside of post-conflict situations. They do so when attempting to form 'close associations' with females in groups into which they are transferring (Furuichi, 1989; Idani, 1991) and during periods of social tension associated with food competition and sharing (Kano, 1980; Kuroda, 1984; Thompson-Handler et al., 1984; de Waal, 1987; Furuichi, 1989). As such, although female Japanese macaques may engage in sexually motivated series mounting and courtship outside of post-conflict contexts, the reconciliation hypothesis predicts that the frequency of these behaviours should peak during post-conflict interactions relative to pre-conflict control periods.

Unlike bonobos, homosexual interactions in Japanese macaques do not cluster around post-conflict periods. During homosexual consortships, dominant female consort partners were as likely to engage in same-sex courtship and mounting before an aggressive interaction as after. In contrast, subordinate consort partners were more likely to perform same-sex sexual solicitations and mounts before an aggressive interaction than after. This suggests that aggressive interactions inhibit, rather than facilitate, the expression of homosexual behaviour among subordinate consort partners (Vasey, 2004). Taken together, the available evidence indicates that female–female courtship and mounting in Japanese macaques is not a socio-sexual adaptation for reconciliation following conflicts.

Tension regulation

In bonobos, homosexual behaviour often occurs between individuals feeding at the same food site where signs of tension over food are apparent (Kano, 1980; Kuroda, 1980, 1984; Thompson-Handler, 1989). Kuroda (1980) interpreted homosexual behaviour among bonobos to be a mechanism to reduce social tension during periods of close proximity in the same food patch. He argued that it 'works to calm anxiety or excitement, to dissolve inter-individual tension', and 'thus to increase tolerance, which makes food sharing smooth' (p. 190). Indeed, observations in support of this interpretation indicate that bonobo individuals entering occupied patches are more likely to acquire food after engaging in homosexual interactions (Kuroda, 1984; de Waal, 1987; Furuichi, 1989). Similarly, observations of other primate species such as baboons (*Papio cynocephalus*; Owen, 1976), pigtail macaques (*M. nemestrina*; Oi, 1990, 1991), and black crested macaques (*M. nigra*; Dixson, 1977) also suggest a close temporal association among homosexual behaviour, increased inter-individual tolerance and reduced aggression during periods of social tension or excitement.

According to the 'social tension hypothesis' homosexual behaviour should occur prior to an aggressive incident and should increase in frequency as social tension mounts; that is, as the time-to-aggression decreases. Contrary to the predictions of the tension reduction hypothesis, homosexual behaviour between female Japanese macaques did not increase in frequency at the impending approach of an aggressive interaction. In fact, homosexual behaviour was no more common immediately before an aggressive interaction than it was during a more distal time period. Taken together, these results suggest that solicitations and mounting between female Japanese macaques do not function as adaptations for reducing social tension associated with incipient aggression.

Is same-sex consortship behaviour in Japanese macaques sexual?

As demonstrated above, a substantial number of studies indicate that the female Japanese macaques do not use same-sex solicitations mounting and courtship to facilitate adaptive socio-sexual goals (Gouzoules and Goy, 1983; Vasey, 1995, 1996, 1998, 2004; Vasey et al., 1998). Despite over 40 years of intensive research on this species, there is not a single study demonstrating any adaptive value for female–female sexual behaviour in Japanese macaques. In the face of these data, it seems increasingly unparsimonious to assert that these behaviours are socio-sexual adaptations. That being said, what evidence is there that same-sex mounting, courtship and consortship activity in female Japanese macaques are really sexual behaviours?

Currently, there are three convergent lines of evidence that indicate that female same-sex courtship, mounting and consortship activity in Japanese macaques are, indeed, sexual (Fedigan and Gouzoules, 1978; Gouzoules and Goy, 1983; Wolfe, 1984; Vasey, 1996; Vasey *et al.*, 1998). To begin with, same-sex sexual interactions in Japanese macaques follow many of the same patterns as heterosexual copulation. For example, like heterosexual behaviour: (a) female–female courtship mounting and courtship are never observed outside of the species' fall–winter mating season; (b) females that engage in same-sex sexual behaviour exhibit a reddening of the face and perineum that is indicative of increased sexual receptivity; (c) female same-sex sexual behaviour occurs within the context of temporary, but exclusive sexual relationships called consortships; (d) female–female courtship behaviours appear to be identical to those observed in heterosexual contexts; (e) certain female–female mounting postures (for example, double foot-clasp mounts with thrusting) are similar to those employed by males in a heterosexual context; (f) female mountees do not try to escape being mounted by pushing or shaking off the mounter. Instead, they typically brace their bodies to support the mounters. Often, the mountees will facilitate mounts-in-progress by reaching back and clasping the mounters while gazing into the mounters' eyes.

When considering the issue of sexual motivation, it also noteworthy that patterns of affiliation during same-sex consortships do not mimic generalized patterns of social affiliation that occur outside these associations. Rather, patterns of affiliation between females during same-sex consortships represent radical departures from normal patterns of affiliation that are widely recognized as social. This suggests that same-sex consortships are not social relationships, but, rather, are sexual ones. First, female Japanese macaques exercise incest avoidance with close female kin ($r \leq 0.25$) (for more details see Chapais *et al.*, 1997). Consequently, mounting, sexual solicitations and consortships are never observed between mothers and daughters, sisters, or grandmothers and grand-daughters (Chapais and Mignault, 1991; Chapais *et al.*, 1997; Wolfe, 1984). In contrast, these close categories of female kin commonly engage in other types of affiliative behaviour with each other, such as grooming, huddling and inter-ventions (Chapais, 1992; Koyama, 1991; Kurland, 1977). Second, female consort partners intervene in conflicts to support each other at consistently high rates during same-sex consortships. Outside the consortship period, these same females rarely support each other, if at all (Vasey, 1996). Third, normal dominance relationships are temporarily destabilized during same-sex consortships because subordinate consort partners receive support from their dominant partners against targets which normally rank above them, or with whom they share ambiguous dominance relationships (Vasey, 1996). Fourth, dominant consort partners groom their subordinate counterparts significantly more often than the reverse (Vasey, 1996). Outside of same-sex consortships, the opposite pattern holds true: female Japanese macaques overwhelmingly direct grooming up the hierarchy, not down (Chapais *et al.*, 1995; Koyama, 1991). Fifth, females are almost exclusively in proximity with their non-kin consort partners during same-sex consortships (Gouzoules and Goy, 1983; Vasey, 1996, 1998; Wolfe, 1984). Outside the consortship period, however, female Japanese macaques spend most of their time in proximity with close kin (Kurland, 1977; Singh, D'Souza and Singh, 1992).

Finally, prolonged and directed vulvar contact and stimulation occur during many same-sex mounts. During sitting mounts, for example, mounters rub their vulvas repeatedly and vigorously against the backs of mountees. During some double foot-clasp and standing mounts, mounters thrust against the mountees' vulvas. In addition, both mounters and mountees often stroke their vulvas with their tails during same-sex mounts.

Taken together, these various lines of evidence suggest that female–female courtship, mounting and genital contact in Japanese macaques are sexual behaviours. Moreover, the overwhelmingly bi-directional flow of sexual solicitations, mounts, clasping, approaches and interventions within consortships (Chapais and Mignault, 1991; Vasey, 1996; Vasey *et al.*, 1998) suggest that these relationships are based on a mutual sexual attraction between the partners. In light of this information, as well as the data bearing on facultative same-sex sexual partner preference among females of this species (Vasey, 1998), it seems reasonable, indeed, even appropriate, to characterize these interactions as 'homosexual behaviour', as many researchers have done in the past.

Evolutionary history

My attempt to account for homosexual behaviour between female Japanese macaques in terms of proximate sexual reward does not address why the behaviour evolved in the first place. If same-sex sexual interactions and the expression of same-sex sexual preference among female Japanese macaques are not adaptive, as the data suggest, then how might we account for these phenomenon within the context of an evolutionary framework? In response to this question, I have suggested, based on the available evidence, that female homosexual behaviour is not an adaptation; rather, it is the *by-product* of an adaptation (Vasey, 1996, 1998, 2002b; Vasey *et al.*, 1998). By-products of adaptations are characteristics that do not evolve to solve adaptive problems, and, thus, do not have a function and are not products of natural selection. Instead, by-products are 'carried along' with particular adaptations because they happened to be coupled with those adaptations (Futuyma and Risch, 1984; Buss *et al.*, 1998). By-products of adaptations cannot be explained in functional terms and, instead, evolutionary history must be invoked to account for their existence. Explanations that invoke evolutionary history focus on reconstructing the evolutionary steps that lead to a current behaviour. This involves understanding the origin of the behaviour and how it changed over time (Alcock, 1989).

I have speculated that female homosexual behaviour in Japanese macaques is a by-product of an adaptation, namely female–male mounting. Female–male mounting is routinely observed in certain populations of this species (Gouzoules and Goy, 1983) and I have suggested that it is an adaptation that females employ to prompt sexually disinterested or sluggish males to copulate with them (Vasey, 2002b). In line with this hypothesis is the observation that proceptive females routinely perform mount prompting and mounts in response to movement by desired males away from their immediate vicinity (Vasey, pers. obs.).

It is by no means apparent, however, why an ability to mount opposite-sex sexual partners would translate into a desire to mount same-sex sexual partners. Why would females mount same-sexual partners in the absence of any fitness-enhancing, socio-sexual pay-off? To understand why it is important to note that female Japanese macaques engage in vulvar stimulation during female–male mounts. They do so, by rubbing their clitorises against the backs of male mountees or by masturbating with their tails during female-male mounts. Once females evolved the capacity to mount, and the capacity to derive sexual gratification during mounts through vulvar stimulation, they could do so just as easily by mounting females, as males. Indeed, at that point, some same-sex sexual partners might be preferred over certain opposite-sex alternatives, simply because females derived more sexual gratification during interactions with them. Within the context of this evolutionary scenario, female–female mounting and occasional same-sex sexual partner preference can be seen as the by-product of an adaptation, namely female–male mounting. Thus, from an ultimate perspective, female–male mounting can be explained in terms of function and evolutionary history, but female–female mounting can only be explained in terms of the latter.

Understanding the ultimate reasons why individual animals sometimes prefer certain same-sex sexual partners relative to certain opposite-sex alternatives may involve abandoning a strictly functional perspective and instead approaching the issue from a historical evolutionary perspective (Futuyma and Risch, 1984; Vasey, 1995; Bagemihl, 1999). The evolutionary scenario outlined above is intended to demonstrate how the perspective of evolutionary history might aid us generate testable hypotheses that can help account for the evolution of homosexual behaviour, particularly in those cases where the behaviour appears to be functionless (Vasey, 2002b). It must be stressed, however, that homosexual behaviour in Japanese macaques is by no means representative of the entire phenomenon of homosexual behaviour in animals. Instead, multiple motivations, developmental pathways and evolutionary histories underlie the phenomenon we call 'homosexual behaviour' and these vary both within and between species (Vasey, 1995; Bagemihl, 1999).

Acknowledgements

I thank Bernard Chapais, Carole Gauthier, Ellen Kapsalis and Volker Sommer. Various stages of this research were funded by the Université de Montréal, the Sigma Xi Society, the Fonds pour la formation des chercheurs et l'aide de recherche (FCAR) Québec, the University of Lethbridge, the L. S. B. Leakey Foundation and the Natural Science and Engineering Research Council (NSERC) of Canada.

References

Alcock, J. (1989) *Animal Behavior*, 4th edn. Sunderland, MA: Sinauer Associates.

Bagemihl, B. (1999) *Biological Exuberance: Animal Homosexuality and Natural Diversity*. New York: St. Martin's.

Buss, D., Haselton, M. G., Shackelford, T. K., Bleske, A. L. and Wakefield, J. C. (1998) Adaptations, exaptations, and spandrels. *Am. Psychol.*, **53**, 533–48.

Chadwick-Jones, J. K. (1989) Presenting and mounting in non-human primates: theoretical developments. *J. Soc. Biol. Struct.*, **12**, 319–33.

Chapais, B. (1992) The role of alliances in the social inheritance of rank among female primates. In *Coalitions in Humans and Other Animals*, eds. A. Harcourt and F. M. B. de Waal, pp. 29–60. Oxford: Oxford University Press.

Chapais, B. and Mignault, C. (1991) Homosexual incest avoidance among females in captive Japanese macaques. *Am. J. Primatol.*, **23**, 171–83.

Chapais, B., Gauthier, C. and Prud'homme, J. (1995) Dominance competition through affiliation and support in Japanese macaques: an experimental study. *Int. J. Primatol.*, **16**, 521–36.

Chapais, B., Gauthier, C., Prud'homme, J. and Vasey, P. L. (1997) Relatedness threshold for nepotism in Japanese macaques. *Anim. Behav.*, **53**, 533–48.

de Waal, F. B. M. (1987) Tension regulation and nonreproductive functions of sex in captive bonobos (*Pan paniscus*). *Nat. Geogr. Res.*, **3**, 318–38.

Diamond, J. M. (1989) Goslings of gay geese. *Nature*, **340**, 101.

Dixson, A. F. (1977) Observations on the displays, menstrual cycles and sexual behavior of the 'Black ape' of Celebes (*Macaca nigra*). *J. Zool.*, **182**, 63–84.

Dixson, A. F. (1998) *Primate Sexuality*. Oxford: Oxford University Press.

Fairbanks, L. A., McGuire, M. T. and Kerber, W. (1977) Sex and aggression during rhesus monkey group formation. *Aggress. Behav.*, **3**, 241–9.

Fedigan, L. M. (1991) History of the Arashiyama West Japanese macaques in Texas. In *The Monkeys of Arashiyama: Thirty-five Years of Research in Japan and the West*, eds. L. M. Fedigan and P. J. Asquith, pp. 54–73. Albany: SUNY Press.

Fedigan, L. M. and Gouzoules, H. (1978) The consort relationship in a troop of Japanese monkeys. In *Recent Advances in Primatology*, Vol. 1, eds. D. J. Chivers and J. Herbert, pp. 493–5. New York: Academic Press.

Furuichi, T. (1989) Social interactions and the life history of female *Pan paniscus* in Wamba, Zaire. *Int. J. Primatol.*, **10**, 173–97.

Futuyma, D. J. and Risch, S. J. (1984) Sexual orientation, sociobiology, and evolution. *J. Homosex.*, **9**, 157–68.

Gouzoules, H. and Goy, R. W. (1983) Physiological and social influences on mounting behavior of troop-living female monkeys (*Macaca fuscata*). *Am. J. Primatol.*, **5**, 39–49.

Huynen, M. C. (1997) Homosexual interactions in female rhesus monkeys, *Macaca mulatta*. *Adv. Ethol.*, **32**, 211.

Idani, G. (1991) Social relationships between immigrant and resident bonobo (*Pan paniscus*) females at Wamba. *Folia Primatol.*, **57**, 83–95.

Izawa, K. and Nishida, T. (1963) Monkeys living in the northern limits of their distribution. *Primates*, **4**, 67–87.

Kano, T. (1980) Social behavior of wild pygmy chimpanzees (*Pan paniscus*) of Wamba: a preliminary report. *J. Hum. Evol.*, **9**, 243–60.

Koyama, N. (1991) Grooming relationships in the Arashiyama group of monkeys. In *The monkeys of Arashiyama: Thirty-five Years of Research in Japan and the West*, eds. L. M. Fedigan and P. Asquith, pp. 211–26. Albany: SUNY Press.

Kurland, J. A. (1977) *Kin Selection in the Japanese Monkey*. Basel: S. Karger.

Kuroda, S. (1980) Social behavior of the pygmy chimpanzee. *Primates*, **21**, 181–97.

Kuroda, S. (1984) Interactions over food among pygmy chimpanzees. In *The Pygmy Chimpanzee*, ed. R. L. Susman, pp. 301–24. New York: Plenum.

McDonald-Pavelka, M. (1993) *Monkeys of the Mesquite: The Social Life of the South Texas Snow Monkeys*. Dubuque, Iowa: Kendall/Hunt.

Nadler, R. D. (1990) Homosexual behavior in nonhuman primates. In *Homosexuality/Heterosexuality: Concepts of Sexual Orientation*, eds. D. P. McWhirter, S. A. Sanders and J. M. Reinisch, pp. 138–70. New York: Oxford University Press.

Nanda, S. (2000) *Gender Diversity: Cross-cultural Variations*. Prospect Heights, IL: Waveland.

O'Neill, A., Fedigan, L., and Ziegler, T. (2004) Ovarian cycle phase and same-sex making behavior in Japanese Macaque females. *Am. J. Primatol.*, **63**, 25–31.

Oi, T. (1990) Patterns of dominance and affiliation in wild pig-tailed macaques (*Macaca nemestrina nemestrina*) in West Sumatra. *Int. J. Primatol.*, **11**, 339–5.

Oi, T. (1991) Non-copulatory mounting in wild pig-tailed macaques (*Macaca nemestrina nemestrina*) in West Sumatra, Indonesia. In *Primatology Today*, eds. A. Ehara, T. Kimura, O. Takenaka and M. Iwamoto, pp. 147–50. Amsterdam: Elsevier.

Owen, N. W. (1976) The development of sociosexual behavior in free-living baboons, *Papio anubis*. *Behavior*, **57**, 241–59.

Parish, A. R. (1994) Sex and food control in the 'uncommon chimpanzee': how bonobo females overcome a phylogenetic legacy of male dominance. *Ethol. Sociobiol.*, **15**, 157–79.

Parker, G. A. and Pearson, R. G. (1976) A possible origin and adaptive significance of the mounting behavior shown by some female mammals in oestrous. *J. Nat. Hist.*, **10**, 241–5.

Singh, M., D'Souza, L. and Singh, M. R. (1992) Hierarchy, kinship and social interaction among Japanese monkeys, *Macaca fuscata*. *J. Biosci.*, **17**, 15–27.

Smuts, B. B. and Watanabe, J. M. (1990) Social relationships and ritualized greetings in adult male baboons (*Papio cynocephalus anubis*). *Int. J. Primatol.*, **11**, 147–72.

Stephenson, G. R. (1973) Testing for group specific communication patterns in Japanese macaques. In *Precultural Primate Behaviour*, ed. E. W. Menzel, pp. 51–75. Basel: S. Karger.

Takahata, Y. (1982) The sociosexual behavior of Japanese macaques. *Z. Tierpsychol.*, **59**, 89–108.

Thompson-Handler, N., Malenky, R. K. and Badrian, N. (1984) Sexual behavior of *Pan paniscus* under natural conditions in the Lomako Forest, Equateur, Zaire. In *The Pygmy Chimpanzee*, ed. R. L. Susman, pp. 347–68. New York: Plenum.

Vasey, P. L. (1995) Homosexual behavior in primates: a review of evidence and theory. *Int. J. Primatol.*, **16**, 173–204.

Vasey, P. L. (1996) Interventions and alliance formation between female Japanese macaques, *Macaca fuscata*, during homosexual consortships. *Anim. Behav.*, **52**, 539–51.

Vasey, P. L. (1998) Female choice and inter-sexual competition for female sexual partners in Japanese macaques. *Behaviour*, **135**, 579–97.

Vasey, P. L. (2002a) Sexual partner preference in female Japanese macaques. *Arch. Sex. Behav.*, **31**, 45–56.

Vasey, P. L. (2002b) Same-sex sexual partner preference in hormonally and neurologically unmanipulated animals. *Ann. Rev. Sex Res.*, **13**, 141–79.

Vasey, P. L. (2004) Pre- and post-conflict interactions between female Japanese macaques during homosexual consortships. *Int. J. Comp. Psych.*, **17**, 351–9.

Vasey, P. L. and Gauthier, C. (2000) Skewed sex ratios and female homosexual activity in Japanese macaques: an experimental analysis. *Primates*, **41**, 17–25.

Vasey, P. L., Chapais, B. and Gauthier, C. (1998) Mounting interactions between female Japanese macaques: testing the influence of dominance and aggression. *Ethology*, **104**, 387–98.

Veenema, H. C. (2000) Methodological progress in postconflict research. In *Natural conflict resolution*, eds. F. Aureli and F. B. M. de Waal, pp. 21–3. Berkeley, CA: University of California Press.

Wallen, K. and Parsons, W. A. (1997) Sexual behavior in same-sexed nonhuman primates: is it relevant to understanding homosexuality? *Ann. Rev. Sex. Res.*, **7**, 195–223.

Wickler, W. (1967) Socio-sexual signals and their intra-specific imitation among primates. In *Primate Ethology*, ed. D. Morris, pp. 69–147. London: Weidenfeld & Nicolson.

Wolfe, L. D. (1984) Japanese macaque female sexual behavior: a comparison of Arashiyama East and West. In *Female Primates: Studies by Women Primatologists*, ed. M. F. Small, pp. 141–57. New York: Alan R. Liss.

Yamagiwa, J. and Hill, D. A. (1998) Intraspecific variation in social organization of Japanese macaques: past and present scope of field studies in natural habitats. *Primates*, **39**, 257–74.

9

Getting to know you: female–female consortships in free-ranging rhesus monkeys

ELLEN KAPSALIS AND RODNEY L. JOHNSON

Introduction

In this paper, we report on the occurrence of female–female (homosexual) consortships among free-ranging, seasonally breeding rhesus macaques (*Macaca mulatta*). Ours is not the first report of homosexual consortships in free-ranging female rhesus. Carpenter (1942b) and Wolfe *et al.* (1991) both observed a small number of such consortships in two other rhesus populations, namely, the macaques on Cayo Santiago, Puerto Rico, and those at Silver Springs, Florida, respectively. Also, Lindburg (1971: Table X) reported isolated instances of female same-sex mounting among the rhesus monkeys resident near Dehra Dun, India, though he did not characterize such behaviour as homosexual (see also Lindburg, 1983). These earlier studies were of limited duration and, therefore, unable to examine interannual variation in the incidence of female homosexual consortships. Here, we demonstrate an increasing incidence of female same-sex consortships over a seven-year period, which paralleled a precipitous decline in the female membership of our study populations.

In contrast to the small number of reports mentioning female homosexual behaviour in free-ranging rhesus monkeys, there have been multiple accounts of such behaviour in captive rhesus (for example, Hinde and Rowell, 1962; Bernstein and Mason, 1963; Fairbanks *et al.*, 1977; Akers and Conaway, 1979; Huynen, 1997; Wallen and Parsons, 1997). We have also observed it in captive rhesus (personal observation), and suspect that same-sex consortships may be uncommon but not rare among captive females. Our previous field experience, however, leads us to conclude that female homosexual consortships are rare to non-existent in most free-ranging rhesus populations. Collectively, we have had opportunities to observe such behaviour (had it been present) in four different free-ranging rhesus macaque populations: behavioural data on more than 80 female rhesus on Cayo Santiago, over four consecutive mating seasons by Ellen Kapsalis both as a field assistant and during her own dissertation research (Kapsalis and Berman, 1996a, b); data on the behaviour of rhesus females were collected by R. L. Johnson both in Nepal during a single mating season, and in India over two consecutive mating seasons, where the animals' sexual activity was a specific focus (Malik *et al.*, 1992; Johnson *et al.*, 1993). He also conducted a ten-month survey (encompassing the 1986–87 mating season) of the same rhesus populations studied for the present paper (Johnson, 1989). Despite this previous field experience, neither of us have ever observed female homosexual consortships prior to the work described here.

Use of the term consortship has been criticized by Manson (1997) since it has sometimes been employed as a proxy for male reproductive success. We use consortship much as Carpenter (1942a, b) did when he defined it as an association between an adult male and female 'with a high degree of reciprocally interactive behavior and "rapport"'. For us, sexual interaction is a necessary but not sufficient part of a consortship. A given pair of animals must also interact in other ways (see Methods) to be considered a consort pair. Unlike Carpenter, we use consortship to embrace female–female as well as male–female dyads, since qualitatively, the behaviour of the same-sex consorts we observed during this study was indistinguishable from that of their heterosexual counterparts (see also Akers and Conaway, 1979; Fedigan and Gouzoules, 1978). Although quantitative differences in the interactions of male–female and female–female consortships likely exist, such differences are not relevant to the present paper since our focus is the incidence of homosexual consortships, not the form of the homosexual behaviour itself.

Our data were collected *ad libitum* during the course of a larger study of reproductive ageing in rhesus females resident on either of two islands situated at the southern tip of the Florida peninsula, namely, Raccoon Key and Key Lois (see Johnson and Kapsalis, 1995, 1998). The interannual changes in homosexual consortships that we observed within both populations are best understood in the context of their unusual histories described below. In our discussion, we argue that our data are supportive of the alliance-formation hypothesis proposed by Fairbanks *et al.* (1977). We also show that the observed interannual differences in the incidence of homosexual consortships on Raccoon Key and Key Lois cannot be accounted for by between-year changes in the adult female–male sex ratio, a proximate mechanism that has been argued as responsible for intraspecific variation in the homosexual behaviour of Japanese macaques (*M. fuscata*).

Methods

Subjects and study sites

Rhesus monkeys are native to a broad swathe of Asia extending from the northeastern tip of Afghanistan westward through Pakistan, India and Nepal, terminating finally at the Pacific coast of China and Vietnam (Wolfheim, 1983). They also occupy a wide variety of habitats including the low-lying mangrove forests of the Sundarbans, West Bengal (Mukherjee and Gupta, 1965) and Himalayan pine-oak forests at elevations as great as 3000 m (Richie *et al.*, 1978). However, many rhesus populations live in close association with humans, residing near or even within villages, towns and cities. Irrespective of where they occur, rhesus are seasonal breeders. Typically, copulations occur in the fall and winter, while births occur in spring and summer. The animals live in social groups containing multiple adult males and females, with the latter outnumbering the former due to sex differences in migration and mortality. Females ordinarily remain throughout their lives in the social groups they are born into. Males, however, typically emigrate from their natal groups around the age of puberty (Lindburg, 1971; Sade, 1972; Colvin, 1986), and are subject to greater mortality at this time (Drickamer, 1974).

In order to make the *M. mulatta* more accessible to primatologists and biomedical researchers in North America, rhesus monkeys from India were used to establish provisioned but free-ranging rhesus populations on six islands in the Caribbean or along the south eastern coast of the United States from 1938 to 1978 (Carpenter, 1942a, b; Drickamer, 1974; Evans, 1989; Taub and Mehlman, 1989). The Raccoon Key and Key Lois monkey populations were two of these six colonies (Pucak *et al.*, 1982). Although monkeys no longer reside on either island, both locations were populated with rhesus macaques for over 20 years.

Key Lois (also known as Loggerhead Key), is a 39-hectare mangrove island located at 24°36′5″ N, 81°28′32″ W in the Atlantic Ocean. Provisioned, free-ranging rhesus monkeys occupied the island from 1973 to 1997. Raccoon Key, an 81-hectare island situated at 24°44′45″ N, 81°29′32″ W in the Gulf of Mexico, was occupied by rhesus Monkeys from 1978 to 1999. The Raccoon and Lois populations were breeding colonies owned and maintained by Charles River Laboratories. Although trails and open areas existed on Lois and Raccoon during data collection, much of the islands were inaccessible owing to the deep mud and sea water that covered a large portion of each (Figure 9.1) especially at high tide (see Sherman, 1980; Johnson and Kapsalis, 1995, 1998, for additional information).

From 1973 to 1976, 1202 female and 176 male rhesus monkeys were released on Key Lois (Sherman, 1980). Before the initial release on Lois, each Indian-born monkey was given a chest tattoo, assigned an estimated year of birth and held in quarantine for up to six months (Sherman, 1980). Given that the animals were juveniles or young adults (that is, two to four years of age) at the time of their entry into the US (Sherman, 1980), we assume the great majority were assigned accurate birth years. From 1978 to 1980, over 500 of the Key Lois rhesus, including both Indian- and island-born animals, were removed and transferred to Raccoon Key. On both islands the monkeys were provided daily with fresh water and a commercial primate diet.

During the 12 years following the transfer of animals from Key Lois to Key Raccoon (that is, 1980–1992), adults were removed from the islands occasionally. However, each year 'yearlings' (that is, animals 9–18 months of age) were captured and removed for sale. Prior to 1986, many island-born animals were not sold; those not slated for sale were given chest tattoos when first captured and released back into the free-ranging population. From 1986 onward, the annual removal of animals included juveniles as well as yearlings and, for both of these age classes, was nearly exhaustive (in accordance with the mandates of the Florida Department of Environmental Protection). With recruitment of immatures into the adult age class severely restricted, there was no meaningful replacement of adults lost due to natural mortality. Hence, the Lois and Raccoon populations were both in decline when this study was begun in February 1992 (Johnson and Kapsalis, 1995). In order to accelerate the decline on Key Lois and hasten the island's retirement as a site for breeding rhesus monkeys (again as mandated by the Florida Department of Environmental Protection), several rhesus social groups on Key Lois were captured during the autumns of 1992–1995, and the adult females transported to Raccoon Key (Johnson and Kapsalis, 1998). Adult males were transferred only in 1992. Live trapping on Key Lois (but not Raccoon Key) became a year-round effort from 1996 onward, and all free-ranging monkeys were removed from the island by the summer of 1997. In contrast, only yearlings and juveniles continued to be targeted for removal from Raccoon Key until the summer 1998 when two social groups were removed almost in their entirety and transferred to captive housing. It was not until the start of 1999 that live trapping of the entire Raccoon Key population became intensive. The retirement of Raccoon Key as a rhesus breeding site occurred in December of that same year.

Population census

The census of the Key Lois and Raccoon Key rhesus monkeys that we began in February 1992 was focused on delineating the survivorship and repro-ductive performance of the islands' adult females (see Johnson and Kapsalis,

1995, 1998). In general, we recorded the identity and reproductive status of each female we encountered during visits that occurred two to four times per week. At first we were able to reliably identify individual females only by means of the animals' chest tattoos. During the latter years of this study, however, we were able to effect identifications of females on the basis of the monkeys' physiognomies alone. Deaths mostly were inferred on the basis of the animals' disappearances; because of the inaccessibility of much of Raccoon Key and Key Lois, the bodies of deceased monkeys were found infrequently, and generally not soon enough after death to permit individual identification. The exact date of birth was not known for either the Indian- or island-born adults on the islands. Therefore, for the purposes of this study, we advanced the age of each female by one year if she survived to 1 January of the next calendar year.

At the beginning of 1992, both the Key Lois and Raccoon Key adult female populations included a large number of aged animals (that is, those $\geq$ 20 years old) and a depauperate number of young adults relative to that found in natural or even in provisioned but unmanipulated macaque populations (Johnson and Kapsalis, 1995). The atypical age structures of the Raccoon and Lois colonies were due to the limited number of island-born animals permitted to stay on the islands long enough to become adults, and to the fact that the founding population of Indian-born animals was comprised of so many similarly aged monkeys. With over 1200 such females released on Lois and Raccoon, it was inevitable that a relatively large number would survive to enter their third decade of life at about the same time. Because the trapping of adults occurred on Raccoon Key much later than on Key Lois (1998 versus 1992), the proportion of aged animals on the former island in particular continued to increase after 1992, producing a concomitant increase in adult mortality (see Results).

Our understanding of the changes in the population of adult males on Raccoon and Lois is more limited than that of the population of females. Individual identification of the adult males on Raccoon and Lois was impractical, especially at the beginning of this study. The reasons were twofold: (a) the animals' dense chest hair and the concomitant unreadability of their tattoos and (b) our inability to closely approach extra-group males in the less accessible parts of each island that they frequented. Lacking a direct measure of the number of adult males on Raccoon and Lois, we had to estimate the mating-season adult sex ratio on the islands by conducting periodic counts of the number of males and females in sight at a given location and time (except during the Lois 1996–97 mating season; see Results). We conducted these counts from September through January, a time period that encompassed most of the annual mating season on the islands (Johnson and Kapsalis, 1998; see below). When making such counts, we tallied the adults seen in successive areas sufficiently removed from one another to minimize the risk of double counting individuals. The data from all counts conducted during each mating season were pooled before calculating that year's colony-wide female–male ratio.

Dominance

The dominance ranks of the Raccoon Key and Key Lois females were based on dyadic agonistic encounters with unambiguous winners and losers. However, the very large number of females belonging to some social groups together with our limited contact time per animal made it impossible for us to observe dominance interactions between every possible pair of females. Although this means that the exact hierarchical position of some animals cannot now be determined, our subjects can be assigned to one of three broad rank categories, that is high, middle or low, depending upon whether their ranks fell within the top, middle or bottom thirds of their respective group's hierarchies.

Consortships

Like all rhesus macaques exposed to seasonal environments, the monkeys on Raccoon Key and Key Lois bred seasonally. Births and conceptions peaked in May and late November respectively (Johnson and Kapsalis, 1995). Nevertheless, copulations were commonly observed from October to March. As indicated above, the reproductive state of individual females was noted as they were encountered, including whether they were participating in consortships. In a typical rhesus heterosexual mating, the male mounts the female by grasping the waist of the latter with his hands, and her lower legs with his feet. This is followed by intromission and pelvic thrusting. Rhesus males are typically multiple mounters with only the last mount in a series culminating in ejaculation. Female–female consortships also include series mounts accompanied by pelvic thrusting (personal observation; see also Carpenter, 1942a). Indeed, we credited a given female as being part of a heterosexual *or* homosexual consortship only if she was observed participating in a mount series. In addition, consortships were only recognized if before or after the series mount, or during the interval between single mounts, the members of a consorting dyad also groomed, maintained close proximity or cooperated in threatening conspecifics (potential sexual competitors) that approached too closely. In many cases, the members of a consort pair moved frequently and furtively such that their series mounts took place out of the view of other animals. In at least two consort pairs, the roles of mounter and mountee were reversed while the animals were being observed. Whether such reversals were a general phenomenon is unknown, since the

Figure 9.1. Free-ranging rhesus monkeys on Raccoon Key at high tide (photo: Rodney Johnson).

relative restlessness of the female–female dyads and difficult observational conditions (Figure 9.1) often precluded prolonged contact with consort pairs.

Due to the similarities in the behaviour of females participating in heterosexual and homosexual consortships, and the fact that all observed female–female consortships occurred during the mating season, we assume that our subjects' motivation to interact sexually with a conspecific, irrespective of the sex of that partner, was linked to their hormonal status. However, each female's choice of partner (for example, male versus female) could be driven by factors other than her reproductive physiology. In contrast to the adult females on Raccoon and Lois, the adult males appear to have expressed their sexual motivation only within the context of heterosexual consortships, since male–male sexual interactions were not observed on either island.

Results

Key Lois

Demography

Not surprisingly, trapping on Key Lois from 1992 to 1997 resulted in a rapid decline in the population of female adults on the island. Before the

Table 9.1 *Mating season ratio of female to male adults on Key Lois and Raccoon Key*

	Mating season						
	1992–93	1993–94	1994–95	1995–96	1996–97	1997–98	1998–99
Key Lois	4.4	4.2	4.5	4.0	5.2[†]	–	–
Raccoon Key	4.6	4.1	4.5	4.8	6.1	4.8	4.4

Note: [†]The Key Lois ratio for 1996–97 was known and not estimated as in other years (see Methods).

large-scale capture of adult animals began in the autumn of 1992, the population of free-ranging adult females on Lois numbered approximately 500 individuals (Johnson and Kapsalis, 1995). The trapping that occurred during the autumns of 1992–94 reduced that number to just 226 females, a decrease of about 55%. By the end of 1996, the total number of free-ranging adult females stood at 94 individuals. Not all social groups were equally affected. In 1992, for example, three groups were removed in their entirety (or nearly so), while all others were left untouched. Even by the autumn of 1994, three of the largest and most visible social groups on Key Lois (groups A, D and C in Table 9.2) had yet to suffer any trapping losses.

The between-year variation in our estimates of the adult sex ratio on Key Lois was minimal (Table 9.1), and may be attributable to measurement error alone. The 5.2 female-to-male ratio for the 1996–97 mating season is higher than that of any previous year (range, 4.5–4.0 females per male). However, this difference probably is a methodological artefact and not real. By the autumn of 1996, all 18 adult males and 94 females still on the island were individually known. Consequently, the sex ratio for the 1996–97 mating season was calculated using these numbers, and not estimated by means of scan samples as in previous years (see Methods). We suspect that the scans overestimated the abundance of males who are larger and more conspicuous than females.

Homosexual consortships

The identities of the females on Key Lois that we observed participating in homosexual consortships, and the mating season during which those consortships occurred are listed in Table 9.2. The animals' ages, group affiliations and, for some, their relative dominance ranks are also shown. Ranks are not provided for those individuals whose respective social groups included fewer than ten adult females; an assignment of high or low rank to an animal belonging to such a small group simply is not comparable with an identical rank assignment given to a female whose group was three to ten times larger. Three of the Key

Table 9.2 *Female–female consortships in free ranging rhesus monkeys*

			Female		Initial social group			
Island	Mating season	Dyad	Age	Rank†	ID	No. females as of 1 Jan†	3-year decline (%)	Group change
Key Lois	1994–95	(1) C-66	23	middle	A	39	35	
		X-357	16	high	A	–	–	
		(2) KL-31	19	high	A	39	35	
		N-295	4	low	A	–	–	
	1995–96	(3) X-289	19		B	6	58	fusion
		X-259	17	middle	C	17	59	
	1996–97	(4) 79–727	16		A1‡	5	62	
		80–645	15		A1	–	–	
		(5) 79–665	17		B	6	76	fusion
		79–524	17	low	C	15	71	
		(6) X-361	18	low	D	17	75	immigration
		X-259	18	middle	C	15	71	
		(7) B-73	25	high	A	20	64	
		80–683	16	low	A	–	–	
Raccoon Key	1996–97	(8) N-296	21	middle	E	59	32	
		T-42	24	high	E	–	–	
		(9) B-88	25		A2‡	8	84	immigration
		X-80	20	low	E	59	32	
	1997–98	(10) X-297	22		F‡	4	90	
		N-464	22	middle	G	44	37	
		(11) B-115	19	low	H	41	58	
		J-349	11	high	H	–	–	
		(12) C-31	26		J	9	25	fusion
		B-149	19	high	I	15	32	
		(13) H-622	13		L	9	61	fusion
		B-8	25	low	G	31	53	
		(14) A-3	25	low	E	44	39	immigration
		79–722	19		M‡	9	31	
		(15) 86–506	12	middle	N	13	57	
		N-464	23	middle	G	28	53	
		(16) B-88	27		§	§	§	
		X-356	20		M‡	9	31	

Notes: †1 January was the approximate midpoint of a given mating season.

‡ Groups originally resident of Key Lois but transferred to Raccoon Key.

§ Group affiliation for B-88 prior to the 1998–99 mating season was unknown. See text.

Lois females assigned relative dominance ranks were high ranking, three were middle ranking and four were low ranking. The relatively advanced age of many of the animals listed in Table 9.2 reflects the relative agedness of the Key Lois population as a whole at the time these data were collected. With the exception of n-295 who was nulliparous at the time of her consortship with KL-31, the Lois females were all multiparous.

The first two female–female consortships we observed on Key Lois occurred during the 1994–95 mating season, the third of our study. These two consortships both involved females belonging to the social group designated as group A in Table 9.2. This group incurred its first trapping losses on 7 November 1994; 11 of 50 adult females (22%) were removed from early November to early December, leaving 39 group-A females on Lois as of 1 January 1995 (Table 9.2). The female membership of this group had been pared previously by mortality and the formation of a small fission group of ten females (group A1) in early 1994 (that is, approximately February). The females captured from group A in November 1994 were relocated to captive housing outside of the Florida Keys. No additional trapping of group A occurred until November 1995 when 14 females were trapped and relocated to Raccoon Key thereby creating a new social group on this island (group A2 of Table 9.2).

Only one homosexual consortship was noted on Key Lois during the 1995–96 mating season. This involved females that originally belonged to two different social groups, B and C. Group C had incurred its first trapping losses in November 1995, while group B females had been removed in the autumns of 1994 and 1995. The participants of this female–female consortship, X-259 and X-289, interacted sexually on 8 February (about three months after the removal of animals from group C) and again approximately one month later on 15 March 1996. Their homosexual behaviour occurred in the context of the fusion of the remaining memberships of groups B and C (Table 9.2). The merger of these groups was underway in early 1996. Members of group B moved with those of C, but mostly remained on the periphery of the combined group. This pattern was little changed by the start of the 1996–97 mating season, a fact that we believe indicates the integration of B and C was still in progress at that time.

It was during the 1996–97 mating season that we observed the greatest number of female homosexual consortships on Key Lois, the same season that the population of free-ranging adult females dropped below 100, as described above. Eight females from five different social groups, all of which had suffered trapping-related losses by 1996, combined to make four consortships. One of the consortships included a female immigrating into group C from group D, while another occurred in the context of the continuing fusion of groups B and C (dyads no. 5 and 6 respectively in Table 9.2). Thus, two of the homosexual pairings of the 1996–97 mating season included females involved in the process of changing their group affiliations. It is particularly noteworthy that X-259, who had participated in a homosexual consortship the previous year during the initial stages of the merger of B and C groups, participated in a second consortship with X-361, the female immigrating from group D. The two other female–female consortships observed during the 1996–97 mating season were comprised of individuals belonging to the same social groups; specifically, two females belonging to the original group A, and two females belonging to the daughter group A1.

Raccoon Key

Demography

The proportion of aged females (that is, those at least 20 years of age) on Raccoon was about 19% in 1992. This proportion reached 26% (199 of 777 females) and 36% (265 of 736) by 1 January 1994 and 1996 respectively due to aging, negligible recruitment of juveniles into the adult age class and the transfer of additional aged animals from Lois to Raccoon from 1992 to 1995. The annual rate of female mortality increased incrementally in tandem with the proportion of aged animals (Figure 9.2). However, mortality increased from just under 10% in 1996 to over 20% in 1997, and remained high in 1998 (Figure 9.2). This acceleration of the colony-wide death rate was expected given that by 1 January 1997, the proportion of females 20 years old and older had increased to 38% (252 of 666), while the proportion of those at least 25 years of age stood at 17% (112 of 666). Like the mortality rate, the adult sex ratio on Raccoon Key each autumn increased incrementally from 1992 to 1996 (Figure 9.2). Unlike the death rate, however, the sex ratio declined after 1996, falling to 4.8 females per male in late 1997 and 4.4 females per male in late 1998. The ratio in 1998 was very similar to the 4.6 females per male present on the island in 1992.

Consortships

The bottom half of Table 9.2 lists the identities of the adult females we observed on Raccoon Key that formed homosexual consortships. As was the case on Key Lois, the ages of the Raccoon Key females in Table 9.2 reflect the relative agedness of the Raccoon population as a whole. All of the Raccoon Key females listed in Table 9.2 were multiparous; and, some were probably of very high parity indeed. Three were high ranking, whereas four were middle ranking and four low ranking. The social groups to which the Raccoon Key females initially belonged all substantially declined in size over the three years previous to their same-sex sexual behaviour. Groups A2, F, M and N originally resided

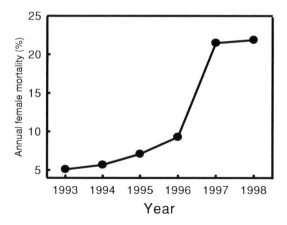

Figure 9.2. Annual rate of mortality among adult female rhesus monkeys on Raccoon Key.

on Key Lois, and were not transferred to Raccoon Key in their entirety; some of their respective members had evaded capture and, therefore, had been left behind on Lois. Thus, the three-year declines for these relocated groups encompass losses due to trapping as well as mortality. The remaining troops listed in Table 9.2, those already resident on Raccoon Key in 1992, eventually incurred trapping losses as well, but not before the occurrence of the nine homosexual consortships we observed on the island. Hence, the three-year declines in the female memberships of these Raccoon-only groups were solely the consequence of natural mortality.

As Table 9.2 shows, the first two female–female consortships observed on Raccoon Key occurred during the 1996–97 mating season, two years after the first Key Lois homosexual pairings. Whereas one of the consortships included two females from the same social group, the other (no. 9 in Table 9.2) was comprised of animals originating from two different groups. One of these females, B-88, immigrated into group E; and, her sexual relations with X-80 occurred in the context of her change in group affiliation. The one consortship observed during the 1997–98 mating season and five of the six observed during the 1998–99 mating season were also comprised of females that initially belonged to different social groups. Three of the intergroup consortships observed during the latter mating season involved females whose group affiliations had recently changed or were changing (that is, nos. 12–14 in Table 9.2), in one case through immigration and in the other two cases through group fusion. Two females, B-88 and N-296, formed consortships twice. After associating with group E for more than a year, B-88 appeared not to belong to any group when she was seen participating in her second homosexual relationship during the 1998–99 mating season, this time with a group M female (Table 9.2). In contrast, the group affiliation of N-296 never changed. However, her 1997–98 and 1998–99 sexual partners both belonged to other social groups.

Discussion

The salient points of the present study are the following: (1) Female homosexual consortships were not observed prior to large-scale losses of adult females on both Key Lois and Raccoon Key. Such consortships were seen only after these losses were well underway. (2) Ten of the 16 homosexual consortships we observed included females initially from different groups. (3) In seven of these ten consortships, one of the interactants had recently joined the group of her sexual partner or was in the process of doing so. (4) The generally increasing incidence of homosexual pairings was not matched by an increasing female:male sex ratio. In the discussion that follows, we consider why we feel our data are supportive of the suggestion of Fairbanks et al. (1977) that female homosexual behaviour aids alliance formation in rhesus. We will also argue that our observations are inconsistent with the notion that the adult sex ratio on Raccoon and Lois was causally related to the expression of female homosexual behaviour on the islands.

Before expanding on these points, however, we need to consider whether the fact that our data were collected ad libitum under difficult observational conditions may have affected our results. It is certain that we witnessed only a minority of the sexual encounters that occurred on Key Lois and Raccoon Key (Johnson and Kapsalis, 1998). Therefore, it is indeed possible that female–female consortships occurred on the islands prior to the 1994–95 Lois and the 1996–97 Raccoon mating seasons and that the incidence of such consortships during all years was higher than our observations would suggest. However, given that neither of us had ever observed a female homosexual consortship prior to the 1994–95 season on Lois (or anywhere else for that matter), we found such pairings to be profoundly 'attention getting' irrespective of when they occurred. Therefore, we are confident that female–female consortships were no more likely to go unobserved during the first two or three mating seasons when we canvassed the islands than they were during any subsequent year, and that the greater frequency of female–female sexual pairings observed during our last field seasons on Key Lois and Raccoon Key reflected a true increase in the incidence of homosexual behaviour among the adult females on both the islands.

Given that the observed interannual differences in female homosexual consortships are unlikely to be observational artefacts, how can they best be accounted for? Certainly, the dramatic changes in the Raccoon Key and Key

Lois populations that occurred during our study implicate demographic factors. Wolfe (1986) showed that interannual differences in female homosexual behaviour among the Arashiyama West Japanese macaques have correlated with interannual changes in the populations' adult sex ratio, with the incidence of such behaviour being greater when the ratio was higher (and males were less available). However, our data do not indicate that the same was true for the rhesus macaques on Key Lois and Raccoon Key. It is true that the incidence of homosexual consortships on Lois was greatest during the 1996–97 mating season, the same season that the number of adult females per male *may* have been greatest (Table 9.1). However, as we indicated, we believe the difference between the sex ratio of 1996–97 and that of the four previous years was the consequence of the between-year difference in the way the sex ratio was determined (see Methods), not a real interannual difference in the ratio itself. Furthermore, the estimated ratio for Raccoon Key was no higher during the 1998–99 mating season (when the greatest number of female–female consortships were observed on that island) than it was during the 1992–93 mating season (when no such pairings were seen). In fact, the ratio peaked during the 1996–97 mating season following the transfer of 70 adult females (but no males) from Lois to Raccoon the previous year, and probably was in decline during our last field season when the incidence of homosexual consortships increased sharply.

Given that homosexual consortships were not observed until after the Lois and Raccoon female populations had declined substantially, we believe the demographic factor that can be most plausibly linked to the frequency of homosexual behaviour is the actual loss of adult females. Certainly, the several cases of immigration and group fusion we observed on Raccoon and Lois were the result of adult loss. Large-scale trapping has been reported to precipitate female migration in Japanese macaques (Sugiyama and Oshawa, 1982), while mortality-driven group fusion has been documented in both Japanese macaques (Sugiura *et al.*, 2002) and vervet monkeys (Hauser *et al.*, 1986). That females attempting to ingratiate themselves into a new group might be motivated to form a homosexual consortship is suggested by the study of Fairbanks *et al.* (1977) on captive rhesus. Fairbanks and colleaques noted high levels of female homosexual behaviour among captive rhesus in newly formed groups, and hypothesized that such behaviour fosters the rapid development of affiliative relationships and alliances. They speculated that, under the unusual circumstances of the laboratory, some rhesus females formed same-sex consortships because they were deprived of the 'normal mechanisms for assuring female–female bonds'. We take a slightly different view in that we believe the predilection to form homosexual consortships is, in fact, part of the 'normal' behavioural repertoire of at least some rhesus females and is manifested whenever there is a premium associated with the rapid formation of an affiliative relationship, as when unfamiliar females are housed together in captivity or a free-ranging female is attempting to gain entry into a new social group. In other words, participating in a homosexual consortship is a means by which a female can put a developing social relationship on the affiliative fast-track.

Reciprocal grooming, maintenance of proximity and cooperative aggression, are all behaviours that occur during female homosexual consortships (Akers and Conaway, 1979; Fairbanks *et al.*, 1977), and that would seem conducive to rapidly forging affiliative relationships, particularly since they are performed repeatedly over a short period of time in the context of what is likely to be a trusting psychological state (Wallen and Parsons, 1997). That a prolonged affiliative relationship can be so formed between animals originally from different groups is indicated by the fact that C-31 and B-149 (Table 9.2, dyad no. 12) were repeatedly observed travelling and feeding in close proximity during the nearly four-month interval between the termination of their consortship and their eventual capture and removal from Raccoon Key. That a sexual liaison may fail at other times to evolve into a longer-term relationship is indicated by the fact that a few Raccoon Key animals (for example, 86–506 and N-464, dyad no. 15) that formed intergroup consortships remained with their social groups after their sexual activity ceased.

The homosexual behaviour of those females that interacted sexually with recent immigrants or with females belonging to their own social groups may have been a mechanism whereby an ambivalent relationship could be quickly fashioned into an affiliative one. The rapid loss of adult females from a given group must have had a profound effect on the network of alliances of at least some of the group's surviving members. For macaque females, alliance formation is thought to be central to the formation and maintenance of dominance relations (Chapais, 1992). Therefore, an abrupt loss of alliance partners, through either trapping or mortality, could conceivably have undermined the hierarchical position of some females (for example, B-73 on Lois and J-349 on Raccoon), thereby motivating them to cultivate new partners quickly. However, the fact that the great majority of females on Raccoon Key and Key Lois did not engage in same-sex sexual behaviour suggests to us that participating in a homosexual consortship was a strategy that only a few were inclined to employ. Some, like N-464 on Raccoon Key (Table 9.2), seemed to have been particularly predisposed to do so.

We conclude by pointing out that our inferences about the importance of adult loss in facilitating female homosexual behaviour are supported by Carpenter's (1942a, b) study of the Cayo Santiago rhesus. Carpenter indicated that the Cayo colony was first populated with 409 monkeys in December 1938. By March 1940, the population stood at 350 animals, a total that included 110 infants born on the island in 1939 and 1940. Thus, the founding population of rhesus had been reduced from 409 to just 240 individuals, a decline of about 41% in a little more than a year. As indicated in the introduction, Carpenter's (1942b) study is just one of the two previous research efforts to document female–female consortships in free-ranging *M. mulatta*. Inasmuch as his observations of female homosexual behaviour, made from February to April 1940, followed a dramatic winnowing of the Cayo population, and inasmuch as female–female consortships were either extremely rare or non-existent on the island during the 1980s (EK, personal observation), a period of rapid growth for the Cayo population (Rawlins and Kessler, 1986), we are confident that the loss of adult females was the demographic factor responsible for the increasing incidence of same-sex sexual behaviour we observed on both Key Lois and Raccoon Key.

Acknowledgements

We wish to thank Paul Schilling for the opportunity to do research on Raccoon Key, colony personnel for their cooperation and assistance, and Anthea Lavallee and Larry Mull for their help in data collection. The Raccoon Key rhesus breeding colony was owned and operated by Charles Rivers Laboratories.

References

Akers, J. S. and Conaway, C. H. (1979) Female homosexual behavior in *Macaca mulatta*. *Arch. Sex. Behav.*, **8**, 63–80.

Bernstein, I. S. and Mason, W. A. (1963) Group formation by rhesus monkeys. *Anim. Behav.*, **11**, 28–31.

Carpenter, C. R. (1942a) Sexual behavior of free ranging rhesus monkeys (*Macaca mulatta*): I. Specimens, procedures and behavioral characteristics of estrus. *J. Comp. Psychol.*, **32**, 143–62.

Carpenter, C. R. (1942b) Sexual behavior of free ranging rhesus monkeys (*Macaca mulatta*): II. Periodicity of estrus, homosexual, autoerotic and non-conformist behavior. *J. Comp. Psychol.*, **33**, 143–62.

Chapais, B. (1992) The role of alliances in the social inheritance of rank among female primates. In *Coalitions in Humans and Other Animals*, eds. A. Harcourt and F. M. B. de Waal, pp. 29–60. Oxford: Oxford University Press.

Colvin, J. D. (1986) Proximate causes of male emigration at puberty in rhesus monkeys. In *The Cayo Santiago Macaques: History, Behavior and Biology*, eds. R. G. Rawlin and M. J. Kessler, pp. 131–57. Albany: State University of New York Press.

Dittus, W. P. J. (1986) Sex differences in fitness following a group take-over among Toque macaques: testing models of social evolution. *Behav. Ecol. Sociobiol.*, **19**, 257–66.

Drickamer, L. C. (1974) A ten-year summary of reproductive data for free-ranging *Macaca mulatta*. *Folia Primatol.*, **21**, 61–80.

Evans, M. A. (1989) Ecology and removal of introduced rhesus monkeys: Desecheo Island National Wildlife Refuge, Puerto Rico. *Puerto Rico Health Sci. J.*, **8**, 139–56.

Fairbanks, L. A., McGuire, M. T. and Kerber, W. (1977) Sex and aggression during rhesus monkey group formation. *Aggress. Behav.*, **3**, 241–9.

Fedigan, L. M. and Gouzoules, H. (1978) The consort relationship in a troop of Japanese monkeys. In *Recent Advances in Primatology, Vol. 1*, eds. D. J. Chivers and J. Herbert, pp. 493–95. New York: Academic Press.

Hauser, M. D., Cheney, D. L. and Seyfarth, R. M. (1986) Group extinction and fusion in free-ranging vervet monkeys. *Am. J. Primatol.*, **11**, 63–77.

Hinde, R. A. and Rowell, T. E. (1962) Communication by postures and facial expressions in the rhesus monkey (*Macaca mulatta*). *Proc. Zool. Soc.*, **138**, 1–21.

Huynen, M. C. (1997) Homosexual interactions in female rhesus monkeys, *Macaca mulatta*. *Adv. Ethol.*, **32**, 211.

Johnson, R. L. (1989) Live birthrates in two free-ranging rhesus breeding colonies in the Florida Keys. *Primates*, **30**, 433–7.

Johnson, R. L. and Kapsalis, E. (1995) Ageing, infecundity, and reproductive senescence in free-ranging female rhesus monkeys. *J. Reprod. Fertil.*, **105**, 271–8.

Johnson, R. L. and Kapsalis, E. (1998) Menopause in free-ranging rhesus macaques: estimated incidence, relation to body condition, and adaptive significance. *Int. J. Primatol.*, **19**, 751–65.

Johnson, R. L., Berman, C. M. and Malik, I. (1993) An integrative model of the lactational and environmental control of mating in female rhesus monkeys. *Anim. Behav.*, **46**, 63–78.

Kapsalis, E. and Berman, C. M. (1996a) Models of affiliative relationships among free-ranging rhesus monkeys (*Macaca mulatta*): I. Criteria for kinship. *Behaviour*, **133**, 1209–34.

Kapsalis, E. and Berman, C. M. (1996b) Models of affiliative relationships among free-ranging rhesus monkeys (*Macaca mulatta*): II. Testing predictions for three hypothesized organizing principles. *Behaviour*, **133**, 1235–63.

Lindburg, D. G. (1971) The rhesus monkey in North India: an ecological and behavioral study. In *Primate Behavior*, ed. L. A. Rosenblum, pp. 1–106. New York: Academic Press.

Lindburg, D. G. (1983) Mating behavior and estrus in the Indian rhesus monkey. In *Perspectives in Primate Biology*, ed. P. K. Seth, pp. 45–61. New Delhi: Today and Tomorrow's Printers and Publishers.

Malik, I., Johnson, R. L. and Berman, C. M. (1992) Control of postpartum mating behavior in free-ranging rhesus monkeys. *Am. J. Primatol.*, **26**, 89–95.

Manson, J. H. (1997) Primate consortships: a critical review. *Curr. Anthropol.*, **38**, 353–73.

Mukherjee, A. K. and Gupta, S. (1965) The Sundarban of India and its biota. *J. Bombay Nat. Hist. Soc.*, **72**, 1–20.

Pucak, G. J., Foster, H. L. and Balk, M. W. (1982) Key Lois and Raccoon Key: Florida islands for free-ranging rhesus monkey breeding programs. *J. Med. Primatol.*, **11**, 199–210.

Rawlins, R. G. and Kessler, M. J. (1986) Demography of the Cayo Santiago macaques. In *The Cayo Santiago Macaques: History, Behavior and Biology*, eds. R. G. Rawlins and M. J. Kessler, pp. 46–72. Albany: State University of New York Press.

Richie, T., Shrestha, J., Teas, J., Taylor, H. and Southwick, C. H. (1978) Rhesus monkeys at high altitudes in northwestern Nepal. *J. Mammal.*, **59**, 443–4.

Sade, D. W. (1972) A longitudinal study of social behavior of rhesus monkeys. In *The Functional and Evolutionary Biology of Primates*, ed. R. Tuttle, pp. 378–98. Chicago: Aldine.

Sherman, B. R. (1980) The formation of relationships based on socio-spatial proximity in a free-ranging colony of rhesus monkeys. *Primates*, **21**, 484–91.

Sugiura, H., Agetsuma, N. and Suzuki, S. (2002) Troop extinction and female fusion in wild Japanese macaques in Yakushima. *Int. J. Primatol.*, **23**, 69–84.

Sugiyama, Y. and Oshawa, H. (1982) Population dynamics of Japanese macaques at Ryozenyama: III. Female desertion of the troop. *Primates*, **23**, 31–44.

Taub, D. M. and Mehlman, P. T. (1989) Development of the Morgan Island rhesus monkey colony. *P. R. Health Sci. J.*, **8**, 159–69.

Wallen, K. and Parsons, W. A. (1997) Sexual behavior in same-sexed nonhuman primates: is it relevant to understanding homosexuality? *Ann. Rev. Sex. Res.*, **7**, 195–223.

Wolfe, L. D. (1986) Sexual strategies of female Japanese macaques (*Macaca fuscata*). *Human Evol.*, **1**, 267–75.

Wolfe, L. D., Kolias, G. V., Collins, B. R. and Hammond, J. A. (1991) Sterilization and its behavioral effects on free-ranging female rhesus monkeys (*Macaca mulatta*). *J. Med. Primatol.*, **20**, 414–18.

Wolfheim, J. H. (1983) *Primates of the World: Distribution, Abundance, and Conservation*. Seattle: University of Washington.

10

A wild mixture of motivations: same-sex mounting in Indian langur monkeys

VOLKER SOMMER, PETER SCHAUER AND DIANA KYRIAZIS

Introduction

Sexual behaviour between members of the same sex seems to be incompatible with the Darwinian view of evolution since natural selection should eliminate traits that interfere with or reduce reproduction. Yet the behaviour is widely distributed amongst humans and other animals, highly variable in frequency of expression, from total absence to levels that approach or even surpass heterosexual activity, and is as likely to be displayed by males as by females (Weinrich, 1980; Vasey, 1995; Bagemihl, 1999).

Explanations which try to reconcile the occurrence of homosexual behaviour with a Darwinian paradigm are as inventive as they are varied. Non-functional explanations view same-sex sexual behaviour as a pathology or an evolutionary by-product, whereas functional explanations link same-sex sexual behaviour with agonistic or affinitive social strategies. Most of these hypotheses are difficult to test, given the multivariate nature of animal societies and the general paucity of specific data from the wild.

Our study examines the evidence for and against such explanations through long-term observations on wild Indian langurs, a colobine monkey which inhabits large parts of the Indian subcontinent. This taxon is particularly well suited for such investigation because of its variable social organization, which includes both bisexual troops with females and males as well as unisexual male bands. Moreover, anecdotal evidence from numerous field studies reveals that same-sex mounting, a typical expression of homosexual behaviour, is ubiquitous (review in Srivastava et al., 1991).

We hope that an analysis of homosexual behaviour of wild animals based on a detailed knowledge of their social and reproductive strategies will contribute to an overall understanding of primate social relationships, including those of human primates.

Methods

Site and study population

Indian langurs (*Presbytis entellus*), also known as Hanuman langurs, have one of the largest geographical ranges of any non-human primate, including parts of Pakistan, India, Nepal, Bhutan, China, Bangladesh and Sri Lanka. They dwell in the Himalayan mountains, foothills and flood planes as well as in semi-arid zones, dense cities and deciduous or evergreen forests.

In some areas, langurs form troops with multiple adult males and females. Elsewhere, a single male lives with multiple females in a so-called 'harem', while surplus males join bachelor bands. In other places, varying proportions of one-male versus multi-male troops are found simultaneously.

Data for this study come from Jodhpur, located in northwest India in the state of Rajasthan at the eastern edge of the Great Indian Desert. Here the climate is dry, with maximum temperatures of up to 50°C during May and June and minimum temperatures around 0°C during December and January. Jodhpur receives 90% of its scanty rainfall (annual average 360 mm) during the monsoon, which lasts from July to September. The town stands on a hilly sandstone plateau which covers approximately 85 km^2.

This plateau is inhabited by a geographically isolated population of langurs, which has been closely monitored by researchers for almost four decades (for example, Mohnot, 1974; Borries *et al.*, 1991; Rajpurohit *et al.*, 1995; Sommer *et al.*, 2002).

Socio-ecology

All langur groups at Jodhpur (Figure 10.1) forage on the natural vegetation, which is xerophytic open scrub, and some groups raid crops. Additionally, most monkeys are fed by local people for religious reasons. In correspondence with an increase in the urban human population, the langur population now stands at close to 3000, whereas it was about 1300 in 1967. The proportions of natural, provisioned and raided food items consumed vary considerably between groups (Sommer and Mendoza-Granados, 1995). Sexual activity and births occur throughout the year, most probably due to provisioning, which offsets shortages in the seasonal supply of natural foods.

The reproductive units are multi-female/one-male troops with an average size of 38.5 members, range 7–93. Each of the 27–29 troops present occupies a home

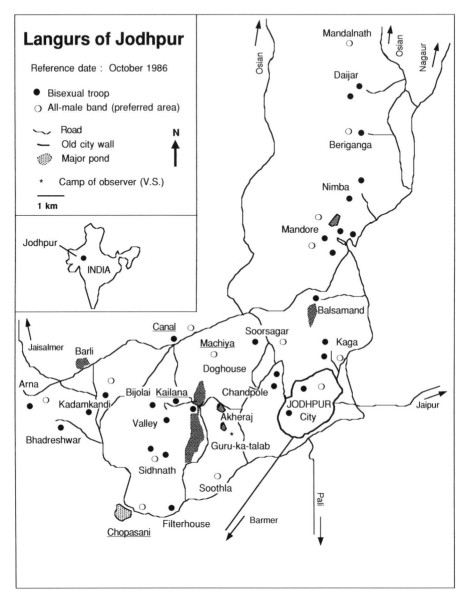

Figure 10.1. Bisexual troops and male bands of langurs around Jodhpur. The study groups are underlined.

range of about 0.5–1.3 km². Females remain in their natal troop for their entire life, while maturing males emigrate and join one of the 12–14 all-male bands in the area. These male bands have an average size of 11.8 members, range 2–47. Band home ranges comprise 1.5–3.5 km², but are not well defined because individual males can move over areas >20 km².

Male bands invade the harem-structured troops in an unpredictable pattern. Harem holders can change during rapid takeovers that occur in days. Alternatively, these changes can be gradual processes that last up to several months with temporary multi-male stages and successive short adult male tenures. Infanticide by immigrant males is common. Residencies of single adult males in bisexual troops range from just a few days to at least 74.0 months, with an average of 26.5 months. Stable one-male situations are typical for 94.5% of a given troop's history (Sommer and Rajpurohit, 1989).

Data collection

The present analysis is based on data collected by V. Sommer, C. Borries and A. Srivastava during several thousand hours of observations between 1981–1987 (Borries *et al.*, 1991, 1994; Srivastava *et al.*, 1991; data on male–male mounting have not been published previously).

The study groups were one harem troop, known as Kailana-I and three all-male bands, known as Machiya, Canal and Chopasani, living under varying ecological conditions (Table 10.1). Study periods lasted between 11 and 17 months and observations were approximately equally distributed throughout the day (06:00--18:00 h) and year. All group members were individually identified and well habituated to close-range observations (see Sommer *et al.*, 2002, for details).

Data were collected *adlibitum* as well as by focal-animal-sampling, with both methods lumped together for analysis. Typically, the observer changed position within the group about every 30 min, while the randomly selected focal animal was changed, thus minimizing the potential bias that particular individuals were overrepresented in the *adlibitum* sample. In a forested area, *adlibitum* records underreported langur interactions by 22% compared with pure focal animal samples (Podzuweit, 1994), but the bias is probably smaller in the open Jodhpur habitat. The *adlibitum* sampling procedure does not easily lend itself to context analysis. It does, however, ensure a data density that allows the reconstruction of solid socio-metric matrices that provide background information about dominance ranks, grooming networks and long-term dynamics of social relationships.

Data analyses

A context analysis of homosexual mounting aimed to detect correlations with variables such as age-class, reproductive state, affinitive behaviour (allo-grooming: 'Coordinated movement through a group mate's fur with one or both hands or mouth. Particles may be removed in the process') and agonistic behaviour (displacement: 'A and B are close to each other or A approaches B.

Table 10.1 *Study groups: methodology, composition and variables of eco-ethology*

	Harem troop			Male band		
	Kailana-I			Machiya	Canal	Chopasani
Focal observations	Oct81–Dec82	Jan–Oct85	Sep86–Dec87	Sep86–Dec87	Sep86–Mar88	Oct86–Feb88
Total days[a]	288	156	200	72	56	59
Hours[b]	942	1019	1272	515	397	420
Animals	All 11 adult females	All 13 adult females	All 12 adult females	All males	All males	All males
Dyads	55	78	66	45	6	42
Mean h / dyad	660	1019	1272	384	316	207
Min h / dyad	660	1019	1272	79	234	10
Max h / dyad	660	1019	1272	515	397	420
Group composition[c]						
Adult (old)	6F	5F	6F	1	2	2
Adult (middle-aged)	2F, 1M	3F, 1M	3F, 1M			3
Adult (young)	3F	5F	3F	2		
Immature (sub-adult)				1	1	1
Immature (juvenile)	4F, 3M	1F, 2M	2F, 1M	6	1	4
Immature (infant)	3F, 4M	3F, 1M	3F, 7M			
Total members	26	27	26	10	4	10
Mean members per month	20.4	ca. 20	23.3	8.8	3.6	7.0

Habitat[d]	75 ha	similar	similar	349 ha	165 ha	142 ha
Range size	75 ha	similar	similar	349 ha	165 ha	142 ha
Open scrub	+++			+++	++	+
Gardens, fields	−			−	++	+++
Human habitations	+, −			+, −	+	++
Water access	+++			+, −	++	+++
Diet (annual feeding time)						
Provisioned food (%)	22.2	16.4	33.5	2.4	3.0	15.6
Natural food (%)	77.8	83.6	66.5	97.7	93.8	64.6
Crop raiding (%)	0.0	0.0	0.0	0.0	3.2	19.8
Source	Sommer, 1985: 57	Borries, 1989: 40	Srivastava, 1989: 21, 28f	Rajpurohit et al., 1995	Rajpurohit et al., 1995	Rajpurohit et al., 1995

Notes: [a] In bands mostly dawn-to-dusk follows.

[b] Continuous focal animal sampling was carried out during most study periods. Focal hours = subsample of present study.

Additional scan and *ad libitum* sampling was carried out during most study periods.

[c] F = female, M = male. The harem troop's adult male (given as 1M) was replaced once during the study period (Jun82). Bands contained only males; not all were present throughout study because of group transfers. Some individuals changed age-classes during the study.

[d] +++ = much / many, ++ = moderate, + = little / few, − = none.

B leaves, often after A exhibits aggression. A then occupies the position or resource previously controlled by B').

Heterosexual sexual behaviour includes female solicitations for copulation ('oestrous behaviour'), which consist of (a) more or less complete lowering of the tail, (b) presentation of the anogenital region and (c) vigorous lateral shaking of the head. The male mounts in response to this 'shake-and-present' posture, usually by clinging with his feet above the female's heels. A complete copulation with pelvic thrusting and intromission might follow. However, males will often ignore female solicitations, presumably because the vast majority of copulations are harassed by other langurs (Sommer, 1989a).

Homosexual behaviour was defined as genital contact between individuals of the same sex (Vasey, 1995), which typically took the form of ventro-dorsal mounting.

The mean basic reproductive parameters for females (Sommer et al., 1992) are a cycle length of 24 days; menstruation two days (clearly visible in Jodhpur langurs), follicular stage nine days, ovulatory phase four days and luteal stage 15 days. Pregnancies last on average 200 days. The interbirth interval was 16.7 months. Sexual solicitations and copulations of cycling females are virtually confined to the ovulatory stage. During pregnancies, sexual behaviour is absent during the first and last phase (days 1–52, days 137-term). However, postconception sexual behaviour is observed during mid pregnancy (days 53–88: frequent sexual solicitations of males and copulations; days 89–136: less frequent solicitations, no copulations.

The following age-classifications applied (see Rajpurohit et al., 1995; mo = months, yr = years):

(a) Females
  • *Infant-1*: 'black coat', from birth to about 5 mo, after completion of fur color change.
  • *Infant-2*: 'white coat', after completion of fur color change to completion of weaning, that is, from about 5 to 15 mo.
  • *Juvenile*: 15 mo to the onset of menarche, at about 2.5 yr.
  • *Sub-adult*: (no sub-adult category in females!)
  • *Adult/young*: from regular cycling until 9 yr; weight 11–12 kg.
  • *Adult/middle-aged*: 10–19 yr.
  • *Adult/old*: 20+ yr.
(b) Males
  • *Infant-1, Infant-2*: see females.
  • *Juvenile-1*: testes internal, glans of penis not visible, about 15–30 mo.
  • *Juvenile-2*: testes descended, glans generally not visible, about 30–48 mo.

- *Sub-adult*: glans usually visible, ischial pads still undeveloped, canine teeth not yet fully erupted, capable of copulatory behaviour; about 4 to 6 yr.
- *Young adult*: glans always visible, ischial pads small and usually pale, canines fully erupted, teeth unworn; not yet full size; about 6–8 yr.
- *Adult*: prime male, full size; generally smooth face with few scars; ischial pads usually pink and puffy; weight 18–19 kg; about 8–13 yr.
- *Old*: teeth conspicuously worn; numerous scars; often wrinkled facial skin; ischial pads usually pink and puffy; 14 + yr.

For calculations of relative age differences, adult females were broken down into old, middle-aged and young individuals. An episode with an old actor interacting with a young recipient would thus be scored as having an age-class difference of +2 and a middle-aged actor interacting with an old recipient would be scored as −1. Males were broken down into six categories (old, adult, young adult, sub-adult, juvenile-2, juvenile-1). Thus, the age difference could maximally be −5 (a juvenile-1 actor interacting with an old recipient) or +5 (an old actor interacting with a juvenile-1 recipient).

With respect to dominance rank, bachelors develop a clear, and for the most part linear, hierarchy. The highest-ranking male is invariably a band's next candidate to gain residency in a troop. Dominance is largely age and weight dependent, and generally decreases from prime adult to young adult to old to sub-adult and finally to juvenile (Rajpurohit *et al.*, 1995). In females, dominance is age inversed. The youngest adult females occupy the alpha positions and the oldest females the omega positions (Borries *et al.*, 1991).

The dominance hierarchy was divided into three classes: high-ranking individuals (first third of the hierarchy, depending on the total number of individuals usually the first three or four positions), middle-ranking individuals (second third) and low-ranking individuals (last third, usually the last four positions). Alternatively, absolute rank differences between individual mounters and mountees were used.

For each category of analyses (for example age-class, dominance rank), interaction rates were calculated for individual dyads, and the observed rates of mounting, grooming and displacement were compared with the expected rates. Expected levels were defined as interactions per hour of observation time, while two animals had the opportunity for dyadic interaction, that is were members of the same group. This procedure controls for monkeys that died, disappeared, immigrated, emigrated or were temporarily absent from the group.

Significance levels for statistical tests were two-tailed.

Table 10.2 *Agonism (displacement), affinity (grooming), and homosexual behaviour (same-sex mounting) in the study groups: proportion of individuals engaging as actors and recipients*

	Females			Males			
	Harem troop Kailana-I (1981/82)	Harem troop Kailana-I (1985)	Harem troop Kailana-I (1986/87)	Male band Machiya	Male band Canal	Male band Chopasani	Average (or sum)
Individuals (n)	11	13	12	10	4	10	60
Displacer (%)	100.0	100.0	100.0	100.0	100.0	90.0	98.3
Displacee (%)	100.0	100.0	100.0	90.0	100.0	100.0	98.3
At least one role (%)	100.0	100.0	100.0	100.0	100.0	100.0	100.0
Groomer (%)	100.0	100.0	100.0	100.0	100.0	100.0	100.0
Groomee (%)	100.0	100.0	100.0	100.0	100.0	100.0	100.0
At least one role (%)	100.0	100.0	100.0	100.0	100.0	100.0	100.0
Mounter (%)	81.8	69.2	83.3	90.0	75.0	70.0	78.3
Mountee (%)	100.0	100.0	100.0	80.0	100.0	80.0	93.3
At least one role (%)	100.0	100.0	100.0	90.0	100.0	90.0	96.7

Results

Agonistic, affinitive and same-sex mounting interactions

The present study explores potential correlations between same-sex mounting and agonistic and affinitive interactions. The data-set is comprised of 12 927 interactions, which includes 3651 displacements, 8660 allo-grooming episodes and 616 mountings, all of which occurred between 60 different individuals. All individuals, be it female or male, were involved in displacements or allo-grooming and 97% were involved in same-sex mounting (100% of females and 83% of males; Table 10.2).

Female–female interactions

A total of 458 episodes of mounting were observed (151 in 1981/82, 189 in 1985 and 118 in 1986/87), in which mounter as well as mountee could be identified, as well as 2930 displacements (486 in 1981/82, 1367 in 1985 and 1077 in 1986/87) and 6655 grooming interactions (2346 in 1981/82, 2578 in 1985 and 1731 in 1986/87).

Behaviours associated with same-sex mounting (Table 10.3) reveal a mixture of affinitive and agonistic motivational states. Female–female mounts closely resembled male–female copulations. Interactions were brief, lasting about 5–10 s. In the majority of cases, mounters applied a double foot-clasp on the mountee's lower legs (Figure 10.2a).

The sequence of female–female mounting was very similar to that of male–female copulations. Most interactions were initiated by the mounter but were often accompanied by the mountee's 'shake-and-present' posture. A langur mounter does not rub her genitals on the rump of the mountee, unlike female rhesus (Akers and Conaway, 1979) or stumptail macaques (Chevalier-Skolnikoff, 1976). In fact, such quadrupedal mounting was observed in only 2.7% of cases. Instead, the mounter would often thrust against the ischial callosities of the mountee (39.3% of all mounts). The mean number of thrusts was 6.0 ($n = 41$ episodes, range 2–11 thrusts). It is unlikely that the clitoris of the mounter is directly stimulated, as would be the case if the mounter rubs her genitals on the mountee's back, although that of the mountee might be. However, indirect mounter stimulation might be achieved due to mechanical pressure on body portions surrounding the clitoral region. There were no visible indications of orgasm for either mounter or mountee, but these are likewise absent in mountees participating in heterosexual copulations. Manual or oral stimulation of the genitalia was not observed.

Mounting could be accompanied by various agonistic behaviours, such as displays (jumps and/or long-distance 'whoop' calls; 2.2%) and displacements (7.5%). Only 1% of unisexual female couples were disturbed, in contrast to heterosexual copulations, of which more than 80% are harassed by non-mating individuals, especially adult females (Sommer, 1989a).

As a reflection of affinitive motivations, females embraced each other, a common gesture of reassurance, in 7.6% of same-sex mountings, and 27.3% were accompanied by allo-grooming. Following copulations, females also often groom the male, whereas males rarely groom females. However, such a clear-cut role of 'mountee = active groomer' vs. 'mounter = passive groomee' could not be found in female–female interactions. The relation was rather reversed, in that 79% of grooming was performed by the mounter.

Male–male interactions

The total of 158 mounts included 44 for male band Machiya, 30 for Canal and 84 for Chopasani. Displacements were recorded 721 times (388 for Machiya, 125 for Canal and 208 for Chopasani) as were 2005 grooming interactions (1002 for Machiya, 476 for Canal, 527 for Chopasani).

Table 10.3 Behaviours connected with mounting. (a) Among adult females in troop Kailana-I (based on 224 episodes during the 1981–82 study period; mounter or mountee was a non-adult individual in 66 cases, usually a perimenarcheal female; after Srivastava et al., 1991). (b) Among males of the all-male bands (based on 212 episodes observed in 3 bands). Sum >100% since a given episode could be accompanied by several variables. na = not applicable; nd = no data

Behaviour connected with same-sex mounting (%)	Female–female (n = 224 episodes)				Male–male (n = 212 episodes)			
	Preceding the mount	During mounting	After the mount	Sum%	Preceding the mount	During mounting	After the mount	Sum%
SEXUAL				84.8				66.5
Female mountee solicits male or copulates	8.5	0.0	4.9	13.4	na	na	na	na
Mountee solicits mounter (present, head-shake)	12.5	0.0	17.0	29.5	18.4	2.4	1.4	22.2
Mounter solicits same-sex mountee	0.0	0.0	0.0	0.0	0.0	0.0	1.4	1.4
Pelvic thrusts of mounter	0.0	39.3	0.0	39.3	0.0	29.7	0.0	29.7
Mounter mounts quadrupedically	0.0	2.7	0.0	2.7	0.0	3.3	0.0	3.3
Penile erection of mounter	na	na	na	na	0.0	4.7	0.0	4.7
Penile erection of mountee	na	na	na	na	0.0	0.5	0.0	0.5
Mounter muzzles / slumps on mountee's back	0.0	0.0	0.0	0.0	0.0	0.0	4.7	4.7
AFFINITIVE	0.0	0.0	0.0	31.3	0.0	0.0	0.0	67.9
Mounter grooms mountee	5.8	0.0	12.9	18.8	0.5	0.0	11.8	12.3
Mountee grooms mounter	0.4	0.0	4.5	4.9	0.0	0.0	8.5	8.5
Embracing, pulling of facial fur	6.3	1.3	0.0	7.6	5.7	0.5	10.8	17.0
Social play (locomotion, rough-and-tumble)	nd	nd	nd	nd	4.2	17.0	9.0	30.2

AGONISTIC	0.0	0.0	0.0	11.2	0.0	0.0	0.0	25.5
Jumping and / or whoop call display of mounter	0.4	0.0	0.4	0.9	7.1	0.0	2.8	9.9
Jumping and / or whoop call display of mountee	0.4	0.0	0.9	1.3	3.8	0.0	5.7	9.4
Mounter displaces (pulls, slaps) mountee	1.8	0.0	1.3	3.1	1.9	0.0	1.9	3.8
Mountee displaces (pulls, slaps) mounter	0.0	1.3	0.0	1.3	0.5	0.5	0.9	1.9
Displacement with 3rd party involvement	2.7	0.0	0.4	3.1	0.0	0.0	0.5	0.5
3rd individual harasses same-sex couple	0.0	0.0	1.3	1.3	0.0	0.0	0.0	0.0
VOCALIZATION	0.0	0.0	0.0	0.9	0.0	0.0	0.0	33.5
Mounter: assertive (grunts, barks, teethgrinding)	0.4	0.0	0.4	0.9	3.3	0.5	0.0	3.8
Mounter: submissive (whimpers, squeals, screams)	nd	nd	nd	nd	9.4	4.2	0.5	14.2
Mountee: assertive (grunts, barks, teethgrinding)	nd	nd	nd	nd	2.8	0.0	0.0	2.8
Mountee: submissive (whimpers, squeals, screams)	nd	nd	nd	nd	6.6	2.8	3.3	12.7

(a) Female 4 (right) of troop Kailana-I abandons her ten-day old infant (left) and solicits troop mate F3 through head shaking. F3 mounts F4, employing a double-foot clasp just below the knee-bend of F4. Her pelvic thrusting is accompanied by grunt-and-grimacing. Actors are often higher ranking than recipients, but the mounter ranks three positions below the mountee in this case (22 April 1982; photo: V. Sommer).

(b) Same-sex mounts might result if heterosexual contact is denied. The harem resident of troop Kailana-I, male 11, is groomed by F13 (right). The male ignores a solicitation for copulation by another female (F11, far left), who is four months pregnant. Instead, F11 is mounted by troopmate F8, who is in her ovulatory phase (6 November 1982; photo: V. Sommer).

Figure 10.2. Homosexual contact among langur monkeys. Same-sex mounts closely resemble male–female copulations.

Male–male mounting was very similar to female–female mounts in both length, position and accompanied variables. Some differences (Table 10.3) are due to the fact that the female sample includes only interactions between adults, while the male sample includes sub-adults and juveniles. As with females, the majority of male–male interactions were initiated by the mounter. Male mountees likewise presented their anogenital area (22.2% of cases) and sometimes even displayed 'female-typical' head-shakes. As with females, quadrupedal mounting was rare (3.3% of cases).

Male mounters would also often thrust against the ischial callosities of the mountee (29.7% of cases). The mean number of thrusts was 3.5 ($n = 53$ episodes, range 1–8 thrusts). Penile erections were observed for mounters in 4.7% of mounters, but only 0.5% of mountees. Sometimes, mounters rubbed their penis under the root of the mountee's tail, obviously aiming for sensory stimulation, but only a single instance of ejaculation was observed. Anal penetration was never recorded, and neither was oral or manual stimulation of another male's genitalia. Mounters would sometimes muzzle or slump on to the back of the mountee (4.7% of cases), a gesture that resembled the sudden relaxation sometimes observed in males at the end of a heterosexual copulation, presumably in connection with an ejaculation.

Displays, including jumps and long-distance 'whoop' calls, were associated with 19.3% of male–male mounts, and displacements were recorded in 6.2%. Mounters groomed mountees before or after 12.3% of all mountings, whereas mountees groomed mounters in 8.5% of cases. Embracing was observed in 17.0% of cases. A large proportion (30.2%) of mounting between sub-adults and juveniles emerged from locomotory or rough-and-tumble play.

Submissive vocalizations, such as high-pitched whimpers, squeals or screams, often accompanied by grimacing, which seemed to indicate ambivalent or fearful emotions, were heard in 26.9% of cases and were typically produced by immature individuals. Assertive grunts or muffled barks from adult males were heard in 6.6% of cases. Both mounters and mountees produced assertive as well as submissive vocalizations.

Annual distribution

Same-sex mounting between both females and males occurred during all months of the year (Figure 10.3), although not equally distributed (chi-square (11) for female–female mounts 78.56, $p < 0.001$, male–male mounts 87.33, $p < 0.001$). Most months were well within the range of what was expected by chance (8.3%). This pattern resembles that of heterosexual copulations which occur year round.

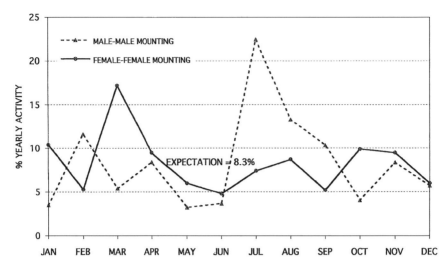

Figure 10.3. Annual distribution of same-sex mounting (female data from Srivastava *et al.*, 1991). Expected value/month = 8.3%.

Female reproductive status

Cycling, pregnant and lactating individuals would all act as mounter as well as mountee (Figure 10.4). However, observed rates deviated, often significantly, from what would be expected by chance. In particular, participation was significantly underrepresented for mounters as well as mountees during menstruation, follicular and luteal periods. Around ovulation, mounter activity was 1.6 times and mountee activity even 4.7 times higher than expected. Mounters had additional significant peaks during the main phase of postconception oestrous and during lactation. Mountees had significant minima during the last anestrous phase of pregnancy as well as during a later stage of lactation. Cases in which mounter as well as mountee were anestrous accounted for only 15% of all episodes.

Agonism and affinity

Several functional explanations view same-sex mounting as either socially bonding (affinitive) or dominance-related (agonistic). It is therefore necessary to investigate the interrelationship of these variables.

A first analysis explores the likelihood that a certain type of interaction (displacements = agonistic; affinitive = grooming) does or does not occur between two animals (Figure 10.5a) and how frequent these interactions were (Figure 10.5b). The direction of interactions was neglected but many if not most

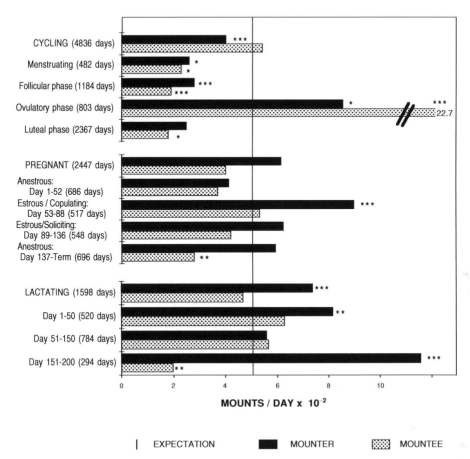

Figure 10.4. Reproductive status of female mounters (black bars) and mountees (hatched bars). Dashed line = expected rate of same-sex mounts. "Days" refer to the total number of observation days of females in a given reproductive status. Observed versus expected: *$p < 0.05$, **$p < 0.01$, ***$p < 0.001$ (chi-square goodness-of-fit test). After Srivastava et al., 1991.

interactions were reciprocal, particularly amongst females (see Borries et al., 1994; Sommer et al., 2002).

Every female in the troop had agonistic as well as affinitive interactions with every other during each observation period. The proportion of observed dyads out of all potential dyads was thus always 100% (Figure 10.5a). Males had fewer social partners within the pool of potential dyads because the values for agonistic and affinitive interactions varied between 60% and 100% for the three study bands. This difference is most likely due to the fact that the two groups with lower values contained several immature individuals, and agonistic as well as affinitive interactions between animals of very different ages, in particular

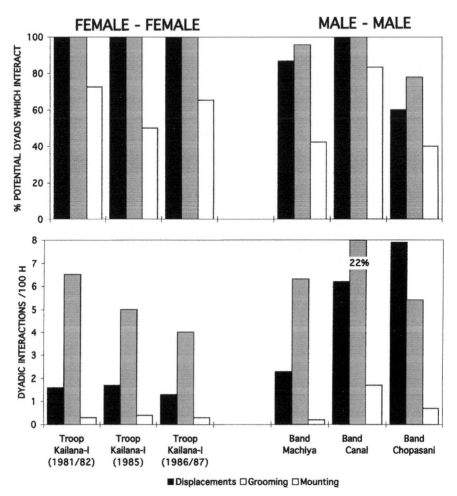

Figure 10.5. Agonism (displacement), affinity (grooming), and homosexual behaviour (same-sex mounting) in the study groups: (a) proportion of potential dyads that are interacting; (b) dyadic interactions/100 h of observation.

juveniles and adults, are less likely to occur than interactions between same-aged individuals. The greatest selectivity was recorded with respect to same-sex mounting, as it occurred in comparatively fewer potential dyads (average 63% for females, 55% for males).

These findings are mirrored when rates are taken into account (Figure 10.5b). Averaged across groups, 3.5 displacements and 8.2 episodes of allo-grooming are observed per 100 h, but only 0.6 episodes of same-sex mounting. Average interaction rates are consistently higher for males, with a factor of 3.7 for displacements, 2.2 for allo-grooming and 3.0 for same-sex mounting.

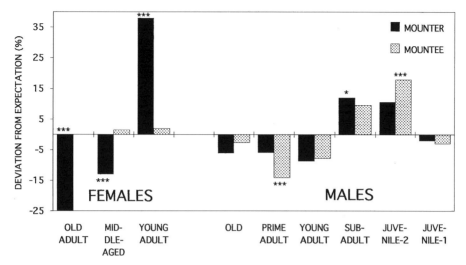

Figure 10.6. Same-sex mounting: distribution over age-classes in (a) females (after Srivastava *et al.*, 1991) and (b) males. Observed versus expected: $*p < 0.05$, $**p < 0.01$, $***p < 0.001$ (chi-square goodness-of-fit test).

The influence of age

Females of the young age class were significantly overrepresented as mounters, whereas middle-aged and old individuals were significantly underrepresented. For mountees, there was no difference between observed and expected values. In male bands, older members were also underrepresented. In particular, prime adult males were significantly less often mounted than expected. Younger males engaged more frequently than expected in same-sex mounting. Sub-adults, for example, were significantly overrepresented as mounters, and juvenile-2 individuals as mountees (Figure 10.6).

With respect to relative age differences, female mounting partners of the same age class were clearly overrepresented (Figure 10.7a), whereas older individuals mounting younger troop mates were underrepresented (Figure 10.7b). Similarly, males preferred partners of a same or very similar age, whereas mounting became less likely with increasing age difference (Figure 10.7b).

The influence of dominance rank

Female mounters of high rank were significantly overrepresented, whereas those of the middle- and low-ranking classes were significantly underrepresented. However, amongst mountees, only high-ranking individuals were overrepresented (Figure 10.8). In contrast, high-ranking males were significantly

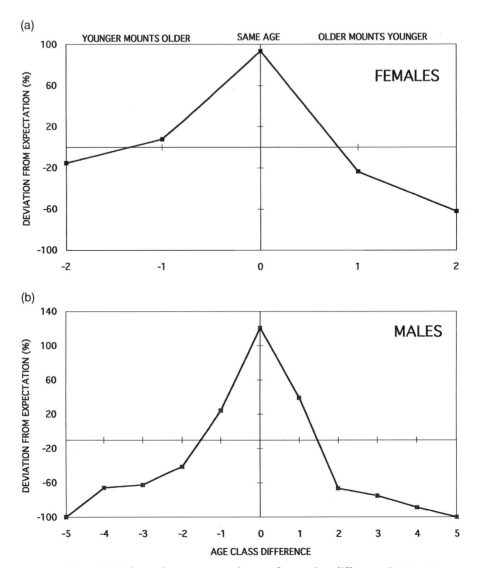

Figure 10.7. Observed versus expected scores for age-class differences in same-sex mounting. (a) Combined data for troop Kailana-I from 1981–1987. (b) Combined data for male bands Machiya, Canal and Chopasani. For calculation of age-class differences, see text.

underrepresented as both mounters and mountees, whereas middle-ranking individuals were significantly overrepresented (Figure 10.8).

At the level of individual dyads, female mounters were higher ranking than mountees in 84.2% of all cases ($n = 436$), with a significant difference found during each period of the study (80.8% of 130 cases during 1981/82, 91.6% of 178 cases during 1985, 77.3% of 128 cases during 1986/87). Male mounters were

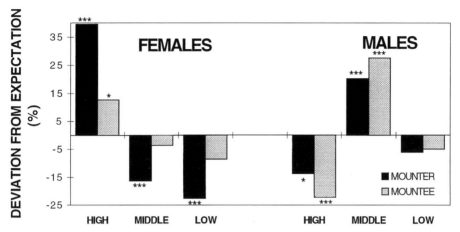

Figure 10.8. Distribution of same-sex mounting over dominance rank-classes. Cut-off points for classes depend on the total number of group members. (a) Females (from Srivastava *et al.*, 1991). (b) Males. Observed versus expected: *$p < 0.05$, **$p < 0.01$, ***$p < 0.001$ (chi-square goodness-of-fit test).

also typically higher ranking than mountees, but the overall proportion was significantly lower (67.3% of 159 cases), and only the individual difference for band Machiya was significant (64.4% of 45 cases for male band Machiya; 83.3% of 30 cases for Canal; 63.1% of 84 cases for Chopasani).

Relative rank differences between actors and recipients (Figure 10.9a) revealed that female mounters tended to be 1–3 ranks higher than mountees, whereas mounters which had a lower rank were always underrepresented. However, an almost inverse picture was observed for males in that mounters (Figure 10.9b), which were 1–2 ranks below the respective mountee tended to be overrepresented. By and large, very broad gaps in the hierarchy were hardly ever bridged.

Relative rank differences influenced the likelihood that individuals of a potential dyad interacted at all, not only with respect to same-sex mounting, but also with respect to displacements and allo-grooming, as shown in Table 10.4. The results for females and males are largely in the same direction.

In 87% of all existing dyads, the higher-ranking individual displaced the lower ranking. In about half of these dyads displacements also occurred in the opposite direction, with a lower-ranking individual displacing a higher ranking. Such cases can be the result of 'reversals', as in 'A displaces B, A displaces B, B displaces A, A displaces B' and so on, or 'rank changes' in which the sequence 'A displaces B, A displaces B. . . .' mutates at one point into 'B displaces A, B displaces A . . .'. Rank changes can feed into the category 'lower rank displaces higher rank' since the calculation was based on average ranks, not rank differences during individual interactions.

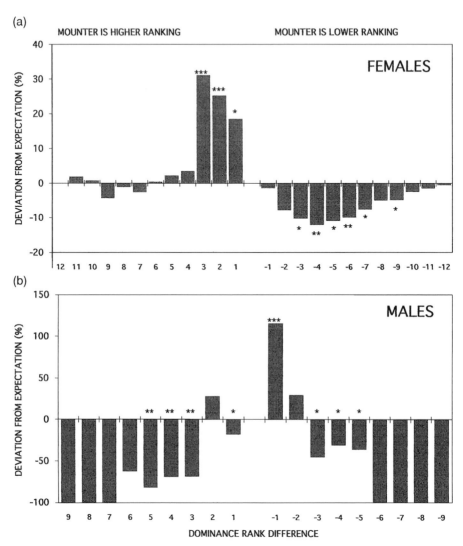

Figure 10.9. Observed versus expected scores for dominance rank differences in same-sex mounting. (a) Combined data for troop Kailana-I from 1981–1987. (b) Combined data for male bands Machiya, Canal and Chopasani. Observed versus expected: *$p < 0.05$, **$p < 0.01$, ***$p < 0.001$ (chi-square goodness-of-fit test).

With respect to allo-grooming, females revealed a very 'promiscuous' grooming style in that rank played no role. Almost every female groomed every other female, independent of dominance direction. Males were more selective, and it is particularly noteworthy that grooming up the hierarchy (87% of potential cases) was more common than grooming down the hierarchy (69% of potential cases).

Table 10.4 *Proportions of potential dyads in which higher ranking actors interact with lower ranking recipients, and vice versa, for displacement, grooming and mounting interactions*

	Troop Kailana-I (1981/82)	Troop Kailana-I (1985)	Troop Kailana-I (1986/87)	Band Machiya	Band Canal	Band Chopasani	Female average	Male average	Total average
Displacements	581	1367	1077	388	125	208	1008	240	624
Higher rank displaces lower rank: % of potential dyads	98.2	100.0	98.5	82.2	100.0	40.0	98.9	74.1	86.5
Lower rank displaces higher rank: % of potential dyads	67.3	39.7	48.5	37.8	66.7	35.6	51.8	46.7	49.3
Grooming	2346	2578	1731	1002	476	527	2218	668	1443
Higher rank grooms lower rank: % of potential dyads	100.0	100.0	98.5	64.4	83.3	60.0	99.5	69.3	84.4
Lower rank grooms higher rank: % of potential dyads	100.0	100.0	92.4	88.9	100.0	71.1	97.5	86.7	92.1
Mounting	157	189	118	44	30	84	155	53	104
Higher rank mounts lower rank: % of potential dyads	58.2	43.6	59.1	31.1	66.7	26.7	53.6	41.5	47.6
Lower rank mounts higher rank: % of potential dyads	36.4	15.4	21.2	20.0	50.0	35.6	24.3	35.2	29.8

Table 10.5 *Correlations of the ratio of grooming/displacement with the dyadic rates of same-sex mounting*

Langur group	Correlation coefficient (rs)	p	n
Harem troop Kailana-I (1981/82)	−0.046	0.741	55
Harem troop Kailana-I (1985)	−0.224	0.048*	78
Harem troop Kailana-I (1986/87)	−0.013	0.919	66
Sum harem troop	−0.031	0.662	199
Male band Machiya	0.054	0.727	44
Male band Canal	0.257	0.623	6
Male band Chopasani	0.054	0.000***	39
Sum male bands	0.356	0.001***	89
Total	0.129	0.028*	288

Note: * $p < 0.05$, ** $p < 0.01$, *** $p < 0.001$.

Mounting interactions occurred in a much lower proportion of mathematically possible dyads than displacements and grooming. Mounting higher in the hierarchy was recorded for only 48% of all potential dyads and mounting lower in the hierarchy even for only 30%.

Relationship quality

Same-sex mounting is not easily categorized as an agonistic or affinitive type of interaction. In order to determine the more likely motivational underpinning, a correlation with the relationship quality of a given dyad was attempted. For this, the ratio of grooming to displacement episodes for any given dyad was taken as an indicator, so that a high value, defined by much grooming and few agonistic interactions, was taken to indicate a 'good', 'relaxed' or 'friendly' relationship, whereas a low value, defined by a great deal of agonism and little affinitive grooming, indicated a 'bad', 'strained' or 'unfriendly' relationship.

Ratios for any potential dyad were then correlated with the rates for same-sex mounting. A resulting positive correlation would indicate that same-sex mounting increases with an increasingly socio-positive quality of relationships. Mounting would thus seem to reflect a more affinitive type of interaction. A resulting negative correlation would indicate that same-sex mounting decreases if the quality of relationships becomes increasingly socio-positive. Mounting would thus seem to reflect a more agonistic interaction, which is reduced when a relationship becomes friendlier.

The data (Table 10.5) for females of all three study periods are slightly negative and reach significance in the 1985 period. The combined values indicate an almost neutral and non-significant relationship ($r_s = -0.031$). The results for males are always in the direction of a positive correlation and reach significance in the male Chopasani band. Overall, a significantly positive correlation is found for males ($r_s = 0.356$), indicating a more affinitive quality of male–male mounting.

Discussion

Our data-set for wild Indian langur monkeys with its almost 13 000 interactions, including 616 same-sex mounts, is the largest for any primate taxon designed to explore explanatory hypotheses about homosexual behaviour. The analysis reveals that same-sex sexual behaviour, as manifested through ventro-dorsal mounting, is an integral part of the social life of both male and female langur monkeys, and expressed by virtually all individuals and throughout all life-history stages.

The lifetime tally of sexual activity of langur monkeys at Jodhpur can be calculated on the basis of the accumulated long-term data (Sommer, 1989b; Sommer and Rajpurohit, 1989; Sommer et al., 1992).

- *Females*. During the roughly 25 years that the average langur female is reproductively mature, she will copulate every 33 hours, amounting to a total of 3370 copulations. Same-sex mounts occur every 38 hours, amounting to a total of 2920 homosexual interactions. Heterosexual mounts therefore account for 54%, homosexual mounts for 46% of her lifetime's sexual interactions.
- *Males*. During the roughly 14 years that a male spends outside his natal troop, he will be a harem holder for an average of 26.5 months; during this tenure, he will copulate every 3.6 hours, amounting to 2686 copulations. Before and after his residency, he will live in a male band, where he will have a homosexual interaction every 82 hours, amounting to a total of 602 same-sex mounts. The ratio of hetero-versus homosexual activity is therefore 82% vs. 18%. However, 23% of all adult males will never gain a residency; their chances to copulate are restricted to brief multi-male stages in the course of harem holder replacements. These males will be involved in a total of 721 homosexual mounts, which represent about 95% of their whole sex life.

We will now attempt to test numerous explanatory hypotheses about homosexual behaviour against our data. Comparative evidence will be largely restricted to Japanese macaques and bonobo chimpanzees, since these are amongst the few other primate taxa for which relatively detailed data about same-sex sexual behaviour exist.

Maladaptation

Traditionally, homosexual behaviour has been interpreted as a pathology, brought about by conditions such as the unnatural circumstances of confinement or 'hormonal imbalance' (reviews in Weinrich, 1980; Bagemihl, 1999). Support for such views could, for example, come from the case of two male rhesus monkeys reared exclusively together which developed an extensive homosexual relationship with each other as adults (Erwin and Maple, 1976). However, three quarters of reports about primate same-sex sexual behaviour come from naturally occurring groups which certainly do not represent abnormal responses to captivity (Vasey, 1995).

The langur data represent another sound refutation of the maladaptation hypothesis. Any attempt to label female mounters as 'hormonally misguided', 'androgenized' or 'masculinized' seems to be futile, because the very endocrinological milieu around mid-cycle that stimulates females to be mounted in heterosexual copulations also induces them to mount other females (see Figure 10.4). However, one does not even need such a sophisticated argument, since a behaviour can hardly be labeled as pathological if it is exhibited by 97% of all study animals (see Table 10.3), throughout all age and sex classes (see Figure 10.6), and throughout the year (see Figure 10.3). Moreover, same-sex mounting did not seem to have any negative effect on reproductive success (cf. Borries et al., 1991, Srivastava et al., 1991).

Homosexual mounting is thus perfectly well integrated into the natural lifestyle of wild monkeys.

By-product

Alternatively, same-sex sex might not be directly selected for but could represent a by-product of selection for other traits (Vasey, 1995). For example, it may be a by-product of selection for sexual gratification. Homosexual behaviour of female Japanese macaques does not seem to facilitate adaptive social or reproductive goals, but as it does not interfere with the participants' reproduction it is assumed to represent a neutral trait (Vasey, 1995).

Components of same-sex mounting in langurs, most notably solicitations, pelvic thrusting and penile erections, clearly reflect sexual excitement (see Table 10.3). A connection with the motivational state that drives heterosexual sex was particularly obvious for females, since the prevalence or paucity of same-sex mounting covaried with heterosexual activity (see Figure 10.4).

On the other hand, gratification gained from same-sex contacts is not identical with that from heterosexual copulations, since clitoral stimulation was unlikely, penile erections were rare and ejaculations almost never occurred. Compared with bonobos, stumptail macaques or Japanese macaques, same-sex-mounting in langurs seemed to be less 'spectacular' and sexually charged. In addition, the 'pleasure' aspect is readily compatible with numerous other functional hypotheses, support for which would correspondingly weaken the by-product argument.

Controlling population size

A first, albeit old-fashioned functional explanation suggests that homosexual behaviour becomes more frequent if population density increases, thus reducing the strain on resource availability since individuals engaging in homosexual sex would leave fewer offspring (review in Kirsch and Weinrich, 1991). This group-selectionist argument is flawed as the genetic material of individuals engaging in same-sex behaviour would not be passed on and the trait should thus be quickly eliminated from the gene pool.

The Jodhpur data also do not support this assumption. The monkey population has more than doubled in the past 40 years, due to increased provisioning by a burgeoning human population. However, nothing indicates a corresponding rise in same-sex mounting.

Allo-parental care

A first functional explanation centered on individual benefits assumes that same-sex sexual behaviour recruits help in rearing offspring, by rewarding the helper with pleasure or simply by exposing a potential caregiver to a partner's offspring. There is some evidence for this in birds (examples cited in Vasey, 1998), but not primates (Vaso, 1998).

For langur monkeys, this hypothesis is easily refuted. Firstly, of course, it is not applicable to male bands since there are no infants present. Secondly, same-sex mounting in females did not depend upon one of the participants lactating (see Figure 10.4). Allo-mothering is common in Indian langurs (Sommer, 1989b) but a contextual link with mounting is absent (see Figure 10.2a).

Training and practice

Homosexual activity, particularly during immature play, could facilitate social and motor development. Indeed, among macaques, adequate opportunity to engage in mounting when young is necessary for competent performance of heterosexual copulation in adulthood, and the actual sex of mounting partners is irrelevant (Goy and Wallen, 1979).

Immature langurs, as young as three months, frequently engage in mounting, with both immature and mature partners of either sex. The combination juvenile female–juvenile female and juvenile male–juvenile male is the most frequent (Srivastava et al., 1991). Immature male band members clearly prefer like-aged partners (Figure 10.7). Mounting was frequently connected with play (see Table 10.3). In accordance with the practice hypothesis, one could also link much of the same-sex sexual behaviour amongst maturing and adult langur males to a need to 'be prepared' for upcoming occasions of heterosexual acts – although this would apply more to mounters than to mountees.

However, the practice hypothesis is not unproblematic. Firstly, there is little evidence that play is in fact practice (Burghardt, 1984). For example, immature males of the Chopasani band, who had easy access to food, played much more frequently than their peers of band Machiya, who were often under food stress. However, the motor development of the latter did not seem to suffer (Sommer and Mendoza-Granados, 1995). Secondly, same-sex mounting was prevalent amongst individuals with much experience in heterosexual sex, such as adult females and ousted resident males, and it is hard to imagine that they would need further training.

Substitute for heterosexual sex

Homosexual activity has been ascribed to have a 'drive-reducing function' (Weber and Vogel, 1970), substituting for a lack of suitable opposite-sex partners. Such correlation has been documented for a population of female Japanese macaques (Wolfe, 1984), but is not found in bonobo females (Parish, 1994) or male stumptail macaques (Chevalier-Skolnikoff, 1976) who engage homosexually even if suitable opposite-sex partners are present.

Langur male bands seem to be a perfect setting for a naturally occurring 'prison situation', in that they are forcefully excluded from females. However, male–male mounts still occur in multi-male/multi-female troops that are formed in other populations (Vogel, 1976), and when females become available, as when harem troops are invaded by male bands. Also, same-sex activity is prevalent in immatures, particularly males, not yet capable of copulation (see Figure 10.6), who therefore have no need for 'substitution'.

Female–female mounting also offers some support for the substitution hypothesis, since competition for access to the one male does clearly exist in harems. However, females also mount other females in multi-male troops (review in Srivastava et al., 1991).

Dominance expression

It is often hypothesized that same-sex mounting reaffirms the dominance hierarchy by expressing social status, thus preventing aggressive physical interactions (Wickler, 1967). While some studies maintain that mounters are generally higher ranking (squirrel monkeys, Talmage-Riggs and Anschel, 1973; stumptail macaques, Chevalier-Skolnikoff, 1976), others have not found such a relationship (Japanese macaques, Vasey, 1996; rhesus monkeys, Kaufmann, 1967; gorillas, Yamagiwa, 1987).

In any case, to equate same-sex mounting and dominance can be a thinly veiled attempt to 'explain away' its sexual nature. Moreover, it ascribes a submissive role to the mountee, and a dominant role to the mounter. This reflects the stereotypical view that females (the quintessential 'mountee') should be viewed as 'submissive', while males (the quintessential 'mounter') are viewed as 'dominant' (see discussion in Srivastava et al., 1991).

On an empirical level, the prediction of a high rank for mounters finds only limited support from the langur data. Indeed, mounters were of higher rank in 84% of all interactions amongst females and 67% of those amongst males. Still, a solid quarter (24.3%) of same-sex mounts amongst langurs contradicts the dominance hypothesis (see Figure 10.8, Table 10.4). Moreover, langurs would often solicit same-sex mounts, including mounts from lower-ranking individuals (Table 10.3).

Conflict management

Homosexual behaviour has also been linked to conflict management. Firstly, such presumably pleasurable sexual interactions could counteract or buffer stress and therefore theoretically function as a 'compensation payment' for agreeing to de-escalate conflicts over resources ('tension regulation', Kuroda, 1984). Secondly, same-sex sex could re-establish social bonds after other mechanisms for regulating aggression have failed ('reconciliation'; de Waal, 1987). A close temporal relationship between feeding and homosexual behaviour was found for female bonobos and post-conflict genital contact exceeded pre-conflict rates (Hohmann and Fruth, 2000). However, this was not true for female Japanese macaques (Vasey, 2004).

At best, a fraction of the langur data could be related to conflict management. For example, 10.9% of male–male mounts were preceded by whoop calls and jumping displays which are normally addressed to other bands or harem residents with whom band members compete for food or females. Few same-sex mounts occurred in the context of conflicts between group mates (females: ca. 7.5%, males: ca. 6.2%). Post-conflict affinity is generally rare, since Jodhpur langurs tend to avoid each other after conflicts, making it likely that the prevalence of 'reconciliation' in other taxa might be a partial artifact of captivity where primates cannot easily retreat (Sommer *et al.*, 2002).

Social bonding

Same-sex behaviour can express mutual affection and affiliation. The psycho-physical rewards it provides could reinforce long-term relationships and may thus promote social bonding. Watanabe and Smuts (1999), in an interesting merger of concepts from biological and social anthropology, pointed out that same-sex behaviour is often a ritualized signal or 'ceremony' (including stereotypical elements of vocalization, solicitation, mounting and mutual grooming) that can be understood as a formalism which conveys the intention to cooperate within a socially complex yet non-linguistic behavioural context. For example, in olive baboon greetings, the males exchange a series of gestures that include presenting and grasping the hindquarters with one or both hands, mounting, touching the scrotum, pulling the penis and on occasion embracing. Virtually all other interactions between male baboons involve antagonistic exchanges and they almost never engage in friendly behaviours outside these greetings (Watanabe and Smuts, 1999).

The social bonding hypothesis suggests that homosexual behaviour can reinforce long-term relationships and therefore predicts same-sex mounting should be more prevalent in 'good relationships', which can be defined by a high ratio of grooming to displacement. However, this was not fulfilled for langur females, indicating that mounts have a more agonistic quality, whereas male mounts seem to reflect affinity (see Table 10.6). Nevertheless, it remains unclear why mounting should reinforce social bonding more effectively than allo-grooming, given that the latter has the added benefit of hygiene. Indeed, in contrast to baboons, grooming is much more common than mounting amongst both female and male langurs (see Figure 10.5b).

Alliance formation

Same-sex contacts allow an individual to intimately explore the physical attributes of a partner, and knowing the strengths and weaknesses of an ally can be advantageous during competitive interactions with third parties. Such hypothesized correlation between alliance formation and homosexual behaviour (Fairbanks *et al.*, 1977) seems to exist in Hamadryas baboons (Colmenares, 1991) and female bonobos (Parish, 1994) but not Japanese macaques (Vasey, 1996).

Table 10.6 *Hypotheses concerning homosexual behaviour and evidence from the langur monkeys of Jodhpur*

Hypothesis	Evidence for langurs			
	Female–female mounting		Male–male mounting	
Maladaptation		−		−
By-product	+	-	+	-
Controlling population size		−		−
Acquisition of allo-parental care		−		−
Training and practice		−	+	−
Substitution	+	−	+	−
Dominance expression	+	-	+	-
Conflict management	+	−	+	−
Social bonding	+	−	++	-
Alliance formation	+	−	+	−
Stimulate sluggish heterosexual partners	+	-		−
Reduce competitor's receptivity	++	-		−

Note: +++ = strong support, ++ = moderate support, + = weak support; − = strong refutation, – = moderate refutation, - = weak refutation.

The hypothesis also draws little support from langurs. It predicts that subordinates should target dominant partners because subordinates would derive more alliance-related benefits from bonding with dominant individuals. However, while it is true that high-ranking individuals are more active mounters, they are also more active mountees. Data for allo-grooming, the other obvious way to recruit allies, do also not support the direction predicted by the alliance hypothesis, since half of all grooming is directed down the hierarchy (Borries *et al.*, 1994). More particularly, alliances of two individuals that gang up against a third are all but absent from the daily langur routine anyway (Table 10.4; Borries *et al.*, 1991). In addition, both males and females prefer same-aged individuals (see Figure 10.7), when in fact, independent of the age difference, they should prefer young adult females or prime adult males, because these are of high rank.

The specific predictions of the alliance formation hypothesis are in any case somewhat unclear. Should lower-ranking individuals mount higher-ranking individuals, or could they not also promote themselves as mountees? Both assumptions seem to have some merit, but this renders the hypothesis so ubiquitous as to be untestable.

Stimulate heterosexual partners

By mimicking a potential rival through display of male copulatory patterns, a female mounter could attract a sluggish male's attention, by either 'arousing him' or 'making him jealous', resulting in copulation (Parker and Pearson, 1976). The mountee, at the same time, would demonstrate her receptivity, indicating her willingness to seek out a male competitor. A problem with this hypothesis is that female mounters do not directly benefit, and are rarely closely related to the mountee; thus, their cooperative participation would need additional explanation. Moreover, female bonobos (Parish, 1994) and Japanese macaques (Vasey, 1998) tend to ignore or even threaten any male that solicits them while they engage in same-sex activity.

Only some female–female mounts were preceded (8.5%) or followed (4.9%) by invitations for sex that mountees directed towards males (see Figure 10.2b). However, homosexual interactions often did occur when the harem resident male was out of sight (Srivastava et al., 1991). Moreover, mounters were potentially fertile in only 14.3% of all episodes and mountees in 40.7% (see Figure 10.4). Finally, the hypothesis cannot explain male–male mounting.

Reduce a competitor's receptivity

A female mounter could use the sexual satisfaction provided by homosexual interactions in order to decrease a mountee's motivation to copulate, thus reducing a same-sex competitor's reproductive success (Srivastava et al., 1991).

This hypothesis has no relevance for langur males. An obvious counterargument with respect to females seems to be that mounters and mountees were fertile in only a fraction of cases (see Figure 10.4). However, many heterosexual copulations also occur during infertile periods, perhaps because females try to deplete sperm otherwise available to competing fertile females (Sommer, 1989a; Sommer and Rajpurohit, 1989). Indeed, the likelihood of conception increases with the number of copulations per day and decreases if several troop mates copulate likewise with the harem holder on a given day (Sommer et al., 1992). Moreover, 81% of all copulations are harassed by group members, including 75% of all copulations of pregnant females. Female–female mounts, on the other hand, were very rarely harassed (see Table 10.3). 'Pseudo-copulations', about a third of which include pelvic thrusting, could thus provide a surrogate sexual satisfaction to the mountee, especially if a male ignores heterosexual solicitations. A female mounter, even if not fertile, might therefore reduce the number of future resource competitors. Fertile female mounters could also, whether or not the mountee is fertile, lower the probability of insemination and corresponding sperm depletion of the male (see Figure 10.2b). The hypothesis does therefore draw some support from the langur data on female–female mounting.

Conclusion

Homosexual behaviour is an integral part of the social life of wild Indian langur monkeys, constituting 46% of the sexual interactions an adult langur female has over her lifetime, and 18–95% of an adult male's sex life. It is not possible to furnish a single functional explanation for this behaviour, given its ubiquity throughout different life-history stages and its occurrence in both bisexual and unisexual social settings. However, numerous hypotheses can be clearly discounted, in particular those that assume same-sex sexual behaviour to be a pathology or an artifact of captivity. With respect to females, same-sex mounting seems to be more linked with agonistic, competitive behaviours, whereas male--male mounting seems to express more affinitive motivations (see Table 10.6).

Same-sex mounts make up almost half of the sexual interactions a female has during her lifetime. Homosexual interactions represent likewise a substantial proportion of a male's sex life, given that they make up almost a fifth of the sexual interactions of males who gain a harem residency, but about 95% of the sex of males who remain band members throughout their adult lifes. It is therefore not surprising that same-sex mounting is not confined to a specific context but appears to be multi-functional. Homosexual interactions contribute to what seems to be a rather 'sexualized nature' of langur societies. This is quite similar to human societies, where social interactions are permeated by mixed motivations, behaviours, displays and linguistic references related to sexuality and reproduction.

Acknowledgements

Thanks are due to Carola Borries and Arun Srivastava who collected parts of the data analysed here. Paul Vasey and Craig Stanford provided critical comments. VS is grateful to those who supported his original fieldwork (C. Vogel and S. M. Mohnot, Universities of Göttingen and Jodhpur; Deutscher Akademischer Austauschdienst; Alexander von Humboldt Stiftung).

References

Akers, J. S. and Conaway, C. H. (1979) Female homosexual behavior in *Macaca mulatta*. *Arch. Sex. Behav.*, **8**, 63–80.

Bagemihl, B. (1999) *Biological Exuberance: Animal Homosexuality and Natural Diversity.* New York: St. Martin's.

Borries, C. (1989) Konkurrenz unter freilebenden Langurenweibchen (*Presbytis entellus*). Ph.D. thesis, Georg-August Universität Göttingen.

Borries, C., Sommer, V. and Srivastava, A. (1991) Dominance, age, and reproductive success in free-ranging female Hanuman langurs (*Presbytis entellus*). *Int. J. Primatol.*, **12**, 231–57.

Borries, C., Sommer, V. and Srivastava, A. (1994) Weaving a tight social net: allo-grooming in free-ranging female langurs (*Presbytis entellus*). *Int. J. Primatol.*, **15**, 421–443.

Burghardt, G. M. (1984) On the origins of play. In *Play in Animals and Humans*, ed. P. K. Smith, pp. 5–42. Oxford: Basil Blackwell.

Chevalier-Skolnikoff, S. (1976) Homosexual behavior in a laboratory group of stumptail monkeys (*Macaca arctoides*): forms, contexts, and possible social functions. *Arch. Sex. Behav.*, **5**, 511–27.

Colmenares, F. (1991) Greeting, aggression, and coalition formation between male baboons: demographic correlates. *Primates*, **32**, 453–63.

de Waal, F. B. M. (1987) Tension regulation and nonreproductive functions of sex in captive bonobos (*Pan paniscus*). *Nat. Geogr. Res.*, **3**, 318–38.

Erwin, J. and Maple, T. (1976) Ambisexual behavior with male-male anal penetration in male rhesus monkeys. *Arch. Sex. Behav.*, **5**, 9–14.

Fairbanks, L. A., McGuire, M. T. and Kerber, W. (1977) Sex and aggression during rhesus monkey group formation. *Aggress. Behav.*, **3**, 241–9.

Goy, R. W. and Wallen, K. (1979) Experimental variables influencing play, footclasp mounting, and adult sexual competence in male rhesus monkeys. *Psychoneuroendocrinol.*, **4**, 1–12.

Hohmann, G. and Fruth, B. (2000) Use and function of genital contacts among female bonobos. *Anim. Behav.*, **60**, 107–20.

Kaufmann, J. H. (1967) Social relations of adult males in a free-ranging band of rhesus monkeys. In *Social Communication Among Primates*, ed. S. A. Altmann, pp. 73–98. Chicago: University of Chicago Press.

Kirsch, J. A. W. and Weinrich, J. D. (1991) Homosexuality, nature, and biology: is homosexuality natural? Does it matter? In *Homosexuality: Research Implications for Public Policy*, eds. J. C. Gonsiorek and J. D. Weinrich, pp. 13–31. Newbury Park: Sage.

Kuroda, S. (1984) Interactions over food among pygmy chimpanzees. In *The Pygmy Chimpanzee*, ed. R. L. Susman, pp. 301–24. New York: Plenum.

Mohnot, S. M. (1974) Ecology and behavior of the common Indian langur, *Presbytis entellus* Dufresne. Ph.D. thesis, University of Jodhpur, Jodhpur.

Parish, A. R. (1994) Sex and food control in the 'uncommon chimpanzee': how bonobo females overcome a phylogenetic legacy of male dominance. *Ethol. Sociobiol.*, **15**, 157–79.

Parker, G. A. and Pearson, R. G. (1976) A possible origin and adaptive significance of the mounting behavior shown by some female mammals in oestrous. *J. Nat. Hist.*, **10**, 241–5.

Podzuweit, D. (1994) Sozio-Ökologie weiblicher Hanuman-Languren (*Presbytis entellus*). Ph.D. thesis, Georg-August Universität Göttingen.

Rajpurohit, L. S., Sommer, V. and Mohnot, S. M. (1995) Wanderers between harems and bachelor bands: male Hanuman langurs (*Presbytis entellus*) at Jodhpur in Rajasthan. *Behaviour*, **132**, 255–99.

Sommer, V. (1985) Weibliche und männliche Reproduktionsstrategien der Hanuman-Languren (*Presbytis entellus*) von Jodhpur, Rajasthan / Indien. Ph.D. thesis, Georg-August Universität Göttingen.

Sommer, V. (1989a) Sexual harassment in langur monkeys (*Presbytis entellus*): competition for nurture, eggs, and sperm? *Ethology*, **80**, 205–17.

Sommer, V. (1989b) Infant mistreatment in langur monkeys – sociobiology tackled from the wrong end? In *The Sociobiology of Sexual and Reproductive Strategies*, eds. A. E. Rasa, C. Vogel and E. Voland, pp. 110–27. London: Chapman & Hall.

Sommer, V. and Rajpurohit, L. S. (1989) Male reproductive success in harem troops of Hanuman langurs (*Presbytis entellus*). *Int. J. Primatol.*, **10**, 293–317.

Sommer, V. and Mendoza-Granados, D. (1995) Play as indicator of habitat quality: a field study of langur monkeys (*Presbytis entellus*). *Ethology*, **99**, 177–92.

Sommer, V., Srivastava, A. and Borries, C. (1992) Cycles, sexuality, and conception in free-ranging langurs (*Presbytis entellus*). *Am. J. Primatol.*, **28**, 1–27.

Sommer, V., Denham, A. and Little, K. (2002) Postconflict behavior of wild Indian langur monkeys: avoidance of opponents but rarely affinity. *Anim. Behav.*, **63**, 637–48.

Srivastava, A. (1989) Feeding ecology and behaviour of Hanuman langur, *Presbytis entellus*. Ph.D. thesis, University of Jodhpur, Jodhpur.

Srivastava, A., Borries, C. and Sommer, V. (1991) Homosexual mounting in free-ranging female Hanuman langurs (*Presbytis entellus*). *Arch. Sex. Behav.*, **20**, 487–512.

Talmage-Riggs, G. and Anschel, S. (1973) Homosexual behavior and dominance hierarchy in a group of captive female squirrel monkeys (*Saimiri sciureus*). *Folia Primatol.*, **19**, 61–72.

Vasey, P. L. (1995) Homosexual behavior in primates: a review of evidence and theory. *Int. J. Primatol.*, **16**, 173–204.

Vasey, P. L. (1996) Interventions and alliance formation between female Japanese macaques, *Macaca fuscata*, during homosexual consortships. *Anim. Behav.*, **52**, 539–51.

Vasey, P. L. (1998) Female choice and inter-sexual competition for female sexual partners in Japanese macaques. *Behaviour*, **135**, 579–97.

Vasey, P. L. (2004) Pre- and post-conflict interactions between female Japanese macaques during homosexual consortships. *Int. J. Comp. Psych.*, **17**, 351–9.

Vogel, C. (1976) Ökologie, Lebensweise und Sozialverhalten der grauen Languren in verschiedenen Biotopen Indiens. Fortschritte der Verhaltensforschung, Suppl. 17, *Zeitschr. Tierpsychol.* Berlin: Parey.

Wallen, K. and Parsons, W. A. (1997) Sexual behavior in same-sexed nonhuman primates: is it relevant to understanding human homosexuality? *Ann. Rev. Sex. Res.*, **8**, 195–223.

Watanabe, J. M. and Smuts, B. B. (1999) Explaining religion without explaining it away: trust, truth, and the evolution of cooperation in Roy A. Rappaport's 'The Obvious Aspects of Ritual'. *Am. Anthropol.*, **101**, 98–112.

Weber, I. and Vogel, C. (1970) Sozialverhalten in ein- und zweigeschlechtlichen Langurengruppen. *Homo*, **21**, 73–80.

Weinrich, J. D. (1980) Homosexual behavior in animals: a new review of observations from the wild, and their relationship to human sexuality. In *Medical Sexology: The Third International Congress*, eds. R. Forleo and W. Pasini, pp. 288–95. Littleton, MA: PSG Publishing.

Wickler, W. (1967) Socio-sexual signals and their intra-specific imitation among primates. In *Primate Ethology*, ed. D. Morris, pp. 69–79. Chicago: Aldine.

Wolfe, L. (1984) Mounting patterns of female Japanese macaques. *Am. J. Physical Anthropol.*, **63**, 235.

Yamagiwa, J. (1987) Intra- and inter-group interactions of an all-male group of Virunga mountain gorillas (*Gorilla gorilla beringei*). *Primates*, **28**, 1–30.

11

Playful encounters: the development of homosexual behaviour in male mountain gorillas

JUICHI YAMAGIWA

Introduction

Homosexual behaviour occurs in various primate species and exists in highly diverse and flexible forms (Vasey, 1995). It often appears in a similar form to heterosexual copulation and shows species-specific patterns in terms of courtship, duration of mounting, form of genital stimulation, positions employed and associated vocalizations. Observations from captivity (Hess, 1973) and in the wild (Harcourt *et al.*, 1980, 1981; Yamagiwa, 1987) indicate that homosexual mounting in gorillas parallels many aspects of heterosexual mounting in this species. Male homosexual behaviour is rarely observed in bisexual reproductive groups, but it occurs frequently in all-male groups of mountain gorillas in the Virunga Volcanoes (Yamagiwa, 1987).

Male homosexual behaviour is often observed among the members of all-male groups (rhesus macaques, Carpenter, 1942; Patas monkeys, Gartlan, 1974; Hanuman langur, Hrdy, 1974). In all-male groups of mountain gorillas, homosexual behaviour occurs in the complete absence of females. One study described homosexual interactions in an all-male group consisting of two silverback (adult) males and four maturing males that usually avoided contact with bisexual groups and traveled independently while keeping stable membership for at least five years (Yamagiwa, 1987). The two silverbacks had different homosexual partners from each other and made courtship approaches frequently to their partners with copulatory vocalizations. Both silverbacks were observed to ejaculate in homosexual interactions. These interactions were purely sexual in context, rather than socio-sexual (that is, a sexual behaviour incorporated into the non-sexual sphere of social communication, such as greeting, to reduce social tension, or to reinforce rank-related interactions), as has been observed in other species (rhesus macaques, Japanese macaques, Hanby, 1974; pig-tail macaques, Oi, 1991; proboscis monkeys, Yeager, 1990; white-handed gibbons, Edwards and Todd, 1991; bonobos, Kuroda, 1984; Furuichi, 1989).

Homosexual behaviour may prevent fertile males from participating in reproduction, and it seems to have negative effects on the male's reproductive fitness. Why have such behavioural tendencies been maintained throughout the evolutionary history of gorillas? How could such behaviour develop among maturing male gorillas? Nadler (1986) reported frequent genital stimulating behaviour among immature mountain gorillas and suggested that sex-related behaviour in immature individuals is a common developmental phenomenon in the great apes. Genital stimulation among immatures develops to provide a tension-reduction function in bonobos in adulthood (Hashimoto, 1997), while this behaviour rarely takes on a socio-sexual function in the adulthood of gorillas (Harcourt *et al.*, 1981). Understanding, the development process of homosexual behaviour in the great apes may provide us with important insights into the sexuality of the great apes in general. Homosexual behaviour in immature gorillas may be a developmental by-product of play copulation, which occurs in a heterosexual context. This behavioural propensity may, in turn, increase the potential for adult males to engage in homosexual behaviour in the context of skewed sex ratios. In this paper, I compare the homosexual behaviour among male gorillas in all-male groups with two types of sexual behaviour seen in reproductive groups: heterosexual copulation and genital stimulation among immature individuals. The occurrence of genital stimulation is compared among the African great apes, and its development patterns are discussed in relation to causational factors associated with homosexual behaviour.

Methods

Study area and subjects

Mountain gorillas are African apes that are terrestrial folivores, which usually form cohesive groups consisting of one or more adult males and several females with their offspring (Schaller, 1963; Fossey, 1983). Both males and females tend to immigrate from their natal groups around the time of sexual maturity, and only females immigrated into other groups (Harcourt, 1978). Males tend to travel alone before attracting females and establishing their own reproductive groups. Such solitary males occasionally form all-male groups with maturing males (Elliott, 1976; Yamagiwa, 1987; Robbins, 1995). Male gorillas were classified into various age classes including: infant (0–3 years), juveniles

(3–6 years), sub-adults (6–8 years), blackbacks (8–12 years) and silverbacks (older than 12 years).

Data were collected during an 11-month field study from December 1980 to December 1982. The study area lies within the Parc des Volcans of Rwanda and the Parc des Virungas of Democratic Republic of Congo (30°E, 1°S). The vegetational zonation has been reported elsewhere (Fossey and Harcourt, 1977; McNeilage, 2001).

The subjects were members of three groups monitored by the Karisoke Research Center. Group 5 was a multi-male group including two silverbacks, one blackback, four adult females and seven immatures. The Nk Group was a one-male group including only one silverback, six adult females and six immatures. The Pn Group was an all-male group including two silverbacks, two blackbacks and two sub-adults. An infant was born in the Nk group in March 1982, and a sub-adult male immigrated into the Pn Group in December 1982. Data on these individuals were excluded because of their short-term presence in the study groups.

These three study groups were well habituated. Dian Fossey and other members of the Karisoke Research Center recorded demographic information on these groups from 1967 onward (Fossey, 1983). The Pn Group, where homosexual behaviour was observed, was formed after the collapse of one of the former Karisoke groups in 1978 (Yamagiwa, 1987). After all of the females transferred to neighbouring groups, a young silverback (Bm), a blackback, a four-year-old juvenile male (Ts) and a three-year-old infant male remained together and moved within the former group's range. The infant died three months later. A mature silverback (Pn) joined the group in 1979, and two sub-adult males (Ah and Pt) and a blackback (Si) subsequently joined them later. The blackback emigrated from the group in 1981 and began to travel alone. When I started to observe the Pn Group in 1981, it consisted of Pn (silverback, c. 25 years old), Bm (silverback, c. 17 years old), Si (blackback, c. 11 years old), Ah (blackback, c. nine years old), Pt (sub-adult male, c. seven years old) and Ts (sub-adult male, seven years old).

Data collection

Observations were made in the daytime. The field assistants and the author tracked the study groups daily by following their trails in the dense vegetation. Data were collected by *ad libitum* sampling. All of the groups were tolerant to the close presence of the author. The author was able to individually identify all of the study animals at the time of data collection. Due to the small group size and high cohesiveness of the group members, the author was able to record almost all agonistic and non-agonistic interactions that occurred. For Group 5, 61 hours of observations were collected during 12 days. For the Nk Group, 66 hours of observations were collected during 11 days. For the Pn Group, 804 hours of observations were collected during 144 days.

Homosexual behaviour is defined as an interaction between same-sex individuals with genital contact and/or stimulation (Vasey, 1995). In gorillas, some homosexual behaviour can involve mounting with pelvic thrusts, similar to that observed during heterosexual copulation. *Play copulation* is the principal form of sex-related behaviour (Nadler, 1986), in which one animal (the actor) executed pelvic thrusts against the body of another (the recipient) during play. A *bout* is defined as a single sexual interaction, and an *episode* is defined as multiple bouts between the same individuals in which successive bouts occur within a few minutes of each other. A *break in mounting* is defined as termination of mounting by leaving of either individual. An episode of social play was counted as a series of play interactions between the same pair that lasted more than two minutes and did not include cessation for at least five minutes. When sexual or play interactions were observed, the author recorded duration of bout, initiation, break, position of their bodies (ventro-dorsal or ventro-ventral), vocalization and intrusion by third parties.

Results

Courtship

Homosexual courtship was usually initiated by the mounter's approach with copulatory pants towards the mountee. Alternatively, the mountee would approach the mounter in a slow and hesitant manner. After a solicitation approach, the approaching male would gaze, in close proximity (within 2 m), at the partner's face (Figure 11.1a). If the male who was approached opened his arms, the approaching male turned to present his back. When the male who was approached ignored the solicitation, the approaching male touched the partner's arm or shoulder and brought his face into closer proximity to the partner's face (Figure 11.1b).

Silverbacks often emitted loud copulatory pants when approaching younger partners (Figure 11.2a) and held them around their waist to mount. When their approaches were avoided, they made chest-beating displays (Figure 11.2b). In most cases, approaches with copulatory pants by silverbacks did not result in mounting. Most of the blackbacks and sub-adults did not emit copulatory pants when approaching. Only Ts gave it twice when he approached Pt.

(a)

(b)

Figure 11.1. Behaviour of seven-year old subadult male Ts. (a) Ts approaches mature silverback Pn and stares at his face. (b) Ts touches Pn's arm, emitting copulatory whimpers.

(a)

(b)

Figure 11.2. Behaviour of young silverback gorilla male Bm. (a) Bm emits copulatory pants when approaching sub-adult male Pt. (b) Chest-beating display by Bm.

Frequency and form of homosexual interactions in the all-male group

Ninety-seven episodes of homosexual interactions were observed among males of the Pn Group in 26 days during the study period (Table 11.1). Out of these, 95 episodes were observed from initiation to termination (break in mount). Unlike heterosexual copulation, the occurrence of homosexual interaction was not cyclic, but concentrated on several consecutive days. For example, homosexual interactions were observed on seven out of 14 days during which observations were made in July 1982. Episodes of homosexual behaviour occurred approximately once every eight hours of observations (0.12 episodes/h), which is higher than that of heterosexual copulations (0.029 heterosexual copulations/h) observed in three groups of mountain gorillas at the Virungas (Harcourt *et al.*, 1980). The median rate of homosexual interactions on days when they were observed (0.53 episodes/h, range: 0.16–1.80) is also higher than that recorded for heterosexual copulations (for example, 0.45 heterosexual copulations/h, Watts, 1991); 0.32–0.38 heterosexual copulations/h, range: 0.16–1.22, Harcourt *et al.*, 1980).

Homosexual mounting occurred in a similar manner as copulation. The mounter sat upright and held the mountee around the waist. The mountee squatted, with one or both hands holding the hands of the mountee. Mounting occurred in both the dorso-ventral (D-V) and ventro-ventral (V-V) positions (Figure 11.3a, b). Partners sometimes adjusted their positions before thrusting began. Fourteen out of 16 cases observed with V-V position involved a silverback and a sub-adult, while the other two were between a blackback and a subadult who subsequently switched to the D-V position. When Ah and Ts engaged in homosexual interactions, they frequently switched positions and sometimes embraced employing complex positions that were neither D-V nor V-V in form. The mounter frequently emitted copulatory pants with pursed lips (Figure 11.4). The mounteee usually emitted growls. Only during mounts between Ah and Ts did both participants simultaneously emit copulatory pants or growl, irrespective of their positions. When dismounting the mounter sometimes emitted a deep sigh, as observed following ejaculation in a heterosexual context. Signs of ejaculation (that is, semen left on the participant's body after separation) were observed once for Pn, who had mounted Ts, and once for Bm, who had mounted Si.

The median duration of homosexual mounting was 75 seconds (range: 5–612), which is a little shorter than the median duration of copulation (96 seconds, $N = 11$, range: 30–310). The duration of some homosexual mounts were outside the range observed for copulations. Those less than 30 seconds involved at least

Table 11.1 *Comparison of heterosexual copulations, homosexual mountings and genital stimulations among immatures*

	Duration (sec)		Initiation (%)			Position (%)			Vocalization (%)				Break mount (%)		
	N	M (Range)	N	A	B	N	V-D	V-V	N	A	E	B	N	A	B
Copulation*	11	96 (30–310)	16	25	75	69	100	0	16	50	25	25	10	20	80
Homosexual (total)	95	75 (5–612)	94	51	49	97	84	16	81	60	35	5	84	6	94
With silverback	40	152 (12–612)	39	18	82	42	67	33	42	95	5	0	42	5	95
Among blackbacks and sub-adults	55	30 (5–210)	55	75	25	55	96	4	39	23	67	10	42	7	93
Genital stimulation among immatures	25	20 (5–263)	25	100	0	28	96	4	16	94	6	0	28	32	68

Notes: M, Median; A, Mounter; B, Mountee.

*Data on copulation from Harcourt *et al.* (1980, 1981). Initiation was calculated from the number of approaches to a partner within 2 m. Vocalization was rated as the proportion of cases in which the mounter emitted more frequently (A), the mountee emitted more frequently (B), or both emitted evenly (E).

(a)

(b)

Figure 11.3. Mounting positions. (a) Blackback gorilla male Si mounts sub-adult male Ts dorso-ventrally; (b) Young silverback gorilla male Bm mounts sub-adult male Pt ventro-ventrally.

Figure 11.4. Young silverback gorilla male Bm emits copulatory pants with pursed lips.

one sub-adult, and those longer than 310 seconds involved at least one silverback. The median duration of homosexual mounting involving silverback males (152 seconds, $N = 40$, range: 12–612) was significantly longer than that involving blackbacks and sub-adults (30 seconds, $N = 55$, range: 5–210) (Mann–Whitney U test, $Z = -5.48$, $p < 0.0001$).

Homosexual mounting was initiated by both mounters and the mountees (Table 11.1). This contrasts with heterosexual copulations during which females (the mountee) initiated more copulations (75%, Harcourt *et al.*, 1980; 63%, Watts, 1991). Homosexual mounting involving silverbacks was initiated by the mountee more frequently ($N = 39$, 82%), and those not involving silverbacks were initiated by the mountee less frequently ($N = 55$, 25%). The difference between the two was significant ($x^2 = 27.03$, df $= 1$, $p < 0.0001$). Most cases of homosexual mounting were broken off by the mountee, as observed during heterosexual copulation (Harcourt *et al.*, 1981). Neither of the two silverbacks was ever mounted.

Some unusual variations in the typical sequence of homosexual interactions were also observed. Si was mounted by Bm, while he mounted Ts. Among Ah, Pt and Ts, one mounted the other in some episodes, while being mounted in other episodes. Ah and Ts frequently exchanged the role of mounter and mountee during the same mounting episode.

Same-sex partner preference and intrusion
on homosexual interactions

The intensity of courtship and the responses of individuals varied depending upon the partners involved. Neither courtship, nor mounting, was observed in four particular pairs (Pn–Bm, Pn–Si, Bm–Ah and Si–Ah). In another three pairs (Pn–Ah, Pn–Pt, and Si–Pt), courtship was observed but not mounting. For example, Ah and Pt totally avoided or ignored Pn's approaches accompanied with copulatory pants.

The two silverbacks seemed to compete for homosexual partners at the beginning of the observation period, but gradually became tolerant of each other's 'ownership' of partners. In December 1981, Bm started to make a courtship approach to Pt, who usually ignored or avoided the solicitations. Pn also made a homosexual approach to Pt, but the frequency was very low (about ten times lower than that of Bm). Bm carried out mounting forcibly at least twice with Pt in the V-V position. Bm succeeded in mounting six times with Pt in December 1981. However, Pt was never observed to solicit Bm, and, instead, frequently avoided Bm and stayed near Pn. Then Bm began chest-beating displays including ground thumping and branch breaking accompanied by hoot vocalizations. Pn also made the same display in response. Intensive fights occurred between Bm and Pn several times, and both were wounded (Yamagiwa, 1987). Pn stopped making courtship approaches to Pt in February 1982. Bm continued to follow Pt while emitting copulatory pants until the end of the study period (December 1982), although no mounting was observed after January 1982.

In contrast, when Pn increased courtship approaches toward Ts in July 1982, Bm did not increase chest-beating displays near Pn. Bm mounted Ts twice only briefly (12 and 13 seconds in duration) in July, but no mounting was observed between them afterwards. No violent fight was observed between Pn and Bm from July until November 1982. It seems likely that Ts's solicitations towards Pn resulted in Bm's tolerance towards homosexual interactions between Pn and Ts. Similarly, when Si showed soliciting behaviour to Bm, Pn did not show courtship to Si or aggression to either Si or Bm. On the other hand, when a new sub-adult male (Ht) immigrated in December 1982 into the all-male study group, both Pn and Bm made intensive courtship approaches to him with copulatory pants. Frequent exchanges of chest-beating displays occurred between the silverbacks in proximity to Ht, and both silverbacks were wounded, although fierce fighting between them was not directly observed. Ht was never observed to show soliciting behaviour to either silverback. In contrast with Ts, the absence of soliciting behaviour by Pt and Ht probably prevented the silverbacks from establishing and recognizing a priority over homosexual partners.

Table 11.2 Number of mounting episodes with pelvic thrusts observed in homosexual interactions

Mounter (A)	Mountee (B)	No. observed	Duration (sec)		Initiation (%)			Position (%)			Vocalization (%)			Break mount (%)		
			N	M (range)	N	A	B	N	V-D	V-V	N	A	B	N	A	B
Pn	Ts	21	20	211 (30–360)	21	0	100	21	48	52	21	95	5	21	0	100
Bm	Si	13	12	170 (48–456)	10	20	80	13	100	0	13	100	0	13	15	85
Bm	Pt	6	6	88 (29–612)	6	83	17	6	50	50	6	100	0	6	0	100
Bm	Ts	2	2	15 (12–18)	2	0	100	2	100	0	2	50	50	2	0	100
Si	Ts	9	9	21 (12–90)	9	78	22	9	100	0	4	0	100	9	0	100
Ah	Pt	1	1	6	1	100	0	1	100	0	0	0	0	0	0	0
Pt	Ah	1	1	48	1	100	0	1	100	0	1	100	0	0	0	0
Ah	Ts	26	26	72 (5–210)	26	65	35	26	92	8	19	0	100	18	11	89
Ts	Ah	5	5	14 (5–209)	5	80	20	5	100	0	5	20	80	5	0	100
Pt	Ts	3	3	21 (15–90)	3	67	33	3	100	0	1	0	100	2	0	100
Ts	Pt	10	10	28 (24–90)	10	90	10	10	100	0	9	78	22	8	13	87

Blackbacks may also compete over the younger homosexual partners. Their intrusions were related to their dominance rank. Si intruded in Ah's homosexual interactions with Pt and Ts nine times, and Ah intruded in homosexual interactions between Pt and Ts four times. Both silverbacks rarely intruded on homosexual interactions between younger males. Pn was never observed to intrude in such interactions, and Bm was observed to do so only once, during an interaction between Ah and Pt. The sub-adult, Ts, who was the youngest male, was the most attractive as the homosexual partner. He was involved in 78% of the total homosexual interactions, in which he mated with all members of the group (Table 11.2). Ts was also involved in most (78%) of play episodes (364) observed during the study period. It seems likely that Ts was the most attractive partner in both homosexual and play interactions.

Play copulation among immatures in bisexual groups

Genital stimulation with pelvic thrusting was also observed among immatures in the two bisexual groups. In its principal form, one individual (the mounter) executes pelvic thrusts against the body of another (the mountee) as defined by Nadler (1986), who observed genital stimulation episodes in the same groups one year before my study. I observed 22 episodes in Group 5 during seven days out of 12 days of observations, and 13 episodes in the Nk Group during three days out of 11 days of observations (Tables 11.3, 11.4). Six episodes between silverbacks and adult females and one episode between a silverback and a sub-adult female were considered copulations. The other 28 episodes involving immatures occurred in the context of play, including chasing, wrestling and embracing (Table 11.1). The frequency of genital stimulation episodes involving immatures recorded during this study (0.22 episodes/h), is similar to that reported by Nadler (1986) for the same troop (0.13 episodes/h). The median rate of genital stimulation episodes among immatures on days that it was observed is 0.51 episodes/h (range: 0.36–0.82).

Genital stimulation episodes among immatures were initiated by the mounter and broken by the mountee in most cases (68%). The mounter was male in all cases, and the mountee was female in 25 out of 28 cases (89%). However, since all five infants were females and all juveniles were males in both groups, data are biased. Mounting occurred between males in three cases: Pb (sub-adult) mounted Sh (juvenile), Ca (juvenile) mounted Sh (juvenile) and Bo (juvenile) mounted Dy (juvenile). Females were never observed to mount others, irrespective of their age. The mounter was usually older than the mountee, except for two cases: Pb (7.9 years old male) mounted Tu (10.2 years old female), and Ca (3.8 years old male) mounted Sh (5.5 years old male) in Group 5. Tu was

Table 11.3 *Frequency of genital stimulation episodes with thrusting in Group 5*

Mounter	Mountee	Bv	Ic	Zz	Ef	Pa	Pu	Tu	Pb	Pp	Sh	Mu	Ca	Mg	Jz
Bv	Silverback														
Ic	Silverback							1		1					
Zz	Blackback							3		1					
Ef	Adult female														
Pa	Adult female														
Pu	Adult female														
Tu	Adult female*														
Pb	Sub-adalt male							1		2	1				1
Pp	Sub-adult female														
Sh	Juvenile male													1	2
Mu	Juvenile female														
Ca	Juvenile male										1			4	3
Mg	Infant female														
Jz	Infant female														

Note: *Nulliparous female.

Table 11.4 *Frequency of genital stimulation episodes with thrusting in Nk Group*

Mounter	Mountee	Nk	Pe	Pd	Fu	Sb	Au	Bo	Su	Dy	BP	Tx	Jn
Nk	Silverback		3	2									
Pe	Adult female												
Pd	Adult female												
Fu	Adult female												
Ps	Adult female												
Sb	Adult female												
Au	Adult female												
Bo	Juvenile male									1	1		
Su	Juvenile male									1			
Dy	Juvenile male										3	2	
BP	Infant female												
Tx	Infant female												
Jn	Infant female												

a nulliparous female who was observed playing frequently with Pb. Sh had been orphaned several years earlier and was relatively small for his age, only as large as Ca. The youngest ages of mounter and mountee were 3.1 years and 1.6 years, respectively.

The mounter emitted copulatory pants or whimpers in 16 cases (57%), while the mountee did not give any vocalization, except for the case between Zz

Table 11.5 *Occurrence of genital stimulation episodes in context of play*

Pair	During	Around	Outside	% in play
Copulation[a]			7	0
Silverback–blackback[b]			13	0
Silverback–sub-adult male[b]			29	0
Blackback–adult or Sub-adult female	3	1		18
Blackback–sub-adult male[b]	22	13	7	15
Sub-adult male–adult female	1			20
Sub-adult male–sub-adult female	2			40
Sub-adult male–sub-adult male[b]	11	2		13
Sub-adult male–juvenile male[b]	1			4
Sub-adult male–infant female	1			25
Juvenile male–juvenile male[b]	2			3
Juvenile male–infant female	17			22

Notes: [a]Silverbacks with adult and sub-adult females.

[b]Homosexual mount.

During play: copulation occurred within 2 minutes before or after play.

Around the time that play occurred: Copulation occurred within 2–5 minutes before or after play.

Outside of play: copulation occurred more than 5 minutes before or after play.

(blackback) and Tu (nulliparous female), who both emitted copulatory pants. The median duration of the genital stimulation episodes among immatures was 20 seconds (range: 5–263), which is significantly shorter than that of copulation (92 seconds, $N = 7$, range: 69–165) observed in the two study groups (Mann–Whitney U test, $Z = -2.74$, $p = 0.0062$). Genital stimulation among immatures occurred in the D-V position with the exception of Ca who mounted Sh once in the V-V position before subsequently taking the D-V position.

Genital stimulation in context of play in all-male and bisexual groups

Silverbacks and blackbacks did not engage in genital stimulation with immature males in the bisexual groups. The sub-adult male (Pb) engaged in genital stimulation on several occasions with immature females (four times), but did so only once with an immature male. These observations suggest that post-pubertal males prefer female sexual partners over males.

Genital stimulation episodes including copulation and homosexual interaction are divided according to its occurrence in the context of play (Table 11.5).

The episodes that did not involve adult males (silverbacks or blackbacks) were observed only during play or around the time that play occurred (within five min before or after play). The episodes that involved silverbacks always occurred in a context other than play (more than five min before or after play). In contrast, most episodes involving blackbacks (85%), and all episodes between sub-adult males occurred during or around the time that play occurred. It seems likely that genital stimulation behaviour may develop out of play in the age class of blackbacks. During play, blackbacks sometimes tried to change the context to sexual interactions. In one case, after a short period of play, Ah (blackback) solicited Ts (sub-adult male) with a stare and pulled Ts down to mount him. Ah may possibly have solicited Ts by staring in order for him to mount (Yamagiwa, 1992). However, Ts gave chuckle vocalization (play pants), instead of presenting with copulatory pants. Ah avoided this contact with Ts and chest beated. In contrast, in another case, Ts interrupted play with Pt (sub-adult male) and solicited him with a stare. Pt presented to Ts, who mounted him with pelvic thrusts, while emitting copulatory pants. The former case suggests that Ah may have tried to change the context from a play interaction to a homosexual interaction, but Ts refused or misunderstood his solicitation made through staring. The latter case shows how the nature of a single interaction can change from play to homosexual courtship or mounting.

Median duration of homosexual interactions between blackbacks and sub-adult males outside of a play context (95 seconds, $N = 7$, range: 33–167) and around the time that play occurred (43 seconds, range: 14–209) were significantly longer than those during play (18.5 seconds, range: 5–210) (Mann–Whitney U test, $Z = -2.88$, $p = 0.004$; $Z = -2.32$, $p = 0.02$, respectively, see Figure 11.3).

Genital stimulation among males occurred throughout the development of gorillas and was not limited to any particular phase of development (Table 11.5). After reaching the blackback age class, genital stimulation gradually occurs more and more outside of a play context and is probably associated with sexual arousal among older participants (that is, blackbacks, silverbacks).

Discussion

Features of homosexual interaction among male gorillas

The frequency of homosexual behaviour by silverbacks is similar to that of heterosexual copulation in bisexual groups (Harcourt *et al.*, 1981). The two silverbacks observed in this study engaged in competition over sub-adult partners for homosexual interactions. They displayed intensively and fought with each other to the point where both were wounded. Strong competition between silverbacks may erupt if same-sex partners do not target a single silverback with solicitations to mount them. In such instances, silverbacks may be uncertain as to who has priority of access to a particular same-sex partner. This appeared to be the case for Pt and Ht, both sub-adults, who did not solicit either silverback. The silverbacks subsequently engaged in severe competition for access to these sub-adult male sexual partners. In contrast, when Ts, another sub-adult, and Si, a blackback, showed solicitation to one of the silverbacks, competition for these sexual partners did not erupt. Neither silverback showed intensive courtship to another silverback's partner. There appeared to be a kind of respect for 'ownership' of other's same-sex partners, as observed in heterosexual copulation (Harcourt, 1978, 1981; Watts, 1996). These observations suggest that homosexual interactions involving silverbacks may be purely sexual in character, because they parallel heterosexual interactions.

Although blackbacks intruded on younger individuals' homosexual mounting interactions, sub-adults never did so. Another difference in homosexual interactions between silverbacks and immatures was the fixed role in mounting for the former. The two silverbacks always mounted the younger individuals, while both blackbacks and sub-adults took the roles of both mounter and mountee. Si was mounted by Bm and mounted Pt. Ah, Pt and Ts mutually changed roles in their interactions. This role switching was also observed in play interactions among immature male gorillas in the all-male group (Yamagiwa, 1987). Most of the homosexual episodes among immatures (87%) occurred in the context of play and thus it seems likely that male homosexual behaviour first develops in the context of play among gorillas.

Genital stimulation in immature African ape males

Genital stimulation occurs in context of play in all of the African great apes (Tutin, 1979; Tutin and McGinnis, 1981; Harcourt et al., 1981; Nadler, 1986; Enomoto, 1990; Kano, 1992; Hashimoto, 1997). Frequent genital stimulation behaviour may be a common feature of the African great apes in the early stage of development. During the first year of life, male infant chimpanzees show an interest in the sexual swellings of adult females, and successfully achieve intromission before reaching two years old (Tutin and McGinnis, 1981). Male infant bonobos begin to show sexual behaviour at less than one year of age and have been observed trying to copulate with their mothers (Kano, 1992). Immature gorillas started to show genital stimulation behaviour during the period of infancy (Nadler, 1986; this study).

Male immatures tend to initiate and participate in play interaction involving genital stimulation more frequently than female immatures (Nadler and Braggio,

1974; Freeman and Alcock, 1973; Hayaki, 1985a; Hashimoto, 1997). However, the age/sex class in which genital stimulation occurs most frequently varies depending on the great ape species under consideration. Juvenile male chimpanzees direct mounting with thrusting to younger individuals of both sexes during play (Tutin and McGinnis, 1981), but adolescent males are attracted to estrous females, who frequently initiate and break-up copulations with them (Hayaki, 1985b; Goodall, 1986). The most frequent partners of immature male bonobos in genital stimulation episodes are adult and adolescent females (in 73% of total episodes, Kano, 1992). Copulation-like genital contacts between immature male bonobos with mature females tend to increase with the male's age and to gradually take on the function of regulating inter-individual relationships between late juvenile and early adolescent periods (Hashimoto, 1997).

In contrast, adult gorillas do not show interest in genital stimulation with juveniles or infants. When copulation occurs, infant gorillas always watch the participants closely, as observed in chimpanzees and bonobos (Tutin and McGinnis, 1981; Kano, 1992). However, they do not interfere with copulating individuals nor do they exhibit any overt interest in the genitals of the adults (Nadler, 1986). Immature gorillas tend to show genital stimulation behaviour with each other. This may be owing to the lack of swelling of the female's sexual skin and, in the absence of such a sexual signal, immature male gorillas may not become sexually aroused in the presence of females. Maturing males gradually lose sexual interest in genital stimulation with immatures. Sub-adults and blackbacks then develop an interest in the soliciting sex from estrous females. However, even adult male gorillas occasionally exhibit interest in genital stimulation with immatures, and solicit them in the absence of females. Since the genital stimulation behaviour of gorillas has no socio-sexual functions, adult males show such behaviour in a purely sexual context. Frequent courtship and chest-beating displays by the two silverbacks around the young partners and intensive fights between them observed in the all-male group show that homosexual behaviour did not have tension-reduction role but, instead, resulted in increased competition between silverbacks over same-sex partners.

It is likely that the homosexual behaviour of immature gorillas is a developmental by-product, which may increase the potential of adult males to engage in homosexual behaviour in the context of skewed sex ratios. After six years of independent travel, the all-male group observed in this study disintegrated. Four males (Si, Ah, Pt and Ht) left the group as solitary males, and two males (Bm and Ts) formed a reproductive group with females. Only the oldest male (Pn) continued to stay in an all-male group, acquiring juvenile and sub-adult males (Robbins, 1995). The formation of all-male groups and homosexual behaviour may possibly contribute to the survival of younger males for future reproduction.

Conclusions

Homosexual interactions occurred in the all-male group more frequently than heterosexual copulations in the reproductive groups. The forms of homosexual interactions were similar to those of heterosexual copulations, in terms of courtship, duration of episode, initiation, break in mounting and vocalizations. In some instances, sub-adults directed sexual solicitations towards silverback males who, in turn, competed for exclusive access to the sub-adults. Most of the homosexual interactions among blackbacks and sub-adults occurred in the context of play. The form and pattern of sexual partner preference observed among immatures in the all-male group resembled genital stimulation among immatures in reproductive groups. Male immatures tended to show genital stimulation more frequently during play than female immatures.

Immature male chimpanzees and bonobos show genital stimulations with immatures and adults in the early development phase. However, the forms of these interactions and choice of partners change around puberty. Post-pubertal male chimpanzees and bonobos may focus their sexual interests on females. In contrast, immature male gorillas usually showed genital stimulation with other immatures. Immature male gorillas may develop same-sex genital stimulation behaviour that is purely sexual during play interactions and may show such behaviour in adulthood under particular demographic conditions, as observed in the all-male group.

References

Carpenter C. R. (1942) Sexual behavior of free ranging rhesus monkeys (*Macaca mulatta*): II. Periodicity of estrus, homosexual, autoerotic and non-conformist behavior. *J. Comp. Psychol.*, **33**, 143–62.

Edwards, A. M. A. R. and Todd, J. D. (1991) Homosexual behavior in wild white-handed gibbons (*Hylobates lar*). *Primates*, **32**, 231–6.

Elliott, R. C. (1976) Observations on a small group of mountain gorillas (*Gorilla gorilla beringei*). *Folia Primatol.*, **25**, 12–24.

Enomoto, T. (1990) Social play and sexual behavior of the bonobo (*Pan paniscus*) with special reference to flexibility. *Primates*, **31**, 469–80.

Fossey, D. (1983) *Gorillas in the Mist*. Boston, MA: Houghton Mifflin.

Fossey, D. and Harcourt, A. H. (1977) Feeding ecology of free-ranging mountain gorilla (*Gorilla gorilla beringei*). In *Primate Ecology*, ed. T. H. Clutton-Brock, pp. 415–47. New York: Academic Press.

Freeman, H. E. and Alcock, J. (1973) Play behaviour of a mixed group of juvenile gorillas and orangutans. *Int. Zoo Yb.*, **13**, 189–94.

Furuichi, T. (1989) Social interactions and the life history of female *Pan paniscus* in Wamba, Zaire, *Int. J. Primatol.*, **10**, 173–97.

Gartlan, J. S. (1974) Adaptive aspects of social structure in *Erythrocebus patas*. In *Proceedings from the Symposia of the Fifth Congress of the International Primatological*

Society, eds. S. Kondo, M. Kawai, A. Ehara and S. Kawamura, pp. 161–71. Tokyo: Japan Science Press.

Goodall, J. (1986) *The Chimpanzees of Gombe: Patterns of Behaviour*. Cambridge, MA: Harvard University Press.

Hanby, J. P. (1974) Male–male mounting in Japanese monkeys (*Macaca fuscata*). *Anim. Behav.*, **22**, 836–49.

Harcourt, A. H. (1978) Strategies of emigration and transfer by primates, with particular reference to gorillas. *Z. Tierpsychol.*, **48**, 401–20.

Harcourt, A. H. (1981) Intermale competition and the reproductive behavior of the great apes. In *Reproductive Biology of the Great Apes*, ed. C. Graham, pp. 301–18. New York: Academic Press.

Harcourt, A. H., Fossey, D. and Stewart K. J. (1980) Reproduction by wild gorillas and some comparisons with the chimpanzee. *J. Reprod. & Fertility*, Supplement, **28**, 59–70.

Harcourt, A. H., Stewart, K. J. and Fossey, D. (1981) Gorilla reproduction in the wild. In *Reproductive Biology of the Great Apes*, ed. C. Graham, pp. 265–79. New York: Academic Press.

Hashimoto, C. (1997) Context and development of sexual behavior of wild Bonobos (*Pan paniscus*) at Wamba, Zaire. *Int. J Primatol.*, **18**, 1–21.

Hayaki, H. (1985a) Social play of juvenile and adolescent chimpanzees in the Mahale Mountain National Park, Tanzania. *Primates*, **26**, 343–60.

Hayaki, H. (1985b) Copulation of adolescent male chimpanzees, with special reference to influence of adult males, in the Mahale National Park, Tanzania. *Folia Primatol.*, **44**, 148–60.

Hess, J. P. (1973) Some observations on the sexual behavior of captive lowland gorillas, *Gorilla g. gorilla* (Savage and Wyman). In *Comparative Ecology and Behavior of Primates*, eds. R. P. Michael and J. H. Crook, pp. 507–81. London: Academic Press.

Hrdy, S. B. (1974) Male–male competition and infanticide among the langurs (*Presbytis entellus*) of Abu, Rajasthan. *Folia Primatol.*, **22**, 19–58.

Kano, T. (1992) *The Last Ape: Pygmy Chimpanzee Behavior and Ecology*. Stanford, CA: Stanford University Press.

Kuroda, S. (1984) Interactions over food among pygmy chimpanzees. In *The Pygmy Chimpanzee*, ed. R. L. Susman, pp. 301–24. New York: Plenum.

McNeilage, A. (2001) Diet and habitat use of two mountain gorilla groups in contrasting habitats in the Virunga. In *Mountain Gorillas: Three Decades of Research at Karisoke*, eds. M. M. Robbins, P. Sicotte and K. J. Stewart, pp. 266–92. Cambridge: Cambridge University Press.

Nadler, R. D. (1986) Sex-related behavior of immature wild mountain gorillas. *Developm. Psychol.*, **19**, 125–37.

Nadler, R. D. and Braggio, J. T. (1974) Sex and species differences in captive-reared juvenile chimpanzees and orangutans. *J. Hum. Evol.*, **3**, 541–50.

Oi, T. (1991) Non-copulatory mounting in wild pig-tailed macaques (*Macaca nemestrina nemestrina*) in West Sumatra, Indonesia. In *Primatology Today*, eds. A. Ehara, T. Kimura, O. Takenaka and M. Iwamoto, M., pp. 147–50. Amsterdam: Elsevier.

Robbins, M. M. (1995) A demographic analysis of male life history and social
    structure of mountain gorillas. *Behaviour*, **132**, 21–47.

Schaller, G. B. (1963) *The Mountain Gorilla: Ecology and Behavior*. Chicago: University of
    Chicago Press.

Tutin, C. E. G. (1979) Responses of chimpanzees to copulation, with special reference
    to interference by immature individuals. *Anim. Behav.*, **27**, 845–54.

Tutin, C. E. G. and McGinnis, P. R. (1981) Chimpanzee reproduction in the wild. In
    Reproductive Biology of the Great Apes, ed. C. Graham, pp. 239–64. New York:
    Academic Press.

Vasey, P. L. (1995) Homosexual behavior in primates: a review of evidence and
    theory. *Int. J. Primatol.*, **16**, 173–204.

Watts, D. P. (1991) Mountain gorilla reproduction and sexual behavior. *Am. J.
    Primatol.*, **24**, 211–25.

Watts, D. P. (1996) Comparative socio-ecology of gorillas. In *Great Ape Societies*, eds.
    W. C. McGrew, L. F. Marchant and T. Nishida, pp. 16–28. Cambridge: Cambridge
    University Press.

Yamagiwa, J. (1987) Intra- and inter-group interactions of an all-male group of
    Virunga mountain gorillas (*Gorilla gorilla beringei*). *Primates*, **28**, 1–30.

Yamagiwa, J. (1992) Functional analysis of social staring behavior in an all-male
    group of mountain gorillas. *Primates*, **33**, 523–44.

Yeager, C. P. (1990) Notes on the sexual behavior of the proboscis monkey (*Nasalis
    larvatus*). *Am. J. Primatol.*, **21**, 223–7.

12

Social grease for females? Same-sex genital contacts in wild bonobos

BARBARA FRUTH AND GOTTFRIED HOHMANN

Introduction

Bonobos are famous for their sexual behaviour (de Waal, 1987, 1995) which is characterized by high frequency, various mating positions, combination of partners and an overall ease with which sex is incorporated into the daily routine. Indeed, a connection has been drawn between the limited use of aggression in bonobo societies and their sex life – including homosexual behaviour (de Waal, 1987, 1995; Wrangham, 1993). The apparently peaceful life of these African apes evoked slogans such as 'make love not war' and is often contrasted with the rather aggressive life style of their sister species, the chimpanzee.

The social structure of a given species is reflected in the dominance relations between its members. Status is either expressed through agonistic behaviour of dominant individuals, or through submissive displays of subordinates (East *et al.*, 1993). In the male dominated chimpanzee society, relative status is determined through fights. Once decided, it is kept up by signals of submission that involve a variety of postures and gestures such as kisses, embraces, presentations and mounts, as well as vocal displays such as pant grunts, squeaks or screams (detailed descriptions in Goodall, 1968). Asymmetries in such displays express status differences (Rowell, 1966; Bygott, 1979; Colmenares, 1990). Records of pant grunts, for example, have allowed scientists to reconstruct a linear dominance hierarchy within a given chimpanzee community.

The degree to which bonobos and chimpanzees differ is still a matter of debate (Boesch *et al.*, 2002; Stanford, 1998). Like chimpanzees, bonobos also present, mount, embrace and vocalize. They do not, however, use pant grunts, which makes it difficult to delineate dominance hierarchies in the same way as for chimpanzees. This has led to descriptions of their society as egalitarian. The lack of formalized signals of dominance (de Waal and van Roosmalen, 1979) was understood to result from low-feeding competition due to superabundant food patches in their habitat (Wrangham, 1993; White and Lanjouw, 1992). Nevertheless, rank differences still exist and can be inferred through differential access to food or initiative for travel and rest.

Another reflection of social status might be the patterning of sexual interactions, in particular sexual contacts between members of the same sex. If invited for a hug by another female, bonobo females add lateral movements which culminate in so-called genito-genital rubbing ('gg-rubbing'; Kuroda, 1980). Same-sex encounters between males include 'french kissing', fellatio and genital massage (de Waal and Lanting, 1997). Homosexual interactions are very rare amongst chimpanzees, but rather common amongst bonobos. A further important difference is that chimpanzees tend to copulate around the time of ovulation, which is associated with large ano-genital swellings, whereas bonobos also mate and engage in homosexual behaviour outside the period of maximum tumescence (Furuichi, 1987).

Numerous hypotheses have been put forward to understand the evolution of homosexual behaviour (see, for example, Vasey, 1995; Kirkpatrick 2000). Data on bonobos seem to be particularly suitable to test explanations related to the expression of social status with its corollaries of tension regulation (Hanby, 1977) and reconciliation (Kappeler and van Schaik, 1992; Silk, 1998). It is known that genital displays and/or elements of mating behaviour may be part of ritualized dominance interactions; in particular, the mounter is often considered to be dominant over the mountee (Kummer *et al.*, 1974; Fox and Cohen, 1977; Nadler, 1990). For bonobos, same-sex genital contacts have been interpreted as an alternative for aggression and a peace-making ritual, particularly in the context of feeding (de Waal, 1987, 1989; Kuroda, 1980).

In a more general way, bonobos have also been thought to use sexual behaviour to regulate their social relations (Wrangham, 1993; de Waal, 1995), for example, in the contexts of greeting (Kano, 1980, 1989) and affiliation (Furuichi, 1989; Kano, 1980; Kuroda, 1984; White; 1988; White and Lanjouw, 1992). Homosexual behaviour has been viewed as a tool to ease immigration of female strangers (Furuichi, 1989) and to function as an expression of female–female cohesion and affiliation (Parish, 1994, 1996). Another line of reasoning proposes that female–female genital contacts facilitate heterosexual mating, in that they arouse males because of their similarity to male–female copulations (Ford and Beach, 1952; Parker and Pearson, 1976).

Most descriptions of bonobo sexual behaviour come from captivity. We will present long-term data from the wild, with a special focus on the potential function of same-sex encounters and a comparison with chimpanzees, the sister species. Bonobos and chimpanzees shared a last common ancestor with humans 6–7 million years ago, it is therefore conceivable that the comparative aspect of our work may ultimately also contribute to an understanding of the evolution of same-sex sexual behaviour in our own species.

Methods

Socio-ecology of bonobos

The genus *Pan* includes chimpanzees (*P. troglodytes*) with subspecies in Central, West and East Africa, and bonobos (*P. paniscus*) which are restricted to the 'cuvette centrale' or central Congo Basin south of the Congo river. Both live on ripe fruit, leaves, flowers, piths and sprouts but enrich their basically vegetarian diet with the consumption of animal protein acquired through hunts of arboreal monkeys (chimpanzees: Boesch and Boesch, 1989, Stanford, 1995) or ground-living duikers (bonobos: Hohmann and Fruth, 1993; Fruth, 1998; Fruth and Hohmann, 2002). Prey may be shared among community members.

Similar to chimpanzees, bonobos live in communities of up to 50 males, females and their offspring which occupy home-ranges of 15–50 km². These communities split up into several parties or subgroups which use diverse parts of the home range independently from each other. Individuals of either sex have almost complete freedom to come and go as they wish ('fission–fusion' social organization, Goodall, 1986). It was previously thought that females emigrate from their natal community (female dispersal), while males stay within it (male philopatry). However, it has become evident that both males and females migrate, and that the average genetic relatedness among males does not differ significantly from that among females (Gerloff *et al.*, 1999; Hohmann, 2001).

In contrast to chimpanzees, bonobos maintain close relationships between females and their adult sons as well as between males and unrelated females (Ihobe, 1992; Hohmann *et al.*, 1999). Moreover, unlike in the male-centered chimpanzee communities, bonobo societies are female dominated. Females cooperate to achieve this, while males tend to compete amongst each other. Even some adolescent females may dominate males, although the social hierarchy is a complex system which depends on sex, age, character and individual social competence.

Study site and subjects

We studied bonobos of the Eyengo community, in the eastern part of the Lomako study site in the Democratic Republic of Congo (Badrian and Badrian,

1984). The site covers about 35 km^2. Most of the habitat is primary, climax, evergreen, polyspecific forest, interspersed with swamp, slope forest and secondary forest.

Over the years, the Eyengo community (for details see Fruth, 1995; Hohmann *et al.*, 1999) has consisted of ten mature (adolescent, adult) and ten immature (infant, juvenile) males, and 20 mature and ten immature females. Three identified mature individuals (one male, two females) of neighbouring communities contributed to the current sample. All individuals were identified by phenotypic traits (disfigured limbs, shape of sexual swellings, pigmentation). The age of animals born before this study was estimated by traits such as body size, body proportions, condition of teeth, length of mamillae and genital swelling patterns.

General behavioural data derived from eight field seasons (1990–1998; total 41 months), while detailed analyses are restricted to six field seasons (1993–1998) lasting between two and nine months (total 27 months). Data collection involved *ad libitum*, focal animal and event sampling (Altmann, 1974). Not all parameters could be recorded for each type of interaction. Thus, sizes of sub-samples vary accordingly.

Results

Sexual behaviour: baseline data

Between 1990 and 1998 we recorded 1201 sexual contacts at Lomako. Of 53 identified individuals of all age sex classes, 49 were observed engaged in sexual activities at some point. A total of 55% of these interactions were homosexual, and 45% heterosexual. Choice of partner, type of sex and age showed some distinct patterning.

Males engage from early infancy in heterosexual activities, while females increase their participation slowly (Figure 12.1a). In contrast, homosexual behaviour of males is rare throughout all ages, while it increases from childhood onwards amongst females (Figure 12.1b).

Focusing on the Eyengo community, homosexual genital contacts were recorded for 44 of 50 known individuals, from which 37 were adult or adolescent respectively. All age classes were involved, although dyads between adults were most common, followed by those between adults and adolescents (Figure 12.2).

Only 4.2% of all homosexual interactions ($n = 661$) occurred between males. Most were mounts with pelvic thrusts; intromission was rare ($n = 22$). One was a copulation (ventro-dorsal mount with intromission), two were genital contacts

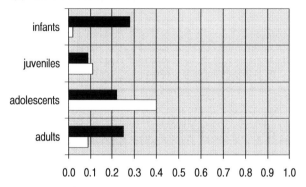

(a) hetero-sexual behaviour

(b) homo-sexual behaviour

Rate of genital contacts

Figure 12.1. Rates of sexual behaviour (genital contacts/observation day) of males (black bars) and females (white bars) of different age classes across all years of observation.

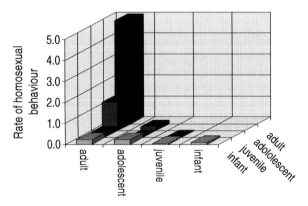

Figure 12.2. Rate of homosexual behaviour (genital contacts/observation day) between different age classes. $n = 635$ dyads including 44 individuals.

with lateral movements comparable to female gg-rubbing, and two were rump–rump contacts.

The vast majority (95.8%) of homosexual interactions occurred between females (Figure 12.3). Most were gg-rubbings in which females embraced ventro-ventrally and rubbed their genital swellings laterally against each other. Only twice did females engage in pelvic thrusts rather than in lateral movements, once ventro-ventrally, once ventro-dorsally.

Gg-rubbing ($n = 470$) occurred in two positions. A small proportion (8%) was performed vertically, with both females exposing their front side to each other, supporting themselves with their hands to the back, or clinging to nearby branches. Most (92%), however, occurred horizontally, with one female's hands behind her back presenting her belly and a second female superposing her body belly-to-belly with both feet and hands on the ground so that the bottom-female embraced the top one and clung to her body, a clinging similar to that of a child in carrying position.

The female who presented her belly and genitals usually initiated the interaction. Females who were invited to superpose ($n = 332$) either mounted (90%) or moved away (10%). When invitations were ignored, solicitors showed signs of frustration such as body rocking, pout face, whimpering or threw tantrums, and sometimes directed aggression against another individual. Occasionally ($n = 27$ times), gg-rubbing appeared to be enforced by the dominant.

Three quarters of all genital contacts ($n = 608$) were silent (75.3%). The rest (24.7%) were accompanied by shrill vocalizations of one or both individuals. Moreover, females often showed an open mouth grin, independent of whether or not they vocalized.

Most genital contacts ($n = 484$) occurred in the context of feeding (79.5%), including contests for food, while 7.0% were related to resting, travel or play. Different contexts overlapped in the remaining cases (13.5%). Female genital contacts averaged one per hour of observation, but striking individual differences were apparent (Figure 12.4).

Homosexual behaviour and social status

If genital contacts serve to demonstrate superiority, one would predict high-ranking individuals to initiate the behaviour more often than low-ranking ones. If the behaviour signals active submission, then the reverse is expected.

Female–female genital contact was usually invited by a ventral presentation, our key criterion to decide which female counted as the initiator. Information was available for 149 records. The two individuals belonged to the same rank category in 35% of cases, and they differed in 65%. In these latter cases, low-ranking

(a)   (b)

Figure 12.3. (a) Bonobo female Vanessa, a member of the Eyengo community of Lomako, with her female infant Virginia (photo: B. Fruth). (b) Genito-genital rubbing in a dorso-ventral position is the most common form of homosexual interaction, in which the female on top rubs her clitoris against the anogenital swelling of a female partner (drawing: B. Fruth; from Hohmann and Fruth, 2000).

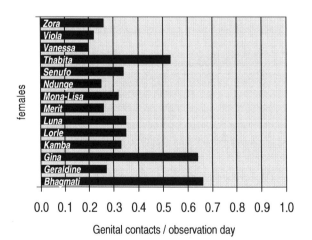

Figure 12.4. Rate of homosexual behaviour (genital contacts/observation day for 14 females of the Eyengo community). Only all-day-follows (8.0 – 12.5 hours of observation) were considered.

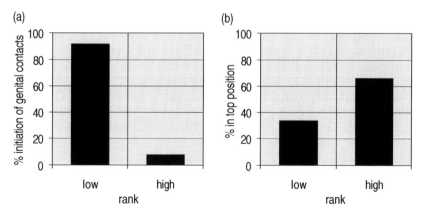

Figure 12.5. Dominance rank and female homosexual interactions: (a) % initiation of genital contacts, (b) % top position.

individuals initiated significantly more often (89 times versus eight times), indicating that genital contacts reflect active submission (Wilcoxon signed rank test, $T = 44$, $n = 97$, $p < 0.001$; Figure 12.5a).

If homosexual behaviour is related to dominance, we should also see status dependant asymmetries with respect to body position. Mounter and mountee belonged to the same rank category in 60.6% of the cases on record ($n = 325$), and they differed in 39.4%. Most same-rank dyads ($n = 13$) performed genital contacts only once or twice. Dyads with more than two contacts ($n = 7$) alternated positions. High-ranking individuals, however, were significantly more often in the top position than low-ranking individuals (84 versus 44 times, Wilcoxon signed rank test, $T = 8$, $n = 128$, $p < 0.0001$; Figure 12.5b).

Homosexual behaviour and tension regulation

Social tension can be expected to increase when inter-individual distances decrease. Proximity and thus tension is expected to vary with party size as well as with size and quality of food patches. If genital contacts serve to regulate tension, they should be more frequent if a food patch can be monopolized than if food is more easily accessible for everybody.

We compared the rates of genital contacts and aggression in situations when food was widely spread with those when access was limited to a few individuals. For this, we analysed the social behaviour of party members at trees of a species that produces many small fruits as well as at trees of a species that produces few and very large fruits. During each of 23 visits to the tree *Irvingia gabonensis* (mean diameter in breast height 113 cm, SD 37 cm) for which we have data, an individual bonobo ate about 10–100 fruits of the about 500–5000 fruits available. A fruit averaged 94 g (SD 22 g). In contrast, during 22 visits to the tree *Treculia africana* (mean diameter in breast hight 92 cm, SD 4 cm), an individual bonobo ate only 1–2 fruits of the about 1–10 fruits available. Such fruits were very heavy, averaging 7600 g (SD 6600 g).

Homosexual behaviour among all party members was significantly more frequent during visits to *Treculia* patches, compared with *Irvingia* patches (mean $\pm$ SD $= 2.82 \pm 3.53$ versus mean $\pm$ SD $= 0.33 \pm 0.77$ genital contacts/h; Mann–Whitney U test, $U = 43$, $N1 = N2 = 45$, $p < 0.0001$). This remained true when only female interactions were considered (*Treculia* mean $\pm$ SD $= 0.31 \pm 0.36$ versus *Irvingia* mean $\pm$ SD $= 0.11 \pm 0.24$, Mann–Whitney U test, $U = 58$, $p < 0.0001$, $n = 45$).

A further test can be conjured from the fact that space is also rather limited in tree crowns which bear ripe fruit. Tension should therefore increase with the number of females feeding in such a crown, and so should the rate of gg-rubbing. We focused on parties feeding in *Polyalthia suaveolens*, a tree with comparatively homogenous dimensions (mean $\pm$ SD for 50 trees: trunk height 23.1 $\pm$ 6.3 m; crown height 9.4 $\pm$ 3.1 m; diameter in breast height 22.4 $\pm$ 9.1 cm; feeding time 21.7 $\pm$ 14.5 min, $n = 113$). Between two and seven females fed in the same tree (mean $\pm$ SD $= 3.42$ females $\pm$ 1.37; $n = 64$ parties). Females had genital contact during the first ten min after entering a crown in 13 cases. The average number of genital contacts per tree was 0.42/10 min (range 0–5, $n = 64$). However, the rate of genital contacts did not depend on how many females fed in the same tree (chi-square $= 5.269$, $p = 0.384$), suggesting that homosexual behaviour was not related to the hypothetical increase in tension.

When the large fruits of *Treculia africana* are available, bystanders badger those who monopolize the fruit with a variety of begging rituals (Fruth and Hohmann, 2002). In addition, they try to get rid of nearby competitors. Consequently, some beggars obtain a share and others do not, despite remarkable patience. In any case, tension is high in a group of food-sharing individuals, and agonism is particularly frequent among bystanders. ($n = 16$; chi-square $= 7.563$, $p < 0.01$; Figure 12.6).

If genital contacts serve to reduce tension, we would expect them to be highest when tension is highest. However, the relationship is the reverse. Significantly more genital contacts occur between bystanders and owners than among bystanders alone ($n = 57$; chi-square $= 8.491$, $p < 0.01$; Figure 12.6).

Homosexual behaviour and reconciliation

The hypothesis that genital contacts serve as reconciliation predicts that homosexual behaviour should be closely preceded by agonistic interactions,

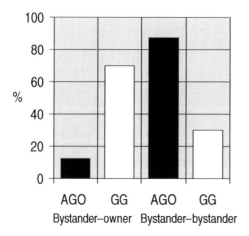

Figure 12.6. Relative proportion (%) of agonistic interactions (AGO: $n = 16$) and genito-genital contacts (GG: $n = 57$) for bystander – owner dyads (left bars) and bystander – bystander dyads (right bars) during the consumption of large fruits (*Treculia africana*).

should increase after agonism and should be more common amongst relatives or individuals with close social ties.

However, only 6.4% of genital contacts between mature individuals ($n = 466$) were preceded by an agonistic interaction in the 15 min interval before, whereas no agonistic encounter occurred in 80.6% (information was incomplete for the remaining 13.0%).Thus, genital contacts were all but independent from agonistic encounters.

We examined all cases of agonism, not just those involving mature individuals, and compared the frequency of genital contacts of 105 pre-conflict with 108 post-conflict intervals. Homosexual interactions were indeed more common after agonistic encounters than before (22% versus 7%, Wilcoxon test, $Z = -3.162$, $p = 0.002$, $n = 105$). Considering only cases for which information on the pre-conflict and corresponding post-conflict interval was complete ($n = 67$), we found that the average rate of pre-conflict genital contacts was four times lower (0.07) than that of post-conflict genital contacts (0.29). However, from the 15 females contributing to the data set, six were seen to be involved in agonistic encounters with other females only once or twice. When we considered only females with more than two agonistic interactions, the overall difference between pre-conflict rates and post-conflict rates of genital contacts was significant (Wilcoxon signed rank test, $p = 0.017$, $T = 0$, $n = 8$; Figure 12.7). This result, however, is again weakened when we look closer at the participating individuals: Out of 26 post-conflict genital contacts, 15 involved both former opponents, one a female and the mother of her former opponent, and four occurred between

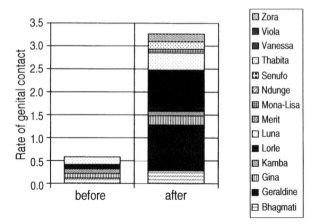

Figure 12.7. Rates of genital contact before and after female–female agonistic interactions.

one of the opponents and a female who was not involved in the conflict. In six cases information was incomplete. These results suggest that genital contacts occur largely independent of agonism. When they were linked, however, rates of genital contacts after agonistic encounters increased in some but not in all females.

Female interactions accounted for 22.7% of all records of agonistic behaviour ($n = 348$ events). We had sufficient information to calculate values of dyadic spatial association for 59 events and relate them to different degrees of overall individual association (see Hohmann and Fruth, 2000). Overall, agonistic encounter frequency among the three classes 'close associates' ($n = 6$), 'non-associates' ($n = 2$) and 'random associates' ($n = 51$) did not differ from expected values (Kruskal–Wallis test, $H = 4.571$, ns). The 26 cases of post-conflict genital contacts among females involved 13 individuals and 14 different dyads. Only one dyad consisted of females who were close associates, whereas they were random associates in all other cases. Non-associates were never seen to use genital contacts to reconcile agonistic conflicts. This shows clearly that the performance of post-agonistic genital contacts was independent of association patterns.

Homosexual behaviour and social bonding

If the exchange of genital contacts reflects close affiliative ties, one would predict a high frequency among kin or individuals who associate and affiliate with each other.

Data on genetic relations were available from a previous study on members of the Eyengo community (Gerloff *et al.*, 1999). Genital contacts between four

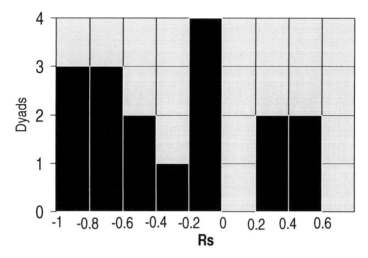

Figure 12.8. Relationship between relative frequencies of social grooming and genital contacts of 17 female dyads. Spearman rank correlation coefficients (Rs) indicate whether the relationship between behaviours was positive (high or low scores for both behaviours, respectively) or negative (high grooming scores correspond with low scores of genital contacts and vice versa). Y axis = dyads which fall in each category.

immature females and mature resident females occurred 34 times. Out of these, 30 involved unrelated females and four involved a mother and her daughter. Considering the average number of mature female party members (mean $\pm$ SD = 4.74 ± 2.05, $n = 310$), genital contacts with unrelated females occurred more often than expected by chance (goodness-of-fit test, $G = 22.178$, $p < 0.001$, $n = 34$).

We compared the proportions of the three association classes for four field seasons (1993–1996) against the relative frequency of genital contacts and found that association classes did not differ (Kruskal–Wallis test, $H = 4.571$, $p = 0.06$, $n = 3$).

Another indicator of friendly behaviour is grooming. Previously we found a positive correlation between grooming and close spatial associations amongst bonobos of the Eyengo community (Hohmann et al., 1999). If we consider homosexual interactions as friendly and affiliative behaviour, we would also expect a positive correlation between genital contacts and grooming.

Four dyads showed a positive relation between both behaviours, that is dyads who groomed also had genital contacts, whereas 13 dyads had a negative relation. Overall, a significant negative correlation existed (Wilcoxon matched pair test, $T = 33.5$, $p < 0.05$, $n = 17$; Figure 12.8).

Homosexual behaviour and mate attraction

If females use homosexual behaviour to attract the attention of a particular male, then we can predict more genital contacts in mixed-sex than in all-female parties.

At Lomako, 76% of parties consisted of adults of both sexes, while all-female parties made up for 20% ($n = 485$; Fruth 1995). Mixed-sex parties had more female members than all-female parties (mean $\pm$ SD $= 4.77 \pm 2.07$, $n = 265$ versus 3.26 ± 1.34, $n = 46$; Mann–Whitney U test, $U = 2980$, $p < 0.0001$). To control for this difference, the hourly rate of genital contacts was divided by the number of mature female party members. It turned out that females had more frequent genital contacts in mixed-sex parties (rate/h/female 0.06 ± 0.1, range 0–2, $n = 260$) than all-female parties (0.03 ± 0.08, range 0–0.4, $n = 45$; Mann-Whitney U test, $U = 4672$, p $= 0.01$).

The mate-attraction hypothesis also predicts that genital contacts should be closely followed by copulations. In wild bonobos, copulations are correlated with tumescence of female genitalia (Furuichi, 1987). We therefore investigated the temporal association of genital contacts between females ($n = 181$) and copulations among mature individuals as recorded during 140 days. However, only one of the two types of sexual interaction occurred during 77 days, whereas both occurred during 63 days. Of all copulations, 73% were independent of a genital contact, while 27% involved a female who had a preceding homosexual interaction. Looking at the time that elapsed between both events, the average interval was more than an hour (73.4 ± 137.5 min; median $= 15$ min, range 0–585 min, $n = 181$). Comparison of the intervals between genital contact followed by copulation with intervals of the other three possible combinations of sexual interaction (Figure 12.9) revealed no difference in length (chi-square $= 3.28$, $p = 0.351$; copulation–copulation $n = 64$; copulation–genital contact $n = 58$; genital contact–genital contact $n = 92$; genital contact–copulation $n = 49$). Pair-wise comparison indicated that intervals between two genital contacts were, at 210 min on average, significantly shorter than intervals between other combinations of sexual behaviour (chi-square test, $p < 0.001$).

If homosexual interactions enhance copulations, then they should occur more often during the phase of tumescence. Out of 181 interactions with complete information about the degree of genital swellings, 128 showed signs of tumescence (swelling stages 3 and 4), while 53 did not (swelling stage 1 and 2). Overall, female homosexual interactions were unequally distributed across the four stages of tumescence duration (chi-square $= 42.59$, $p < 0.001$). However, durations of cycles and swelling stages are known to vary (Dahl, 1986, Vervaecke et al., 1999). Therefore, we used only data from complete cycles, which revealed a large

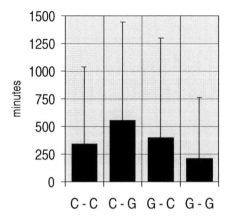

Figure 12.9. Duration of intervals in minutes (average, SD) in the four follow-up combinations of sexual behaviour. Copulation = male/female. Genital contact = female/female. C-C = copulation follows copulation ($n = 64$); C-G = copulation follows genital contact ($n = 58$); G-C = genital contact follows copulation ($n = 49$); G-G = genital contact follows genital contact ($n = 92$).

inter-individual variation and fewer than expected contacts during de-tumescence (stages 1 and 2), while more than expected during tumescence (stages 3 and 4). However, this deviation from expectation was significant only at stage 4 (maximum tumescence; Wilcoxon sign test, $T = 4.5$, $n = 10$, $p < 0.02$).

It is also predicted that copulations which are closely preceded by genital contacts should involve more tumescent females. For 166 copulations between mature individuals, we knew the swelling stages for 67 of the females. We compared females who copulated without preceding homosexual interactions ($n = 50$) with those who mated after same-sex genital contacts ($n = 17$). Neither the proportions of de-tumescent females (chi-square = 0.457, ns) nor those with maximum tumescence differed (chi-square = 0.120, ns; Figure 12.10).

Discussion

Same-sex sexual interactions are common amongst wild bonobos, in particular amongst females. We tested five of the hypotheses commonly invoked to ascribe a function to homosexual behaviour and were able to find support for some, but not all of them.

Social status

The hypothesis that female homosexual interactions reflect social status is fully supported by our data. We found rank-related asymmetries in that subordinate females initiated homosexual behaviour more often than dominant

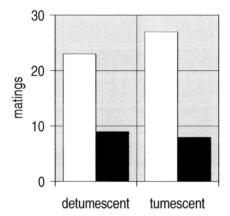

Figure 12.10. Frequency of 'enhanced' matings (black bars) versus 'non-enhanced' matings (white bars) in relation to the phases of tumescence. Detumescent = from less swollen to unswollen; tumescent = fully swollen. Mating was considered 'enhanced' when preceded by homosexual activity.

ones. Moreover, dominants took up the top position more often than subordinates, which is in line with findings from captive studies (Parish, 1994, 1996; de Waal, 1987). Genital contacts were particularly frequent when access to food was limited and conflicts were likely. An immediate pay-off for subordinates who are accepted by the dominant individual is obvious. However, what benefit dominants may gain from genital contacts is less clear, apart from a confirmation of status.

Sexual pleasure may be a factor. Bonobos have external labiae and a pronounced clitoris, which are swollen for extended periods of a female's cycle. When engaging in genital contacts, the clitoris is involved and sexual stimulation is very likely to occur. Frantic lateral movements, screams and grimacing will often accompany genito-genital rubbings, which suggests that an orgasm might occur. Thus, on the proximate level, dominant individuals can be considered to gain pleasure. However, it remains unclear why close associates do not practise this pleasurable event more frequently than those not particularly fond of each other, and why most genital contacts occur among individuals of different rank.

Tension regulation

Several predictions of the hypothesis that same-sex sexual interactions regulate tension are supported by our data. When some individuals monopolized food against others, rates of genital contacts were high, confirming an earlier captive study (Parish, 1994). However, White and Lanjouw (1992) reported more frequent contacts when food was abundant and lower rates when it was limited. Our study measured patch quality via crown size of food trees, while distribution and quality of food, as well as type and intensity of feeding competition were ignored. These variables contribute to the differences in social tension, as is confirmed when they are kept constant. We chose *Polyalthia suaveolens* trees, which are rather homogenous in quality and dimensions, and did not find a positive correlation between party size and rate of genital contacts. This demonstrates that tension is regulated by food quality, accessibility and differences in the type of competition. Future hormonal studies may be able to demonstrate the immediate effect of genital contacts on individual stress levels (Sapolsky, 1982; Aureli *et al.*, 1999).

Reconciliation

The majority of genital contacts occurred independent of agonistic encounters, which seems to contradict reconciliation as a functional explanation. However, rates rose after agonistic encounters, which speaks in favour of the hypothesis. Nevertheless, a large proportion of genital contact was independent from agonistic behaviour.

De Waal (1987, 1995) found an overall increase of socio-sexual behaviour after conflicts, but the number of contacts before and after remained unspecified. Data from another captive group (Parish, 1994; Manson *et al.*, 1997) and from a long-term field study (Kano, 1989; Furuichi, 1989) relate genital contacts to social tension rather than to agonistic conflicts. Our proportion of genital contacts linked to agonistic encounters is small. However, for captive individuals, reconciliation may be the only option for conflict resolution, while, in the wild, opponents can separate and travel apart.

Moreover, reconciliation is most likely to occur among close kin (reviewed by Kappeler and van Schaik, 1992). In captivity it is not uncommon for groups to consist of related individuals. At San Diego, for example, all adult females studied by de Waal (1987) were full siblings. This is different from wild conditions, where close genetic ties among resident females are absent (Hashimoto *et al.*, 1996; Gerloff *et al.*, 1999), which may explain the different reconciliatory potential.

Conflicts arise from ambiguities about access to resources and are likely to be followed by agonistic disputes. Reconciliation implies that an agonistic interaction has taken place before the exchange of friendly behaviour (de Waal and van Roosmalen, 1979). Tension regulation, on the other hand, does not necessarily require the emergence of agonistic behaviour. Bonobos seem to employ both types of conflict resolution. When feeding on fruits of *Treculia*, for example, most agonistic interactions occurred between individuals who competed for food, while genital contacts were almost exclusively exchanged between a bystander and an owner of food.

Social bonding

Predictions of the social bonding hypothesis were not supported by our data. Furuichi (1989) proposed that immigrants use genital contacts to form close associations with residents. However, the data required to show that genital contacts do actually affect the social relations between two individuals in a predictable way are not yet available. Moreover, the hypothesis does not explain why high-ranking resident females who are well integrated in the social network of the community should still engage in such a behaviour. Apart from that, if genital contacts really served bonding, why should the largest proportion occur in the context of feeding? Woolly spider monkeys do embrace each other in tense situations (for example intergroup encounter) but also during long resting periods (Strier, 1992). Spotted hyenas perform greeting ceremonies not only when they feed but also when they meet at the communal den (East *et al.*, 1993). Male squirrel monkeys also engage in coalitionary genital displays in different situations (Mitchell, 1994). If the main function of genital contacts is to develop or maintain affiliative ties, one would expect them to be less associated with feeding.

Mate attraction

Although some predictions of the mate attraction hypothesis were met, we cannot support this explanation. While females had genital contacts more often in the presence of males, and maximally swollen females did so with a higher proportion than females with reduced swellings, there was no evidence that same-sex sexual behaviour induced copulation. Moreover, higher rates of genital contacts in mixed-sex parties are explained by party size rather than party composition, since mixed-sex parties are larger than all-female parties (Fruth, 1995; Hohmann and Fruth, 2002).

Bonobos: chimpanzees with a difference?

Recently the use of genital contacts among captive female chimpanzees has been reported by Anestis (2004) but similar observations from wild chimpanzees are lacking. This raises the question of why do bonobos, but not chimpanzees, use socio-sexual interactions in a prominent fashion to mitigate social relations? Ecological as well as social scenarios have been put forward, which sometimes assume almost paradisiacal environmental conditions that allow female bonobos to socialize freely (Wrangham, 1980, 1993; de Waal and Lanting, 1997). A prolonged period of female sexual attractiveness is thought to foster paternity confusion, thus reducing the threat of infanticidal males. Homosexual contacts are considered to cement female relationships, allowing females to form a so-called second sisterhood, which enables them to dominate males.

However, such a scenario is not without problems. Extended length of genital swellings seems to indeed conceal ovulation to a certain degree, because in a recent captive study, maximum tumescence correlated with fertile cycle stages in only 60% of phases (Reichert et al., 2002).

Nevertheless, paternity confusion might not be so effective as to completely forestall the risk of infanticide by males. In the wild, resident males were sometimes seen to charge females with small infants for unknown reasons. Such male attacks provoked intense counter-aggression by females and males. In one case, a young adult male on the verge of gaining the alpha position attacked a female with a recently born baby, apparently in an attempt at infanticide. Females and also other males beat him badly and he vanished thereafter. He may have fled or his injuries may have even led to his death. (Hohmann and Fruth, 2002, 2003).

Such observations indicate that infanticide might indeed be a factor in bonobo societies. However, while female coalitions may be an effective counterstrategy, female–male associations may also function in this way. Moreover, female relationships are far from salient, as illustrated by the death of a newborn, kidnapped by a resident female in the wild (Hohmann and Fruth, 2002), and several cases of infant kidnapping and cruelty by captive females (Vervaecke et al., 2003).

Competition is certainly not absent in bonobo societies, particularly not where they use open savannah to forage (Thompson-Myers, 1997). And, while the more typical dense forest habitat provides at times rich and abundant resources, it is not free from competitors either. Here, bonobos compete seriously with wild boars, elephants and several species of monkey over fruit.

Intra-specific competition over ripe fruit occurs with neighbouring communities (Fruth, 1995), where smaller parties are left standing. This may be one reason why female bonobos travel together more often (Hohmann and Fruth, 2002) than do female chimpanzees, although dyadic associations of females do not exceed that of some chimpanzee populations (Hohmann et al., 1999).

Morphology might also have much to do with the high incidence of genitogenital rubbings observed in female bonobos. Chimpanzees mount each other ventro-dorsally; they touch, embrace and kiss. Bonobo mounting is sometimes ventro-dorsal as well. However, their genitals are positioned more ventrally, so it is not surprising that they interact preferentially ventro-ventrally. The clinging individual's position is reminiscent of that of an infant, and the associated lateral 'rubbing' movements may simply occur because they are easier to execute in such a close embrace than pelvic thrusts.

We thus conclude that genital contacts among bonobos are not a sign of affiliation, friendship or 'sisterhood' per se, instead, female homosexual behaviour is an instrument that allows distantly related individuals to coexist closely and nevertheless peacefully, by functioning as a method of status acknowledgement, tension regulation and, to some extent, reconciliation. From this, bonobo female coalitions might derive ultimate benefits related to resource defence, infanticide avoidance and dominance over males.

Individuals who are not close relatives are not expected to get along with each other easily, but bonobo females seem to have found a way to circumvent this problem. And a pleasurable one too. It would certainly not be detrimental if humans would do more of the same.

References

Altmann, J. (1974) Observational study of behavior: sampling methods, *Behaviour*, **49**, 227–67.

Anestis, S. F. (2004) Female genito-genital rubbing in a group of captive chimpanzees. *Int. J. Primatol*, **25**, 477–88.

Aureli, F., Preston, S. D. and de Waal, F. B. M. (1999) Heart rate responses to social interactions in free-moving rhesus macaques (*Macaca mulatta*): a pilot study. *J. Comp. Psychol.*, **113**, 59–65.

Badrian, A. and Badrian, N. (1984) Social organization of *Pan paniscus* in the Lomako Forest, Zaire. In *The Pygmy Chimpanzee: Evolutionary Biology and Behavior*, ed. L. S. Susman, pp. 325–46. New York: Plenum.

Boesch, C. and Boesch, H. (1989) Hunting behavior of wild chimpanzees in the Taï National Park. *Am. J. Physic. Anthropol.*, **78**, 547–73.

Boesch, C., Hohmann, G. and Marchant, L. F. (eds.) (2002) *Behavioural Diversity in Chimpanzees and Bonobos*. Cambridge: Cambridge University Press.

Bygott, D. (1979) Agonistic behavior, dominance, and social structure in wild chimpanzees of the Gombe National Park. In *The Great Apes*, eds. D. A. Hamburg and E. R. McCrown, pp. 405–27. Menlo Park: Benjamin/Cummings.

Colmenares F. (1990) Greeting behaviour in male baboons: I. Communication, reciprocity and symmetry. *Behaviour*, **113**, 81–114.

Dahl, J. F. (1986) Cyclic perineal swelling during the intermenstrual intervals of captive female pygmy chimpanzees (*Pan paniscus*). *J. Human Evol.*, **15**, 369–85.

de Waal, F. B. M. (1987) Tension regulation and nonreproductive functions of sex in captive bonobos (*Pan paniscus*). *Nat. Geogr. Res.*, **3**, 318–38.

de Waal, F. B. M. (1989) *Peacemaking Among Primates*. Cambridge, MA: Harvard University Press.

de Waal, F. B. M. (1995) Sex as an alternative to aggression in the bonobo. In *Sexual Nature, Sexual Culture*, eds. P. R. Abramson and S. D. Pinkerton, pp. 37–56. Chicago: The University of Chicago Press.

de Waal, F. B. M. and van Roosmalen, A. (1979) Reconciliation and consolation among chimpanzees. *Behav. Ecol. Sociobiol.*, **5**, 55–66.

de Waal, F. B. M. and Lanting, F. (1997) *Bonobo: The Forgotten Ape*. Berkeley, Los Angeles: University of California Press.

East, M. L., Hofer, H. and Wickler, W. (1993) The erect 'penis' is a flag of submission in a female dominated society: greetings in Serengeti spotted hyenas. *Behav. Ecol. Sociobiol.*, **33**, 355–70.

Ford, C. S. and Beach, F. A. (1952) *Patterns of Sexual Behavior*. London: Eyre & Spottiswoode.

Fox, M. W. and Cohen, J. A. (1977) Canid communication. In *How Animals Communicate*, ed. T. A. Sebeok, pp. 728–48. London: Indiana University Press.

Fruth, B. (1995) *Nests and Nest Group in Wild Bonobos (Pan paniscus): Ecological and Behavioural Correlates*. Aachen: Shaker.

Fruth, B. (1998) Invited comment on 'The social behavior of chimpanzees and bonobos: empirical evidence and shifting assumptions' by Craig B. Stanford. *Curr. Anthropol.*, **39**, 408–09.

Fruth, B. and Hohmann, G. (2002) How bonobos handle hunts and harvests: Why share food? In *Behavioural Diversity in Chimpanzees and Bonobos*, eds. C. Boesch, G. Hohmann and L. F. Marchant, pp. 231–76. Cambridge: Cambridge University Press.

Fruth, B., Hohmann, G. and McGrew, W. C. (1999) The *Pan* species. In *The Nonhuman Primates*, eds. P. Dolhinow and A. Fuentes, pp. 64–72. Mountain View, CA: Mayfield.

Furuichi, T. (1987) Sexual swelling, receptivity, and grouping of wild pygmy chimpanzee females at Wamba, Zaire. *Primates*, **28**, 309–18.

Furuichi, T. (1989) Social interactions and the life history of female *Pan paniscus* in Wamba, Zaire. *Int. J. Primatol.*, **10**, 173–97.

Gerloff, U., Hartung, B., Fruth, B., Hohmann, G. and Tautz, D. (1999) Intracommunity relationships, dispersal pattern and paternity success in a wild living community of bonobos (*Pan paniscus*) determined from DNA analysis of faecal samples. *Proc. Roy. Soc. London (B)*, **266**, 1189–95.

Goodall, J.v.L. (1968) The behaviour of free-living chimpanzees in the Gombe Stream Reserve. *Anim. Behav. Monogr.*, **1**, 163–311.

Goodall, J. (1986) *The Chimpanzees of Gombe: Patterns of Behaviour*. Cambridge, MA: Harvard University Press.

Hamilton, W. D. (1964) The genetical evolution of social behaviour. *J. Theoret. Biol.*, **7**, 1–52.

Hanby, J. P. (1977) Social factors affecting primate reproduction. In *Handbook of Sexology*, eds. J. Money and H. Mustaph, pp. 461–84. Amsterdam: Excerpta Medica.

Hashimoto, C., Furuichi, T. and Takenaka, O. (1996) Matrilineal kin relationship and social behavior of wild bonobos (*Pan paniscus*): sequencing the D-loop region of mitochondrial DNA. *Primates*, **37**, 305–18.

Hohmann, G. (2001) Association and social interactions between strangers and residents in bonobos (*Pan paniscus*). *Primates*, **42**, 91–9.

Hohmann, G. and Fruth, B. (1993) Field observations on meat sharing among bonobos (*Pan paniscus*). *Folia Primatol.*, **60**, 225–9.

Hohmann, G. and Fruth, B. (2000) Use and function of genital contacts among female bonobos. *Anim. Behav.*, **60**, 107–20.

Hohmann, G. and Fruth, B. (2002) Dynamics in social organization of bonobos (*Pan paniscus*). In *Behavioural Diversity in Chimpanzees and Bonobos*, eds. C. Boesch, G. Hohmann and L. F. Marchant. pp. *–*. Cambridge: Cambridge University Press.

Hohmann, G. and Fruth, B. (2003) Intra- and intersexual aggression by bonobos in the context of mating. *Behaviour*, **140**, 1389–413.

Hohmann, G., Gerloff, U., Tautz, D. and Fruth, B. (1999) Social bonds and genetic ties: kinship, association and affiliation in a community of bonobos (*Pan paniscus*). *Behaviour*, **136**, 1219–35.

Ihobe, H. (1992) Observations on the meat-eating behavior of wild bonobos (*Pan paniscus*) at Wamba, Republic of Zaïre. *Primates*, **33**, 247–50.

Kano, T. (1980) Social behavior of wild pygmy chimpanzees, *Pan paniscus*, of Wamba: a preliminary report. *J. Hum. Evol.*, **9**, 243–60.

Kano, T. (1989) The sexual behavior of pygmy chimpanzees. In *Understanding Chimpanzees*, eds. P. G. Heltne and L. A. Marquard, pp. 176–83. Cambridge, MA: Harvard University Press.

Kappeler, P. M. and van Schaik, C. P. (1992) Methodological and evolutionary aspects of reconciliation among primates. *Ethology*, **92**, 51–69.

Kirkpatrick, R. C. (2000) The evolution of human homosexual behavior. *Curr. Anthropol.*, **41**, 385–413.

Kummer, H., Götz, W. and Angst, W. (1974) Triadic differentiation: an inhibitory process protecting pair bonds in baboons. *Behaviour*, **49**, 62–87.

Kuroda, S. (1980) Social behavior of the pygmy chimpanzee. *Primates*, **21**, 181–97.

Kuroda, S. (1984) Interactions over food among pygmy chimpanzees. In *The Pygmy Chimpanzee*, ed. R. L. Susman, pp. 301–24. New York: Plenum.

Manson, J. H., Perry, S. and Parish, A. R. (1997) Nonconceptive sexual behavior in bonobos and capuchins. *Int. J. Primatol.*, **18**, 767–86.

Mitchell, C. A. (1994) Migration alliances and coalitions among adult male South American squirrel monkeys (*Saimiri sciureus*). *Behaviour*, **130**, 169–90.

Nadler, R. D. (1990) Homosexual behavior in nonhuman primates. In *Homosexuality/Heterosexuality: Concepts of Sexual Orientation*, eds. D. P. McWhirter, S. A. Sanders and J. M. Reinisch, pp. 138–70. New York: Oxford University Press.

Parish, A. R. (1994) Sex and food control in the 'uncommon chimpanzee': how bonobo females overcome a phylogenetic legacy of male dominance. *Ethol. Sociobiol.*, **15**, 157–79.

Parish, A. R. (1996) Female relationships in bonobos (*Pan paniscus*): evidence for bonding, cooperation, and female dominance in a male-philopatric species. *Human Nature*, **7**, 61–96.

Parker, G. A. and Pearson, R. G. (1976) A possible origin and adaptive significance of the mounting behavior shown by some female mammals in estrous. *J. Nat. Hist.*, **10**, 241–5.

Reichert, K., Heistermann, M., Hodges, J. K., Boesch, C. and Hohmann, G. (2002) What females tell males about their reproductive status: are morphological and behavioural cues reliable signals of ovulation in bonobos (*Pan paniscus*). *Ethology*, **108**, 583–600.

Rowell, T. E. (1966) Hierarchy in the organization of a captive baboon group. *Anim. Behav.*, **14**, 430–43.

Sapolsky, R. M. (1982) The endocrine stress response and social status in the wild baboon. *Hormon. & Behav.*, **16**, 279–92.

Silk, J. B. (1998) Making amends: adaptive perspectives on conflict remediation in monkeys, apes, and humans. *Human Nature*, **9**, 341–68.

Stanford, C. B. (1995) Chimpanzee hunting behaviour and human evolution. *Am. Scient.*, **83**, 256–61.

Stanford, C. B. (1998) The social behavior of chimpanzees and bonobos: Empirical evidence and shifting assumptions. *Curr. Anthropol.*, **39**, 399–420.

Strier, K. B. (1992) Causes and consequences of nonaggression in the woolly spider monkey or muriqui (*Brachyteles arachnoides*). In *Aggression and Peacefulness in Humans and Other Primates*, eds. J. Silverberg and J. P. Gray, pp. 100–16. New York: Oxford University Press.

Thompson-Myers, J. A. (1997) The history, taxonomy and ecology of the bonobo (*Pan paniscus*, Schwarz 1929) with a first description of a wild population living in a forest/savannah mosaic habitat. Ph.D. thesis, University of Oxford.

Vasey, P. L. (1995) Homosexual behavior in primates: a review of evidence and theory. *Int. J. Primatol.*, **16**, 173–204.

Vervaecke, H., van Elsacker, L., Möhle, U., Heistermann, M. and Verheyen, R. F. (1999) Intermenstrual intervals in captive bonobos (*Pan paniscus*). *Primates*, **40**, 283–9.

Vervaecke, H., Stevens, J. and van Elsacker, L. (2003) Interfering with others: female–female reproductive competition in *Pan paniscus*. In *Sexual Selection and Reproductive Competition in Primates: New Perspectives and Directions*, eds. C. B. Jones, pp. 231–54. New York: Alan R. Liss.

White, F. J. (1988) Party composition and dynamics in *Pan paniscus*. *Int. J. Primatol.*, **9**, 179–93.

White, F. J. and Lanjouw, A. (1992) Feeding competition in Lomako bonobos: variation in social cohesion. In *Topics in Primatology*, Vol. 1, *Human Origins*, eds. T. Nishida, W. C. McGrew, P. Marler, M. Pickford and F. B. M. de Waal, pp. 67–79. Tokyo: University of Tokyo Press.

Wrangham, R. W. (1980) An ecological model of female-bonded primate groups. *Behaviour*, **75**, 262–300.

Wrangham, R. W. (1993) The evolution of sexuality in chimpanzees and bonobos. *Human Nature*, **4**, 47–79.

13

The evolution of male homosexuality and its implications for human psychological and cultural variations

DENNIS WERNER

Evolutionary theories for male homosexuality

In the past few decades, evolutionary psychology has demonstrated its value in accounting for human behaviours and in inspiring new ideas to be tested. An evolutionary view of male homosexuality may also contribute to many current debates regarding the nature (and culture) of human homosexuality. As with all evolutionary theories, there are at least two questions that must be answered. First, we must explain how homosexual orientations could survive and/or reproduce in light of selection pressures. This is the question of adaptation. Second, any evolutionary explanation must also clarify how a given trait could have arisen. No matter how adaptive a trait may be, it must have a traceable past. This is the question of phylogeny. I will examine separately each of these questions.

Adaptation and male homosexuality

There are at least three different levels at which homosexuality may be seen as adaptive, and each of these levels has its own implications with regard to how homosexual behaviours manage to continue among humans. At the most abstract level, homosexuality might be unrelated to genetic differences or to universal genetic programs that regulate ontogeny. Instead, homosexuality may be culturally determined in a manner far removed from direct genetic influences. Following Dawkins (1976), we might call this the 'memic' level. At a less 'abstract' level, we might assume that homosexuals and heterosexuals have the same genes, but that, during ontogeny, universal genetically determined 'programs' get 'switched on' or 'switched off', depending on environmental influences. Thus, people may have the same genotypes, but these genotypes may produce different phenotypes in different situations. This might be called the 'epigenetic' level. Many theories (including Freudian ideas) about the psychological dynamics behind homosexuality are of this type. Finally, at the most concrete level, we might posit genetic differences between homosexuals and heterosexuals. I will call this the 'genetic' level. Many biological studies seem to support this view.

Evolutionary arguments have been offered for all of these levels. For example, Symons (1979) argues that homosexuality results from a meme that takes advantage of the inborn male propensity to find sexual variety interesting. Normally this propensity would be adaptive because it would encourage males to attempt sexual relations with many women, which would result in more offspring. But a meme could parasitize this propensity, redirecting sexual interest to non-adaptive objects, as in fetishism or homosexuality. Epigenetic arguments have been inspired by work on pre-natal hormones. Dörner and his colleagues (Dörner *et al.*, 1980; LeVay, 1994) link stress during pregnancy to hormonal effects on the fetus that would lead to homosexuality. This could be part of a mother's adaptive reproductive strategy. That is, in times of stress, when it is difficult to raise children, it may be adaptive to have some homosexual children who could help their siblings raise offspring instead of having offspring of their own. Finally, studies of gene linkages have pointed to genetic differences between male homosexuals and heterosexuals (Hamer *et al.*, 1993; Mustanski *et al.*, 2005). Various evolutionary theories have been proposed to account for genetic arguments.

Whether genes lead directly to homosexuality or simply program for homosexuality under certain environmental conditions, they would still appear to undercut reproduction. Thus, the basic question for genetic and epigenetic arguments is to account for how such genes could survive the pressures of natural selection. Several hypotheses have been proposed. First, a maladaptive trait might re-occur repeatedly in a population if it results from the frequent mutation of a gene that is normally adaptive. However, people with maladaptive genes rarely exceed 1% of the population, while male homosexuality apparently occurs much more frequently (Gadpaille, 1980; Whitam and Mathy, 1986; Diamond, 1993). Another possibility is that genes that lead to homosexuality might have hidden advantages. For example, homosexuals could directly help their relatives (who share the 'homosexual' genes) raise more children. Nevertheless, cross-cultural research shows that, although male homosexuality is somewhat more common in patrilocal societies, it is not more common in societies with endogamy or extended families where homosexuals live closest to their relatives and so could presumably most help them (Werner, 1979).

A second theory argues that a gene that is especially advantageous for females might inadvertently cause homosexuality in males. According to this argument, the female relatives of male homosexuals should be especially successful in their reproduction. Recent research lends support to this argument by indicating that female maternal relatives of homosexuals have higher fecundity than female maternal relatives of heterosexuals (Camperio-Ciani *et al.*, 2004).

A third theory argues that genes 'for homosexuality' might be maladaptive by themselves, but might be advantageous when combined with other genes. This latter scenario is sometimes known as the 'heterozygous', 'hybrid vigor' or 'heterosis' hypothesis (Sommer, 1990; LeVay, 1994). What, then, might be the advantage of homosexual genes when combined with heterosexual genes? Kirsch and Rodman (1982, cited in Sommer, 1990) suggest that this advantage may have something to do with dominance hierarchies. The maintenance of these hierarchies presumably helps animals, including humans, live peacefully together, and this peaceful living provides advantages to the group, and to the individuals in the group. The heterosis argument suggests that homosexuality results from genes for submissive behaviour. But the key to the argument is the disadvantage of an animal that has only dominance genes. While an animal possessing only submissive genes would fail to reproduce for lack of trying, one possessing only dominance genes might also fail. After all, an animal that 'fights and runs away, lives to fight another day', but an animal that never gives in often dies young. In this light, consider Chagnon's (1988) discovery that Yanomami men who had killed more enemies had more offspring than milder men. Chagnon used this correlation to argue that aggressivity really does enhance reproductive success. Yet Chagnon's study suffers from a sampling problem – it included only living males. Very possibly the more aggressive males also had a greater probability of dying before ever being able to reproduce at all! Thus, on average, milder men may have more offspring than very aggressive men.

According to the heterosis argument, it is the males with a mixture of dominant and submissive genes who would most likely reproduce. By genetic laws, this would leave every generation with a certain percentage of non-reproducing individuals at the extremes. For example, if each of a pair of chromosomes had only one 'homosexual' locus with only two possible alleles (one for dominance and one for submissiveness), this would leave a heterozygous couple with 25% of their offspring homozygous for submissiveness, 50% heterozygous and 25% homozygous for dominance. If, however, homosexuality were the result of the interaction of many genes, possibly spread across different chromosomes, then fewer individuals would have only submissive genes. Of course there are many possible genetic scenarios for this argument. Indeed, different genes may affect different aspects of homosexuality, as new research is beginning to suggest (Mustanski *et al.*, 2005).

The heterosis argument still needs to be tested. It predicts, for example, that the relatives of exclusive homosexuals should have less dominant personalities than the general population, in addition to predicting that homosexuals themselves would be more likely than heterosexuals to avoid fights (at least physical ones), for which there is already abundant evidence (Whitam and Mathey, 1986; Cardoso, 2004).

Phylogeny and male homosexuality

Many evolutionists (Gould, 1977b; Antinucci, 1990; Reichholf, 1992; Rieppel, 1992) have complained that sociobiologists concentrate too much on the fitness part of the natural selection algorithm and neglect the question of where the variations to be selected for come from – the question of phylogeny. According to these authors, sociobiologists have an overly 'atomistic' approach to evolution, acting as if specific biological traits can evolve all by themselves, without regard to structural constraints on design or to the possibility that given changes may or may not be possible in phylogenetic history. Tracing phylogeny is important because natural selection does not operate like an engineer, drawing up a blueprint, and then constructing a machine from scratch in the most efficient way possible. Rather, natural selection operates more like a 'tinkerer' taking advantage of materials already available to produce new forms that 'work' at the moment. Elsewhere (Werner, 1999b) I argued that adaptational arguments (like functional arguments in general) are more common in the Anglo-Saxon academic world, while phylogenetic arguments (like structural arguments in general) are more common among continental academics. A more balanced perspective is in order.

The heterosis theory of homosexuality proposes that genes that affect submissiveness also affect homosexuality. This suggests that the key to understanding homosexuality may lie in the evolution of submissive behaviour. We are unlikely to find many clues in the fossil record, but a search for homosexual analogues and homologues in other animals may be informative.

Sexual relations between males can be found even in very simple species. In his review of animal homosexuality, Sommer (1990) mentions the case of 'homosexual rape' in a parasitic worm (*Moniliformis dubius*). In this case, the raped male's genital opening is blocked off with a spermless semen plug donated by the rapist. The raped male cannot then fertilize females. A variation on this theme is the homosexual rape of the bedbug *Xylocaris maculipennis*. In this case, instead of blocking up the rival male's genital opening, the raping male inserts his own sperm into the rival male's semen ducts. This also occurs with fresh water snails of the genus *Biomphalaria*, which are vectors for schistosomiasis (Forsyth, 1991). When the raped male copulates with a female, he fertilizes her with his rival's sperm.

Very different is the 'homosexual' behaviour found in the ten-spined stickleback fish. During spawning activities the male must court a female in order to get her to lay her eggs. Sometimes a second female is invited into the courting arena, and then the male may succeed in fertilizing the eggs of two females. But sometimes the second 'female' is really a 'transvestite' subordinate male, who, because of his female-like morphology, succeeds in entering the rival's arena and fertilizing the female's eggs himself. Alternatively, the 'transvestite' male may simply eat any fertilized eggs in the mating arena and thereby reduce the territorial male's reproductive success (reviewed in Sommer, 1990). A variation on this theme is found in the bluegill sunfish. In this species there are three different types of males. First is the larger 'parental' male, who defends a nest or courting arena. Second, is the much smaller 'sneaker' who sometimes succeeds in darting into a parental's courting arena and fertilizing the female's eggs before the larger male perceives him. Third are the 'transvestite' males who resemble female bluegill sunfish, and manage to gain access to the mating arena (Wilson, 1994). As they grow, the 'sneakers' gradually turn into 'transvestites', but the 'parental' males are genetically different. In both these examples, the territorial male directs sexual behaviours in the form of courtship to the 'transvestite' male fish.

The 'transvestite' behaviour demonstrated by males in these fish species may be structurally related to homosexual activities in the lizard *Anolis garmani*. In this species, the most visible difference between males and females is body size. Small males are usually driven out of the territories of larger males, but much smaller males are sometimes confused for females. On entering a more dominant male's territory, these small males behave like females and will even serve as sexual partners to the larger males. This behaviour not only gives the subordinates access to the dominant's territory and to his females, but also succeeds in making the dominant male 'waste' his sperm. A similar ploy is used in forest salamanders – subordinates succeed in tricking dominant males into giving up their sperm packets (reviewed in Sommer, 1990). In Manitoba garter snakes (*Thamnophis sirtalis parietalis*) 'transvestite' males give off female scents that attract other males, who then give up their sperm to these female mimics.

Many of these examples of 'homosexuality' in phylogenetically distant animals may be analogous rather than homologous to human homosexuality, but as we move closer to humans the likelihood of homologous behaviours increases. In mammals many different behaviours have been observed that might be associated with male homosexuality. Among primates homosexual behaviours are particularly diverse (reviewed in Sommer 1990; Vasey, 1995). These include such practices as the mounting of one male by another (for example, langurs, pig-tailed macaques, baboons, orangutans, chimpanzees, bonobos), including mounting with anal penetration (for example, stump-tailed macaques, squirrel monkeys – Chevalier and Skolnikoff, 1976; Maple, 1977), and mounting with anal penetration and ejaculation (Japanese macaques, rhesus macaques, gorillas – Gadpaille, 1980; Edwards and Todd, 1991; Bagemihl, 1999). Masturbation of other males has also been reported, including mutual masturbation (for example, stump-tailed macaques – Chevalier and Skolnikoff, 1976) as well as genital–genital contacts (for example, bonobos – Enomoto, 1990), at times leading to ejaculation (for example, gibbons – Edwards and Todd, 1991). Fellatio has also been reported for stump-tailed macaques.

Other perhaps related behaviours include sniffing/inspecting the genitals/anal region of other males (for example, stump-tailed macaques – Chevalier and Skolnikoff, 1976), 'displaying' an erect penis to other males (for example, vervet monkeys – Henzi, 1985), and urinating a few drops on the other male during the display (for example, squirrel monkey – Castell, 1969). In some cases, males have shown a preference for their homosexual partners over heterosexual partners (for example, rhesus macaques).

These behaviours have been reported in a variety of situations, including those involving displays of dominance and submission, in cases of general excitation, and in more playful situations between adolescent and/or adult males who demonstrate special affective relationships with each other.

At first glance the relationships between the different behaviours and the situations in which they occur seem arbitrary. For example, in some cases it is the dominant or older male who mounts the younger or subordinate (for example, observations among gorillas – Yamagiwa, 1992), while in other cases the reverse is true (for example, pig-tailed macaque – Oi, 1990). Presenting the anal region to another animal may indicate one's dominance (for example, squirrel monkey – Ploog *et al.*, 1963), or it may be a submissive gesture (for example, baboon – Lorenz, 1963). It may be the dominant who sniffs/licks the subordinate adolescent's genitals (for example, howler monkeys – Young, 1983), or the subordinate who sniffs/licks the dominant (for example, Hapalidae – Epple, 1967). It may be the dominant who ejaculates (for example, squirrel monkey – Ploog *et al.*, 1963) or the subordinate adolescent (for example, gibbons – Edwards and Todd, 1991). In some cases it is the dominant individual who urinates (for example, squirrel monkey – Castell, 1969), while in others the reverse occurs (for example, a subordinate wolf may urinate on itself – Lorenz, 1963).

The different sexual activities and their social associations seem so diverse that it is tempting to conclude that more complex animals have simply evolved a 'flexible' sexuality that allows them arbitrarily to find virtually anything sexy. Indeed, phenomena like the occasional sexual imprinting on different species or even on objects would seem to confirm this view (Lorenz, 1963; Maple, 1977). In light of this diversity Bagemihl (1999, pp. 261–2) simply eschews all attempts at explanation, and prefers instead to exalt in a 'biological exuberance', which 'embraces paradox' and 'is about the unspeakable inexplicability of earth's mysteries'. Still, I think it possible and profitable to search for order behind this apparent chaos. We just need to look a little more closely at the situations in which these behaviours occur.

I propose that male homosexuality evolved through various stages. The most primitive stage occurs in animals without multi-male groupings, where adult males are generally intolerant of each other's presence. In this stage, homosexuality basically occurs in response to a deception tactic on the part of 'transvestite' males who seek to gain access to a courting male's territory. Note that the territorial male is not actually aware that he is even engaging in homosexual courtship. This is the stage in the evolution of male homosexuality characteristic of the 'transvestite' lizards and snakes described above.

A more complex stage of homosexuality occurs in situations where multi-male groups are adaptive, perhaps in avoiding predators or for other reasons, but where males do not cooperate with each other on specific tasks. In this situation, more powerful males may have reasons to expel other males from the group, but they also have an incentive for allowing them to stay. The 'solution' to this dilemma is the maintenance of a clear dominance hierarchy. The subordinates may stay in the group, but must periodically acknowledge the superior positions of dominant individuals in the hierarchy. The rituals used to define these hierarchies may have derived from two sources. First, some of the 'transvestite' deceptive behaviour described above may have been exapted (changed function) to express submission. The connection may have something to do with how males succeed in imitating females. Lorenz (1963) gives the example of *Cichlid* fish. In this species males recognize females only by the fact that females can mix the emotions of fear and sexual excitement, while males cannot. Lorenz argues that many vertebrates distinguish between the sexes only by the different ways males and females mix basic emotions. It is behaviour, not morphology, which is recognized as distinguishing males from females. By allowing itself to be sexually mounted, a phylogenetically simpler animal may simply be mimicking a female to gain access to the dominant's territory. In later phylogenetic history, this same tactic may no longer deceive the dominant male, but may be used instead to clarify the subordinate's acceptance of his lower status. As in most, if not all, evolutionary changes in function, this change may have been gradual. At one point in phylogeny the same behaviour may very well have served both functions: deceiving some males, while pacifying others by communicating acceptance of a subordinate status.

Another source of later dominance rituals may have been the behaviour used to mark off territories. Many animals have scent glands (often near the genital-anal region) that incorporate urine or faeces secretions to deposit scents on a territory's borders or elsewhere. When subordinate males are allowed to stay near a dominant male, part of the price may be having these territorial markers literally 'rubbed in their faces'. For example, Epple (1967) describes the genital displays (perhaps best described as 'mooning') among dominant males in different *Callitrichidae* species. In silvery marmosets the dominant male shows his anal-genital region to a subordinate while threatening him. The subordinate, in turn, crawls towards the dominant, performing submissive gestures and emitting submissive sounds, before smelling the dominant's genitals. When extremely fearful, submissive males may only smell the dominant's tail. Sometimes the dominant male backs into the submissive animal and rubs his genitals in the submissive's fur, thus marking the subordinate directly with his own odor. Often the dominant withdraws his testicles, but shows an erection during these displays. The dominant's erection may serve to clarify that he is allowed to have an erection (and sexual relations with females), and the subordinate must get used to seeing this. The subordinate is not permitted this license.

Squirrel monkeys have similar, but somewhat more complex dominance rituals. The dominant male approaches a subordinate, opens his thighs (sometimes while touching the subordinate with his hands and/or penis) and 'displays' his erect penis in the subordinate's face, sometimes emitting a few spurts of urine in the process. The subordinate usually hunches over and may succeed in turning his face away from the display, but if he does not remain sufficiently passive or if tries to move away, the dominant may become aggressive (Ploog *et al.*, 1963; Castell, 1969).

The relationship between these displays and sexuality involves more than just the fact that an animal is being mounted and that erect penises are being displayed – the reproductive success of subordinate individuals may actually be compromised. Among mouse lemurs, for example, exposure to the urine of an active dominant male reduces the testosterone level of adult males (Stoddart, 1990) and this may, in turn, reduce their sexual activity with females. Male prairie voles experience a delay in their sexual maturation if exposed to male urine (Forsyth, 1991). In squirrel monkeys, there is clear evidence that the subordinate animals engage in less sexual activity (Ploog *et al.*, 1963).

The squirrel monkey's genital display is very similar to the *Callitrichidae*, but the squirrel monkey has a more complex repertoire. In addition to genital displays, dominants also demonstrate their status by actually mounting submissives. These male–male mounts sometimes involve anal penetration (Ploog *et al.*, 1963). Possibly, an ancestral 'transvestite deception' system was combined with a 'scent-marking' system to produce the squirrel monkeys' dominance rituals. Or perhaps the ancestral display of erect penises by dominant males simply became more explicit. In any case, the squirrel monkey's dominance system also has other novelties. Dominant males sometimes 'ask' other males to stay. They do this by rolling on their backs and exposing their bellies, sometimes showing an erect penis. Animals that would normally run away stay around if the dominant performs this ritual. This permits more non-related males to join together more cohesively among the squirrel monkeys than occurs among the *Callitrichidae* (Ploog *et al.*, 1963; Castell, 1969).

A particularly dramatic variant of the genital display is found in vervets (Henzi, 1985). Vervets have bright red penises with powder blue scrota. During submissive gestures, males retract both their penises and their scrota into their bodies, while the dominants leave both extended. In one display, the dominant walks or runs toward the submissive and then turns perpendicular to him so the submissive can observe the dominants' genitals. In another, the dominant circles around the submissive showing his anal/genital region, sometimes holding on to the submissive during the display. In addition to these non-solicited displays, the submissives sometimes seek out the dominants in order to demonstrate their subordinate status. They do this by submissively running after them, or creeping up to them, and at times cupping the dominant's testicles in their hands and tugging. Both dominants and submissives have been observed with erections during these latter rituals. Other multi-male species also engage in these sorts of ritualized interactions involving rank recapitulation. Lorenz (1963) describes an incident in which a defeated baboon chased after his conqueror 'presenting' his behind until the dominant finally mounted him. This may help explain why, among white-handed gibbons, an adolescent was observed soliciting sexual contact with the adult male in his family group. As Edwards and Todd (1991) suggest, the youth may have needed 'reassurance' that he was still welcome in the family.

I think the different rituals of the *Callitrichidae* versus the squirrel monkeys, and vervets reflect different levels of male/male cooperation, especially cooperation between unrelated males. In the *Callitrichidae* small family groups, all the males are closely related and cooperate in rearing the dominant male's offspring. In the larger squirrel monkey and vervet monkey groups, unrelated males are more numerous and they do not cooperate in rearing young. The reproductive competition inherent in such social situations may require more elaborate rituals to confirm dominance hierarchies just to guarantee peace.

One aspect of dominance hierarchies that deserves special attention is their stability. There are unstable moments when hierarchies are challenged, and stable moments when they are quietly accepted. Some of the apparently 'contradictory' dominance rituals reported by ethologists may be explained by whether the hierarchies were relatively stable or unstable at the time of observation. For example, among pig-tailed macaques submissives have been observed mounting the dominants more than the reverse (although it is the dominant who solicits this behaviour). This may be because dominants, when secure in their positions, may profit by 'ceding' to submissives at times. This lets the subordinates know that they are welcome. Yet immediately after a challenging pig-tailed macaque assumes the alpha rank in its group, it refuses to let others mount it. This may be because this is an especially vulnerable moment, when doubts about hierarchical positions need to be allayed (Oi, 1990). However, since there have been some observations of subordinates mounting dominates immediately after agonistic displays, this cannot be the whole story (Vasey, 1995). Mounting may serve to 'negotiate status' among primates, but the details of just how these negotiations work still needs clarification.

In many primates immature males seem to 'practice' these dominance rituals, displaying or presenting to each other and allowing their partners to mount them (for example, vervets, rhesus, gorillas, orangutans). 'Neoteny' is a common way for 'new' traits to appear in a species (Gould, 1977a). So this youthful play may be the phylogenetic and developmental precursor of adult homosexual behaviour, in which neither partner is clearly dominant to the other. Adult bonobos, for example, often cement alliances with homosexual activities (Waal, 1989); Smuts and Watanabe (1990, cited in Vasey, 1995) showed that cooperating *Papio cynocephalus* baboons often cemented alliances by engaging in homosexual activities with each other just prior to challenging a rival.

By way of conclusion, there are a few important points to be made about primate homosexuality. The first is that, in most cases, the homosexual behaviour has a pacifying effect – averting aggression, reassuring subordinates and dominants of their place in the hierarchy, and/or cementing alliances – although a homosexual display may also represent a challenge during a hierarchical dispute. Moreover, the different forms of homosexuality seem to 'scale', that is the behaviours and associations found in more phylogenetically primitive species are also found, under certain circumstances, in species that are more phylogenetically derived and socially complex. However, the latter have some additional behavioural complexities not present in their more primitive counterparts. In line with this reasoning, Vasey (1995) pointed out that in platyrrhine primates homosexual behaviours have been observed in conjunction with play and with dominance displays, but not in conjunction with alliance formations. Catarrhine species seem to have elaborated on this repertoire including more homosexual relations involved in reconciliation and alliance formation. If the scaling argument is correct, then this suggests that homosexuality is closely tied to the evolution of more complex social behaviour – probably due to its effect in reducing hostilities between males.

Although some authors (for example, Kirsch and Rodman, cited in Sommer, 1990) emphasize only dominance hierarchies in the origins of proto-homosexual behaviours, these hierarchies should probably be seen as only one act in a longer play that begins with territoriality and deceptive 'transvestite' tactics and ends with alliance formation and affection. It is an argument about how exaptations occurred – not miraculous adaptations in which new functions appear out of the blue. Rather it seeks to elaborate on how small changes and additions resulted in gradual functional changes, in which new functions co-occur with older functions. What changes over time is the gradual reduction in importance of one function and the increase in importance of another. Eventually one function may disappear entirely (for example, the deceptive function of female mimicry), but other more primitive functions continue.

For this theory, homosexuality is not an evolutionary 'spandrel' – a coincidental by-product with no adaptive significance at its origin (Gould, 1979). On the contrary, homosexuality is central to understanding how male animals came to cooperate with each other. As a basic element for the origin of society, it is one of the most important issues in evolution. In thinking about cooperation most evolutionary theorists have concentrated either on 'altruism' (usually analyzed in terms of kin selection, parental selection, etc.) or on 'reciprocity' (usually analyzed with game theory and usually assuming 'equal' players). I find it astonishing that (except for ethologists) few theorists have dealt with the evolution of personal hierarchies. Yet the animal literature makes it clear that reciprocal cooperation between equals is much rarer than cooperation arising out of hierarchical disputes.

Human homosexuality fits well within this scenario for the evolution of society. Humans are certainly much more cooperative than other animals, and our societies are much more complex. A closer look at the more 'egalitarian' human societies studied by anthropologists also reveals the tremendous importance of status hierarchies (Werner, 1996, 1981). It is to be expected, then, that humans should exhibit more elaborated forms of homosexuality. What, then, does all of this imply about human homosexuality?

Implications of the hierarchy/cooperation theory for human homosexuality

Before seeking evidence for the hierarchy/cooperation theory among humans, it is necessary to clarify questions surrounding one of the major academic debates of the last two decades – the 'essentialists vs. constructivists' (see Cardoso and Werner, 2004 for a brief historical review of this debate.) 'Essentialists' postulate more universal biological and psychological bases for homosexuality, while 'constructivists' see 'homosexuality' as having no meaning whatsoever outside of its cultural context. Supporting an 'essentialist' viewpoint, most biologists cite evidence for genetic and epigenetic causes of homosexuality, while psychologists point to the common childhood precursors and common personality and cognitive traits of homosexuals. To support a 'constructivist' view, anthropologists and social historians cite the great variation across human cultures in the ways homosexuality is organized or defined. A brief summary of the evidence for these views may help clarify the debate.

'Essentialists': biologists and psychologists

Biological studies have documented differences between male homosexuals and heterosexuals in their exposure to pre-natal hormones (LeVay, 1994; Reinisch et al., 1991), brain structures (Swaab and Hoffman, 1990; LeVay, 1994), genetic markers (Hamer et al., 1993; Mustanski et al., 2005) and possibly other characteristics such as fingerprint patterns (see Downtown, 1995 with regard to a University of Western Ontario study) that may be related to testosterone exposure (Jamison et al., 1993). In addition, effeminate boys (who have a strong tendency to become adult homosexuals, see Bailey and Zucker, 1995) are judged more attractive than other boys (Zucker et al., 1993), which agrees with Green's (1987) finding that parents of effeminate boys rated them as more 'beautiful' babies than their other children.

Blanchard and Sheridan (1992) found that homosexual men had more older siblings than non-homosexual men, and other studies found homosexual males had more older brothers than heterosexual males (Blanchard and Klassen, 1997; Cardoso, 2004; Blanchard, 2004), possibly reflecting a progressive immunization of some mothers to H-Y histocompatibility antigen. Other researchers (Dörner et al., 1980; Cardoso, 2004) found that the mothers of male homosexuals may have suffered more maternal stress during pregnancy.

Finally, support for genetic arguments has come from studies of twins and of homosexuality in family histories (Whitam, 1983; Pillard and Weinrich, 1986; Bailey and Pillard, 1991; Buhrich et al., 1991; Eckert et al., 1993; Flores, 1994; Cardoso, 2004). Anecdotal evidence from tribal level societies also suggests inheritability of homosexuality. For example, Wilbert (1972, p. 101) reported that the Warao Indians of Venezuela think that effeminate males are more common in some of their families than in others.

The most complete and careful study of family relationships and homosexuality is Green's (1987) 15-year comparison of effeminate and masculine boys beginning from the time the boys were four to 12 years old, and continuing into adulthood. The boys and parents were interviewed and observed regularly over this time period, and psychological tests were administered at various points. Of 30 effeminate boys accompanied throughout this period who had had sexual experiences, 24 were 'more than incidentally homosexual' as adults. Of the 25 'masculine' controls, only one was more than incidentally homosexual as an adult. Many of the effeminate boys were subjected to behaviourist or other therapies during their childhood, without any apparent effect on their later homosexual behaviours or fantasies. In addition, although a good deal of attention has been given to the role of parental child-rearing behaviour, neither Green nor others (Greenstein, 1966; Siegelman, 1974; Green, 1987) found much support for these arguments.

However, childhood gender non-conformity consistently predicts adult homosexual orientations in North America and Europe, as well as other cultures (Whitam and Zent, 1984; Whitam and Mathy, 1986; Green, 1987; Phillips and Over, 1992; Cardoso, 1994; Bailey and Zucker, 1995). While homosexuals are more likely to have been effeminate as boys, there are still many homosexual males who have reported more normal childhoods (Phillips and Over, 1992). Weinrich and his colleagues (cited in LeVay, 1994) showed that it is the homosexuals who prefer a more 'passive' role (as 'insertee') who are most likely to have been effeminate boys. These findings have led researchers like Green to propose causal models for homosexuality that begin with the influences of genes and pre-natal hormones. Characteristics of parents (like the desire for a girl or a boy) might affect acceptance or tolerance of feminine behaviour in boys, which in turn might affect their adult femininity, but have less effect on their homosexuality.

'Constructivists': anthropologists and historians

Most anthropologists and historians feel frustrated with the biological and psychological work on male homosexuality because it seems to account so poorly for the cross-cultural variation in male–male sexual relations. This variation is so great that it seems impossible even to *define* homosexuality in a cross-culturally meaningful way. To find *causes* that are cross-culturally valid seems preposterous. For example, Dickemann (1993) cites the case of homosexuality in medieval Europe. She argues that during the period of Charlemagne parents simply decided that their last-born son should adopt homosexuality, and apparently the parental decisions were followed. When we consider the even more 'exotic' cultures of New Guinea (Kelly, 1974; Herdt, 1993), the arbitrariness of cultural definitions of homosexuality seems even clearer. In societies like the Sambia or Etoro, all boys are expected to have sexual relations with older males.

Indeed, people believe that the boys' maturation would be impossible if they did not receive semen from the older males. Among the Etoro, sexual relations between men and women are taboo most days of the year, although homosexual relations are constantly encouraged.

Compatibility of the essentialist and constructivist views

If there is no such thing as 'homosexuality' in a meaningful cross-cultural sense, then what should we make of the biological findings? I think there are two possibilities. First, the biological findings may be peculiarly Western. That is, there may be no gene for 'homosexuality', but rather genes for other characteristics that our particular culture associates with 'homosexuality'. For example, parents may define certain inborn facial features as 'beautiful' in their babies. This may lead them to treat these babies as more delicate and 'feminine' than other babies. It is the later 'femininity' of these boys which then gets defined as 'homosexual'.

I think a more likely possibility is that the cultural differences are not really so great. The apparent incompatibility between the biological and cultural arguments may simply be a question of their dealing with different phenomena. Biologists are more interested in sexual orientation – what individuals are sexually attracted to and/or the nature of their sexual fantasies. On the other hand, anthropologists are more interested in sexual practices – that is, who has sex with whom and what people actually do in these sexual relations. Anthropologists are also extremely interested in sexual identity – how cultures define individuals and how individuals classify themselves. There is certainly a great deal of cultural variation pertaining to sexual practices and identity, but we know much less about the cross-cultural variation in orientation as defined here. The differences among these concepts are not always clear in the work of different scholars. For example, Herdt (1993, p. xlvii) states that, 'Identity includes feelings, ideas, goals and sense of self.' Money and Ehrhardt (1972) often speak of sexual 'identity' without distinguishing this from 'orientation' and without considering the influence of culture in forming these identities.

How different are cultures? Evolutionary psychologists tend to emphasize what is universal to all cultures, while cultural anthropologists generally concentrate on what is unique. To resolve the discrepancies between 'essentialists' and 'constructivists', I think the most productive approach is to examine those cultural features that are common to *some* cultures but not to others. One typology, originally suggested almost 40 years ago (Gorer, 1966), groups cultures into one of three male homosexual systems, now labelled 'gender-stratified systems', 'age-stratified systems', and 'egalitarian systems'.

By far the most common homosexual system in the ethnographic record is the 'gender-stratified system' in which gender-typical males may engage in active (inserter) sexual relations with distinctly identified 'pathics' who generally prefer a passive (insertee) role. That typical males may engage in sexual behaviour with pathics without receiving a 'homosexual' identity is clarified by Cardoso's (2004, 2005) study of homosexual behaviour in a Brazilian fishing village. Pathics were distinguished with a specific term in the local vocabulary, and informants agreed very highly on who the pathics were. However there was no term to distinguish men who had sex with the pathics from the men who did not, and informants could not even agree on just which men had these relationships. A culturally differentiated pathic (often a cross-dressing male) might take on an honored religious role as among the South African Zulu or the Patagonian Tehuelche, or might be mistreated as among the Angolan Mbundo or the Bolivian Chiriguana, although his active partner received no rebuke. He might eventually assume the role of a second or third wife to a gender-typical male, as among the Warao of Venezuela, the Tanala of Madagascar, the Karen of Burma or the Chukchee of Siberia, or he might serve as a sexual partner to various men in the community throughout his lifetime as among the Brazilian Tupinamba. Just how often typical men in these societies participate together in sexual escapades with the pathics is unknown, but judging from a few ethnographic reports and drawings of the 'dance to the berdache' (see Williams, 1992, Plate 6) such 'partying' may not be rare (Katz, 1976; Werner, 1999a). Murray (2000) coded 120 societies as 'gendered-stratified'.

'Age-stratified' homosexual systems include 'mentorship' or 'ritualized homosexuality' systems, in which an older male takes on a younger male as an 'apprentice', as well as 'catamite' systems in which younger males serve simply as sexual objects to a powerful, older male. 'Mentorship' systems have been found among groups as diverse as the ancient Greeks, Australian Aborigines, numerous New Guinea cultures, Tibetan monks, Japanese samurai, Egyptian Siwans and the Zairean Azande. They are remarkably similar in many respects. Boys may begin their initiation/apprenticeship as young as seven years, as among the New Guinean Sambia, or as old as 12 among the Azande, and may continue their apprenticeship until as old as 20 or 25, as among the New Guinean Etoro. After this age, young men generally switch to more active roles with younger boys. It is very common for a father to choose carefully his son's tutor, and in most, if not all cases, the relationship between the two is monogamous, and involves instruction in practical matters as well as moral courage and discipline (Werner, 1999a). 'Catamite' systems were quite common in the Hellenistic empire as well as other societies, such as ancient Rome, China, Korea, Japan, Egypt, Turkey and the African Ashanti. Among the West African Mossi, chiefs had sex with boys on

Fridays when sex with women was taboo. In some cases the boys were simply kept as slaves, while in other cases they traveled with theatrical troupes and served as prostitutes as well. Murray (2000) coded 53 societies as 'age-stratified'.

'Egalitarian' homosexual systems include societies: (a) where homosexual relations occur among typical males only during adolescence, (b) where 'blood-brotherhoods' may formalize homosexual ties between gender-typical males throughout life and (c) where homosexually identified males have sex primarily with other homosexually identified males (the modern 'gay' system). 'Egalitarian' systems are relatively rare in the anthropological record. Murray (2000) coded only 30 societies as 'egalitarian'. And 'gay' systems may be limited to Northern Europeans and their descendants of the past few centuries, which is probably what most social constructivists (for example, Foucault, 1978) are talking about when they refer to the recent cultural construction of 'homosexuality'. Still, it is the 'gay' system, which most non-anthropologists have in mind when they talk about homosexuality. The 'gay' system seems to be gaining ground in many countries of the world. For example, Murray and Arboleda (1995) noted changes over time from 'gender-stratified' to 'gay' systems in Guatemala, Mexico and Peru. In the 1970s, only 50% of their informants had heard of the term 'gay', and only 23% thought it referred to both 'passive' and 'active' partners. In the 1980s, 76% had heard of the term and 58% applied it to both 'passives' and 'actives'. Cardoso's (2004) data from Thailand, Turkey and Brazil showed that the 'gender-stratified' system is most common among the lower classes in each of these cultures, while the professional classes generally adhere to the 'gay' system.[1]

Most societies can be classified as adopting one or the other of these homosexual systems, but different sectors of a culture may adhere to different systems, and at times even the same sector may recognize different systems. For example, the ancient Greeks had 'mentorship' systems of homosexuality, but also recognized pathics with distinct terms and prohibited them from holding public office (Murray, 2000). The Australian Murngin also recognized pathics as distinct from other males, although a 'mentorship' form of homosexuality also characterized their society. In addition, homosexual behaviour may also occur within a society in very context specific social situations such as among prisoners, in street gangs or to humiliate enemies in warfare (Duerr, 1993).

It is important to recognize that most of the cultural variation in homosexuality refers to the homosexual behaviour of typical males, not to the sexual orientations of pathics. It is entirely possible for researchers to acknowledge the cultural 'construction' of homosexual behaviours among typical males, while still recognizing that pathics share 'essentially' the same sexual orientations in all human societies.

At this point it is important to clarify two points. First, 'essential' traits are found in 'orientations/desires', not behaviours. For example, many Brazilian cross-dressers (especially those working as prostitutes) comment that they often assume the active (inserter) role in sexual relations. Yet this behaviour may tell us more about the sexual desires of their partners than of their own. Cardoso's study of Brazilian, Turkish and Thai homosexuality showed that the vast majority of homosexuals (coded as those who preferred sex with males) found passive anal intercourse exciting (90% of Turks, 87% of Thais and 87% of Brazilians), but of the gender-typical (culturally non-differentiated) men who had sex with the pathics, far fewer liked the passive role (17%, 31% and 30% respectively for Turks, Thais and Brazilians). In all three cultures pathics somewhat preferred the passive roles, while among the men who had sex with them, far more preferred the active role (92% of Turks, 62% of Thais and 92% of Brazilians).

Second, it may also be important to clarify a few questions regarding differences in the cultural categories for homosexually identified males. For example, Lakota Sioux informants noted the similarities of their traditional *winktes* with modern 'gays'. But they commented that *winktes* had sex with men, not with other *winktes* like gays do, and one Indian complained: 'It makes me mad when I hear someone insult *winktes*. A lot of the younger gays, though, don't fulfill their spiritual role as *winktes*, and that's sad too' (Williams, 1985, p. 195). Perhaps not all modern gays have 'essentially' the same orientations as pathics. As mentioned above, Weinrich and his colleagues (cited in LeVay, 1994) found that it was the gays who had gender atypical childhoods who most preferred the passive roles. These are the Western 'homosexuals' who most resemble pathics elsewhere. I suspect that if the other Western-style 'gays' had been brought up in another cultural context they might have been identified with the culturally undifferentiated 'typical' males.

The case of modern day Iran illustrates dramatically just how important differences in how homosexuals are identified can be. In Iran homosexual behaviour is punishable with the death sentence. Yet the country's health system provides for sex-change operations and permits transsexuals to marry men. Neither the transsexual's nor her partner's behaviour are considered 'homosexual'. Most anthropologists would likely cite this as evidence for the enormous impact of arbitrary cultural categories. But there may be a less 'arbitrary', more practical question behind the Iranian distinction. During a discussion of gender-stratified systems, a Palestinian friend once commented to me that the Brazilian system seemed very similar to the one he was familiar with. He added, however, that Islamic law looks down on homosexuality because it sees the 'typical' man's behaviour toward the pathic as 'abusive' of the pathic. One might wonder whether this view stems from historical experience of truly abusive homosexual behaviours in the treatment of enemies or prisoners (Duerr, 1993).

Implications for humans of the dominance hierarchy/cooperation theory

What kinds of evidence might support or refute the 'hierarchy/ cooperation' argument for homosexuality? There are three areas of inquiry for which evidence may be forthcoming. First, the theory makes predictions about deeply rooted feelings in all humans. Second it may account for individual variation in all cultures. Third, it may help explain why cultures vary.

Universals

If the association between cooperation and homosexuality is correct, then we should expect humans to show more homosexuality than most other complex animals, with the possible exception of bonobos, who, as Wrangham and Peterson (1996) argue, may be even more cooperative and more homosexual. I believe that humans do practice more homosexuality, although an over-concentration on North European cultures and an over-concentration on genital-sex may sometimes confuse the issue.

The review of animal homosexuality suggests that animals living in socially complex societies add new dimensions to the already existing repertoires of simpler species, yet they still retain the older repertoires. This makes sense in terms of the 'tinkerish' economy of natural selection. One of the first predictions of the hierarchy/cooperation theory is that we should find evidence of these older repertoires in humans. I think such evidence is available for the cognitive associations all humans share regarding homosexual behaviours.

As modern neuroscientists have pointed out (Damasio, 1994; LeDoux, 1996), the human brain is constructed in 'layers'. The phylogenetically inner layers are more conservative, varying less from one species to the next, in conformity with the great dictum of natural selection: 'If it ain't broke, don't fix it.' The outer layers are more recent, and vary more from one species to the next and from one individual to the next (Cairns-Smith, 1996). Thoughts or perceptions organized in the outer layers need to produce their effects by acting through the inner layers. For example we may reason our way (using the outer brain layers) into a fearful state (produced in the inner layers), or may similarly reason our way out of a fearful state. However, when the different regions of the brain are in disagreement, feelings may be less 'convincing' (Damasio, 1994; LeDoux, 1996). When they all agree, they may be perceived as especially powerful or 'raw'.

These sorts of differences are seen even in language. As etymologies show, our abstract concepts are almost always constructed from analogies with more concrete phenomena, and children learn concrete concepts long before they are able to make abstractions. Very likely, the concrete concepts are phylogenetically more primitive. Swear words are especially concrete. Unlike other language capacities, swearing has its origins in the phylogenetically older sub-cortical parts of the brain. Brain-damaged individuals may lose their capacity for virtually all language, but still retain the ability to swear (Pinker, 1994). Psychoanalysts (for example, Arango, 1989) and anthropologists (Duerr, 1993) have taken advantage of this phenomenon to analyze the sources of our most powerful emotions. In the case of dominant/submissive relations our swearing vocabulary is rich and revealing. Primate mounting behavior is illustrated in expressions like 'he wants your ass', or 'up yours!'. References to submissive individuals in different languages are consistently related to primate gestures of submission. In English we call overly submissive individuals 'ass kissers' or 'brownies'. In German and Vietnamese, the expressions are only slightly different: 'Arschkriecher', and 'NimBa', respectively, expressing the act of crawling up to the dominant's behind. Spanish speakers say 'lame culo', southern Slavs 'Dupolizac', and Russians 'Podliza' all referring to licking the dominant's behind. The Brazilian term, puxa-saco (scrotum tugger), refers to another primate gesture, while Turks use 'Kiçimi yala' referring to ass licking, although they also possess the milder 'Dalkavuk', referring simply to the subordinate's bowed back. Uruguayans say 'chupa medias' (literally 'sock sucker'), referring to what may be another scent marker. Other English expressions based on primate gestures include 'rubbing it in', 'smearing it in your face' and 'sucker'. The reader can probably fill in more examples, including some only slightly more abstract expressions.[2]

These cognitive associations may also have their behavioural counterparts. Especially in dramatic situations where we are controlled by 'raw' emotions the more primitive associations of homosexual behaviours with dominance and submission may resurface. This may be what is behind the wartime humiliation of enemies through homosexual rape or other homosexual behaviours, or prison rapes during hierarchical disputes (Duerr, 1993).

Individual differences

'Dirty words' are helpful in understanding attitudes that may be deeply rooted in all individuals. But in all cultures there may be important differences between people as well. Most anthropological studies have concentrated on general cultural norms, categories and symbolic systems, not on individual differences. This is unfortunate, because it tells us little about individual variation in homosexual activities for these societies. Often it is difficult enough just to confirm that exclusive homosexuals are absent. For example, Werner (1984) describes a traditional myth about a transvestite among Brazil's Kayapó Indians, but the Indians reported having never heard of a case. On the other hand, Crocker (1990) reports the presence of transvestites among the Kanela Indians of Brazil, but suggests the culture had no tradition of transvestites. One wonders whether cultures simply adjust their behaviours or attitudes to accommodate transvestites or homosexuals in their midst when they happen to appear.

Anecdotal accounts of pathics in many tribal societies suggest similar childhood backgrounds to Western homosexuals, but more systematic data are extremely rare. Still, statistical studies in a few more complex non-Western societies have confirmed US/European findings with regard to the correlations between exclusive homosexuality (and especially preference for passive roles) on the one hand and childhood behaviours or family histories on the other (Whitam, 1983; Cardoso, 2004). I strongly suspect that similar correlations would be found in all human societies.

Even rarer are studies of the psychological and social differences between gender-typical men who do or do not have occasional homosexual relations. Such studies are uncommon even for the USA. These studies may be particularly revealing with regard to the evolutionary theory postulated in this paper. For example, McConaghy and Blaszcynski (1991) looked at homosexual feelings among males with predominant heterosexual attractions and discovered that homosexual feelings are most correlated with having disliked outdoor and contact sports. There were no correlations of homosexual feelings with activities like cooking or playing with dolls as a child. Likewise, in his study of 41 men from a Brazilian fishing community, Cardoso (1994, 2002, 2005) found that the men most interested in sex with the community's exclusive homosexuals were significantly more likely to have avoided playing soccer as children than were other men who had fewer relationships with the homosexuals. But these men were not different from other men in cross-gender behaviours (like playing with dolls). These findings suggest that submissiveness may be more central to homosexual feelings among culturally undifferentiated 'typical' males than is femininity.

Further clarification of the characteristics of 'bisexuals' in different cultures comes from a larger study in which Cardoso (2004) gathered interview and questionnaire data from 880 men between the ages of 20 and 30 in cities in Brazil, Turkey and Thailand. Cardoso chose these societies in order to maximize diversity, including predominantly Christian, Muslim and Buddhist cultures. His sample was stratified to include homosexual, bisexual and heterosexual men from both working and professional classes in each country. Comparisons of 'bisexuals' (men who had sex with both males and females) with exclusive heterosexuals were revealing. Bisexuals were intermediate between heterosexuals and homosexuals on some childhood precursors of homosexuality (like not playing soccer, preferring girl's tasks, playing with girls or wanting to be a girl), but they were much closer to the heterosexuals than to the homosexuals on these variables. Perhaps more telling, they rated higher than both homosexuals and heterosexuals in 'liking to dominate' their sex partner and 'receiving sexual invitations from women'. Some of the class distinctions help clarify the meaning of these results. Compared with professional-class bisexuals, the working-class bisexuals were less likely to have preferred girls' tasks or have wanted to be a girl, and they were actually more likely than working-class heterosexuals to have bullied others when young (Cardoso, 2004). Very likely there are two quite distinct profiles of 'bisexuals' in the working classes. In Cardoso's sample of working-class bisexuals, those who liked to bully also liked to dominate their sex partner ($r = 0.50$, $p < 0.001$), but they were not more likely to have 'girlish' childhoods (all correlations with bullying were negative, although not statistically significant). On the other hand, those who showed one 'girlish' childhood precursor also tended to have the other girlish precursors (correlations between 0.21 and 0.48, all significant) (Cardoso, personal communication). It is perhaps the presence of a non-intermediate type of 'bisexual' among the working classes that explains the apparently greater frequency of homosexual behaviours found in his haphazard samples of typical working-class males.

These cross-cultural studies suggest a few factors that might be behind the cultural variations in homosexual behaviours reported by anthropologists. To what extent, are cultural variations in line with the hierarchy/cooperation theory?

Cultural variation

The 'hierarchy/cooperation' theory suggests that variations in the ways men cooperate or form hierarchies might explain some of the cultural variation in male homosexual systems. What is the evidence?

Egalitarian, gender-stratified and age-stratified systems. Elsewhere, Cardoso and Werner (2004) reviewed the results of statistical cross-cultural studies on homosexuality. Of particular interest were the results of Murray's (2000) and Crapo's (1995) comparisons of 'gender-stratified', 'age-stratified', and 'egalitarian' systems of homosexuality. In a nutshell, these studies showed that 'egalitarian' systems are most common where males are more involved with infant care, and where there is a generally more egalitarian social structure. 'Gender-stratified' systems are most common where males and females do similar tasks, where women have more power and where there is no adolescent segregation of the sexes. 'Age-stratified' systems occur where there is more patrilocality and patrilineality and greater segregation of the sexes. This suggests that cultural variations in homosexuality may reflect different strategies for male/male cooperation. Where men must invest more in children they eschew homosexual relations, limiting homosexuality to adolescence, to rare 'blood-brotherhood' ties or to 'gays'. Where they have more time to invest in cooperation they may engage in more homosexual behaviour, perhaps to better cement relations. If the society is sexually segregated, this takes the 'mentorship' form of homosexuality. Otherwise the 'gender-stratified' system permits sexual escapades with pathics.

The modern 'gay' system. Historical changes in male/male cooperation may also help account for the origin of the modern 'gay' system in which culturally undifferentiated 'typical' males do not ordinarily engage in sex with culturally identified 'homosexuals' or 'gays'. In brief, I think the gay system may have its roots in one of the major historical changes of the past century or so – the increasing importance of specialized knowledge and skills in performing jobs. When there was a vast pool of labourers with sufficient skills for a job, the skills themselves were not the decisive criterion for hiring decisions. Instead, characteristics like trust, personal ties or familial obligations carried more weight. But over the years specialized knowledge and skills have become increasingly important, and the number of years needed to acquire knowledge and skills has been rising steadily. Employers have been forced to make hiring decisions based on *what* candidates know, rather than on *whom* they know. Personal loyalty has lost its value in the market place and may be downright discouraged, criticized as 'nepotism' or 'favoritism'.

Even in contemporary societies, the value given to personal loyalties may be very different among unskilled workers versus professionals. For example, in all three societies studied by Cardoso (2004) lower-class males were more likely than professional class males to agree that 'getting ahead depends more on *whom* you know than on *what* you know'.

If male homosexuality is related to male/male cooperation, we should expect more homosexual behaviour where loyalties (and personal cooperation) are more important. Indeed, in Cardoso's haphazard samples of presumably exclusive heterosexuals more 'bisexuals' appeared 'by chance' in the working classes than in the professional classes for all three societies, suggesting that homosexual activities may indeed be more common among undifferentiated 'typical' working-class males. More importantly, within the lower classes of all three societies, the bisexuals agreed more than did the exclusive heterosexuals that 'getting ahead depends more on whom you know than on what you know', illustrating a direct link between bisexual behaviour and the importance of personal loyalties. These findings, then, suggest that the increasing importance of professionalism as societies become more industrialized throughout the world may help explain the expansion of the 'gay' system at the expense of traditional 'gender-stratified' systems.

Concern about one's place in the hierarchy. Concern with one's place in the hierarchy may also affect the types of homosexual behaviour men perform. For example, where men are more preoccupied about their position in a dominance hierarchy, the distinctions between what the submissives do and what the dominants do should be accented. Indeed, in his study of prison rape, Silva (1998) showed that, despite their verbal 'justification' of prison rape 'to protect women and families' against convicted rapists, it was actually those prisoners most preoccupied about their personal status who advocated raping fellow prisoners.

Concern about one's hierarchical position may be especially high when hierarchies are threatened. Unstable hierarchies (which change frequently) represent constant threats. As suggested by the pigtail macaque example cited above, a dominant secure in his position may be able to demonstrate his 'generosity' by engaging in 'submissive' homosexual gestures, while one unsure of his position may be more likely to perform only 'dominant' gestures. Could the stability of hierarchies affect the types of homosexual activity carried out by humans? To examine this idea Mendes (1997) carried out a series of experiments in which she asked male university students to complete a short comic strip story about a prison. Mendes varied the stories slightly to ascertain the effects of these variations on student responses. Stories that included 'intimate visits with women' significantly reduced the likelihood of ending stories with a homosexual rape scene, which might support the idea that 'lack of women' leads to prison homosexuality. Yet even with these visits, 40% of the subjects still ended their stories with rape (other choices were physical fighting, non-sexual friendship, and friendly sex). When Mendes contrasted prisons organized on the basis of personal loyalties, versus an abstract evaluation system, there was a slight (non-significant) tendency to cite rape more often in the 'personal loyalties' situation (53.3% versus 33.3% of 60 students), which might support the 'professionalism' argument about homosexuality. Still, the statistically most significant difference occurred when Mendes combined the 'personal loyalties' condition with either 'stable' or 'unstable hierarchies' (few or frequent changes of cellmates). Where hierarchies were more unstable, respondents were much more likely to end their stories with a rape scene (68.5% versus 26.7% of respondents). In sum, policies that make personal loyalties important, but unstable, may create conditions especially conducive to prison rapes.

Hierarchy and cooperation in other studies of homosexuality. These findings suggest it may be useful to include questions of hierarchy/cooperation in general studies of homosexuality. This may help explain a few apparently 'anomalous' findings reported in the literature. For example, consider the finding of Adams *et al.* (1996) that homophobic men are more sexually excited by homosexual pornography than are non-homophobes. The standard psychoanalytic interpretation is that homophobia results from repressed homosexual desires. But another possibility is that the 'homophobes' were not at all like the pathics they humiliated. They may simply have found it exciting to use primitive gestures of dominance to affirm their own status. Finding such gestures exciting may have little to do with the standard psychoanalytical arguments, but rather result from some of the other factors discussed here – like greater concern with one's status in a personal, rather than professional hierarchy. Or 'homophobes' may simply be more prone to 'stimulus-seeking' behaviour. This might explain Cardoso's (2004) findings that many bisexuals were not intermediate between gays and straights, but rather more 'masculine' when it came to childhood bullying behaviour. Future studies that address these other factors might help to clarify these issues.

Many other relationships between forms of male/male cooperation/hierarchy and homosexuality might be hypothesized. But I hope these examples are sufficient to show the value of an evolutionary perspective in suggesting what kinds of relationships to look for.

Biological adaptation, psychological adjustment and morality

The ideas I have been discussing here have been around for many years, but they have been neglected, perhaps in part because many find them politically distasteful. I think this attitude results from misunderstandings about the relationships between biological adaptation, psychological adjustment and morality. In short, many people seem to have confused these really unrelated concepts. Gadpaille (1980, p. 354), for example, argues that 'homosexuality as a preferential or obligatory mode must by definition be biologically deviant', and implies that preferential homosexuality is pathological. Similarly, the psychoanalyst, Arango (1989), in proposing a close tie between dominance hierarchies and homosexuality, argues that homosexuality is not 'love' but 'masochism'.

But let us be clear here. Biological adaptation is not the same as 'psychological adjustment'. Biological adaptation refers to the passing on of genes, not to psychological well-being. The heterosis argument has often been presented as the 'sickle-cell' argument, in analogy with the well-known case of sickle-cell anemia. In malaria areas, individuals homozygous for the sickle-cell die of sickle-cell anemia, and individuals homozygous for the absence of sickle-cell are more likely to die of malaria – so mostly heterozygous individuals pass on genes. Now, in the case of sickle-cell anemia we really are talking about an illness! No one wants to get sickle-cell anemia, and people die from it. 'Illness' and 'health' are defined in terms of individual well-being, and perhaps at times (for example, psychopathic killers) in terms of social well-being. People do not need to pass on genes to be considered healthy. They need to feel healthy and happy, and to not cause harm to others. Arango's argument that homosexuality is 'masochism' is also off the mark, because it makes it sound as if our 'real' selves are what we find in the innermost regions of the brain. But human nature is based on our whole brains. And of course all the different forms of human 'love' (not just homosexual love) have their evolutionary history. I doubt very much whether Arango would reduce these forms of love to their homologues in ancestral fish!

Biological adaptation also tells us nothing about whether something is moral or not. Many adaptive traits are evil – for example the killing of another male's offspring when a new male overtakes a former dominant. Sommer (1990, chapter 15) has shown the absurdity of using the criteria of 'natural' (adaptive) or 'unnatural' (maladaptive) to decide whether homosexuality is 'good' or 'bad'. He found historical examples of scholars who argued for all the different possibilities: (a) that homosexuality is natural (found in animals), therefore it is good, (b) that homosexuality is natural, therefore it is bad, (c) that homosexuality is unnatural, therefore it is good and (d) that homosexuality is unnatural, therefore it is bad.

Still, there may be a tie between the notion of morality we actually have (not necessarily what we *ought to* have) and homosexuality. Elsewhere I have argued that the evolution of hierarchies is also behind much of what we sense to be 'moral'. In particular, psychological experiments show that people are more likely to feel moral indignation when they feel their place in the hierarchy has been threatened than when some 'unfair' act has been committed (Werner, 2003). Also, the correlations between the ways one behaves towards one's gods and the ways one behaves towards institutions that have power over us also suggest that the evolution of hierarchies may be behind religious belief systems (Swanson, 1960).

Morality refers to the surrendering of one's own interests to the well-being of others. Humans are capable of such surrendering because in their evolutionary past they learned to yield at times rather than aggressively defend their own interests. If the hierarchy/cooperation argument is right, then the evolution of morality depended on the evolution of homosexuality. This may sound bizarre. If homosexuality is at the base of morality, why are exclusive male homosexuals so defiled in so many places? I think the answer is simply that they are easy to mistreat because they generally yield more easily than others.

This contradiction between what we define as moral, and how we treat those most likely to yield to others' interests may well be one of the major conflicts in human society. It deserves a name at least as catchy as the Oedipus complex, although it is not an individual psychological complex, but rather a social complex. If it is really as important as my argument suggests, then I imagine this complex must appear in human myths. There are several possibilities. For example, the Kayapó Indians have a story about a boy who shunned men's work, and was sexually abused by a bat man, which caused him to giggle – the very first laugh ever, unworthy of a warrior, but necessary for life (Werner, 1984). Among the Cashinuaha there is a story about a great transvestite artist who showed the Indians how to draw, but who died because he was impregnated by a lover, and the baby could not be born (Lagrou, 1996). But the best-fitting story is closer to home. The story of Jesus is about a man who 'turned the other cheek' instead of fighting, who did not compete with other men for women, and who, in the end, was easily mistreated. Perhaps someday humans will learn to recognize this 'Jesus Complex' and things will change. Then maybe Jesus' prophecy will be born out: 'Blessed are the meek, for they shall inherit the earth.'

Notes

1 The differences between the North American/North European criteria and Brazilian lower-class criteria were most dramatically clarified for me in observing Brazilian reactions to American pornographic films. At one point in an American film a woman inserted her finger into her male partner's anus. The Brazilian audience went wild, crying out '*viado*' (queer) to refer to the male actor. On the other hand, in Brazilian *porno-chanchada* films of the 1970s it was not uncommon for a man to have (active-role) sexual relationships with several women, and also with a *bicha* (effeminate homosexual). No one referred to the inserter male as *viado* in these film sequences.

2 The use of more abstract 'symbolic' expressions for dominance may be found in primates as well. Enomoto (1990) reports the case of one male bonobo expressing its dominance over another by using a gesture normally used to solicit sex from an estrous female.

References

Adams, H. E., Wright, L. W. and Lohr, B. A. (1996) Is homophobia associated with homosexual arousal? *J. Ab. Psychol.*, **105**, 440–5.

Antinucci, F. (1990) The comparative study of cognitive ontogeny in four primate species, In '*Language*': *and Intelligence in Monkeys and Apes: Comparative Developmental Perspectives*, eds. S. T. Parker and K. R. Gibson, pp. 157–71. New York: Cambridge University Press.

Arango, A. C. (1989) *Dirty Words: Psychoanalytic Insights*. Northvale, NJ: Jason Aronson.

Bagemihl, B. (1999) *Biological Exuberance: Animal Homosexuality and Natural Diversity*. New York: St. Martins.

Bailey, J. M. and Pillard, R. C. (1991) A genetic study of male sexual orientation. *Arch. Gen. Psychiat.*, **48**, 1089–96.

Bailey, J. M. and Zucker, K. J. (1995) Childhood sex-typed behavior and sexual orientation: a conceptual analysis and quantitative review. *Dev. Psychol.*, **31**, 43–55.

Blanchard, R. (2004) Quantitative and theoretical analyses of the relation between older brothers and homosexuality in men. *J. Theor. Biol.*, **230**, 173–87.

Blanchard, R. and Klassen, P. (1997) H-Y antigen and homosexuality in men. *J. Theor. Biol.*, **185**, 373–8.

Blanchard, R. and Sheridan, P. (1992) Sibship size, sibling sex ratio, birth order, and parental age in homosexual and nonhomosexual gender dysphorics. *J. Nerv. Ment. Dis.*, **18**, 40–7.

Buhrich, N., Bailey, J. M. and Martin, N. G. (1991) Sexual orientation, sexual identity and sex dimorphic behavior. *Behav. Genetics.* **21**, 75–96.

Cairns-Smith, A. G. (1996) *Evolving the Mind: On the Nature of Matter and the Origin of Consciousness.* Cambridge: Cambridge University Press.

Camperio-Ciani, A., Corna, F. and Capiluppi, C. (2004) Evidence for maternally inherited factors favouring male homosexuality and promoting female fecundity. *Proc. Biol. Sci.*, **271**, 2217–21.

Cardoso, F. L. (1994) *Orientação sexual numa comunidade pesqueira* (Sexual orientation in a fishing village). Master's thesis in Anthropology, Universidade Federal de Santa Catarina, Brazil.

Cardoso, F. L. (2002) 'Fishermen': masculinity and sexuality in a Brazilian fishing community. *Sexuality & Culture*, **6**, 45–72.

Cardoso, F. L. (2004) Male sexual behavior in Brazil, Turkey and Thailand among middle and working social classes. Ph.D. dissertation, The Institute for Advanced Study of Human Sexuality, San Francisco, CA.

Cardoso, F. L. (2005) Cultural universals and differences in male homosexuality: the case of a Brazilian fishing village. *Arch. Sex. Behav.*, **34**, 105–11.

Cardoso, F. L. and Werner, D. (2004) Homosexuality. In *Encyclopedia of Sex and Gender: Men and Women in the Worlds Cultures: Vol. I. Topics and Cultures A–K*, eds. C. R. Ember and M. Ember, pp. 204–15. New York: Kluwer Academic.

Castell, R. (1969) Communication during initial contact: a comparison of squirrel and rhesus monkeys. *Folia Primatol.* **11**, 206–14.

Chagnon, N. (1988) Life histories, blood revenge, and warfare in a tribal population. *Science*, **239**, 985–92.

Chevalier-Skolnikoff, S. (1976) Homosexual behavior in a laboratory group of stumptail monkeys (*Macaca arctoides*): forms, context, and possible social functions. *Arch. Sex. Behav.*, **5**, 511–27.

Crapo, R. H. (1995) Factors in the cross-cultural patterning of male homosexuality: a reappraisal of the literature. *Cross-Cultural Res.*, **29**, 178–202.

Crocker, W. (1990) *The Canela (Eastern Timbira), I: An Ethnographic Introduction.* Washington: Smithsonian Contributions to Anthropology, Number 33.

Damasio, A. R. (1994) *Descartes' Error: Emotion, Reason, and the Human Brain.* New York: G. P Putnam's Sons.

Dawkins, R. (1976) *The Selfish Gene.* New York: Oxford University Press.

Diamond, M. (1993) Homosexuality and bisexuality in different populations. *Arch. Sex Behav.*, **22**, 291–310.

Dickemann, M. (1993) Reproductive strategies and gender construction: an evolutionary view of homosexualities. In *If You Seduce a Straight Person Can You Make Them Gay?: Biological Essentialism versus Social Constructionism in Gay and Lesbian Identities*, eds. J. P. De Cecco and J. P. Elia, pp. 55–71. New York: Haworth Press.

Dörner, G, Geier, T., Ahrens, L. and Krell, L. (1980) Prenatal stress as possible aetiogenetic factor of homosexuality in human males. *Endokrinologie.*, **75**, 365–8.

Downtown. (1995) *Fingerabdrücke*. Cologne: Michael Sürth Verlag.

Duerr, H. P. (1993) *Obszönität und Gewalt: Der Mythos vom Zivilisationsprozess*. Frankfurt a.M.: Suhrkamp Verlag.

Eckert, E. D., Bouchard, T. J., Bohlen, J. and Heston, L. L. (1993) Homosexuality in monozygotic twins reared apart. *Brit. J. Psychiat.*, **148**, 421–25.

Edwards, A. R. and Todd, J. D. (1991) Homosexual behaviour in wild white-handed gibbons (*Hylobates lar*). *Primates*, **32**, 231–6.

Enomoto, T. (1990) Social play and sexual behavior of the bonobo (*Pan paniscus*) with special reference to flexibility. *Primates*, **31**, 469–80.

Epple, G. (1967) Vergleichende Untersuchungen über Sexual- und Sozialverhalten der Krallenaffen (*Hapalidae*). *Folia Primatol.*, **7**, 37–65.

Flores, R. Z. (1994) *Cultura, família e genética: um estudo das causas do homossexualismo em uma população de Porto Alegre*. Relatório de Projeto de Pesquisa, Depto. de Genética, UFRGS.

Forsyth, A. (1991) *Die Sexualität in der Natur*. Munich: Deutscher Taschenbuch Verlag.

Foucault, M. (1978) *Histoire da la sexualité*. Paris: Librairie Plon.

Gadpaille, W. J. (1980) Cross-species and cross-cultural contributions to understanding homosexual activity. *Arch. Gen. Psychiat.*, **37**, 349–57.

Gorer, G. (1966) *The Danger of Equality*. London: Cresset.

Gould, S. J. (1977a) *Ontogeny and Phylogeny*. Cambridge, MA: Harvard University Press.

Gould, S. J. (1977b) *Ever since Darwin: Reflections in Natural History*. New York: W. W. Norton.

Gould, S. J. and Lewontin R. C. (1979 The Spandrels of San Marco and the Panglossian Paradigm: a critique of the adaptationist programme. *Proc. R. Soc, London B*, **205**, 581–98.

Green, R. (1987) *The 'Sissy Boy Syndrome' and the development of homosexuality*. New Haven: Yale University Press.

Greenstein, J. M. (1966) Father characteristics and sex typing. *J. Pers. Soc. Psych.*, **3**, 271–7.

Hamer, D. H., Hu, S., Magnuson, V. L., Hu, N. and Pattatucci, A. M. L. (1993) A linkage between DNA markers on the X chromosome and male sexual orientation. *Science*, **261**, 321–27.

Hayaki, H., Huffman, M. A. and Nishida, T. (1989) Dominance among male chimpanzees in the Mahale Mountains National Park, Tanzania: a preliminary study. *Primates*, **30**, 187–97.

Henzi, S. P. (1985) Genital signalling and the coexistence of male vervet monkeys (*Cercopithecus aethiops pygerythrus*). *Folia Primatol.*, **45**, 129–47.

Herdt, G. (1993) *Ritualized homosexuality in Melanesia*. Berkeley, CA: University of California Press.

Jamison, C. S., Meier, R. J. and Campbell, B. C. (1993) Dermatoglyphic asymmetry and testosterone levels in normal males. *Am. J. Phys. Anthro.* **90**, 185–98.

Katz, J. (1976) *Gay American History: Lesbians and Gay Men in the USA*. New York: Thomas Y. Crowell.

Kelly, R. (1974) Etoro social structure: a study in structural contradiction. Ph.D. dissertation, University of Michigan.

Lagrou, E. (1996) Xamanismo e representação entre os Kaxinawá. In *Xamanismo no Brasil: novas perspectivas*, eds. E. J. M. Langdon, pp. 197–232. Florianópolis: Editora da UFSC.

LeDoux, J. (1996) *The Emotional Brain: The Mysterious Underpinnings of Emotional Life.* New York: Simon & Schuster.

LeVay, S. (1994) *Keimzellen der Lust: Die Natur der Menschlichen Sexualität* (German edition of *The Sexual Brain*). Heidelberg: Spektrum Akademischer Verlag.

Lorenz, K. (1963) *Das Sogenannte Böse*. Munich: Deutsches Taschenbuch Verlag.

Maple, T. (1977) Unusual sexual behavior of nonhuman primates. In *Handbook of Sexology*, eds. J. Money and H. Musaph, pp. 1167–87. Elsevier: North-Holland Biomedical Press.

McConaghy, N. and Blaszcynski, A. (1991) Initial stages of validation by penile volume assessment that sexual orientation is distributed dimensionally. *Comprehen. Psychiat.*, **32**, 52–8.

Mendes, J. C. (1997) Hetero e homo: uma relação entre homens. Bachelor's thesis, Social Sciences, Univ. Federal de Santa Catarina. Florianópolis.

Money, J. and Ehrhardt, A. A. (1972) *Man and Woman, Boy and Girl*. Baltimore: Johns Hopkins University Press.

Murray, S. O. (2000) *Homosexualities*. Chicago: University of Chicago Press.

Murray, S. O. and Arboleda, M. G. (1995) Stigma transformation and relexification: gay in Latin America. In *Latin American male homosexualities*, ed. S. O. Murray, pp. 138–44. Albuquerque: University of New Mexico Press.

Mustanski, B. S., DuPree, M. G., Nievergelt, C. M., Bocklandt, S., Schork, N. J. and Hamer, D. H. (2005) A genomewide scan of male sexual orientation. *Human Genetics online.* http://springerlink.metapress.com/ DOI: 10.1007/s00439-004-1241-4.

Oi, T. (1990) Patterns of dominance and affiliation in wild pig-tailed macaques (*Macaca nemestrina nemestrina*) in West Sumatra. *Int. J. Primatol.*, **11**, 339–56.

Phillips, G. and Over, R. (1992) Adult sexual orientation in relation to memories of childhood gender conforming and gender nonconforming behaviors. *Arch. Sex. Behav.*, **21**, 543–58.

Pillard, R. C. and Weinrich, J. D. (1986) Evidence of familial nature of male homosexuality. *Arch. Gen. Psychiat.*, **43**, 808–12.

Pinker, S. (1994) *The Language Instinct: How the Mind Creates Language*. New York: William & Morrow Co.

Ploog, D. W., Blitz, J. and Ploog, F. (1963) Studies on social and sexual behavior of the squirrel monkey (*Saimiri sciureus*). *Folia Primatol.*, **1**, 29–66.

Reichholf, J. H. (1992) *Der schöpferische Impuls: eine neue Sicht der Evolution*. Stuttgart: Deutsch-Verlags Anstalt.

Reinisch, J., Ziemba-Davis, M. and Sanders, S. A. (1991) Hormonal contributions to sexually dimorphic behavioral development in humans. *Psychoneuroendocrinol.*, **16**, 213–78.

Rieppel, O. (1992) *Unterwegs zum Anfang: Geschichte und Konsequenzen der Evolutionstheorie*. Munich: Deutscher Taschenbuch Verlag.

Siegelman, M. (1974) Parental background of male homosexuals and heterosexuals. *Arch. Sex. Behav.*, **3**, 3–17.

Silva, E. A. (1998) A natureza cultural da justiça: por uma teoria multidisciplinar da justiça, vista através do ritual de violência sexual no presídio masculino de Florianópolis. Unpublished master's dissertation, Universidade Federal de Santa Catarina, Brazil.

Sommer, V. (1990) *Wider die Natur? Homosexualität und Evolution*. Munich: C. H. Beck Verlag.

Stack, M. K. (2005) Iran warms up to sex changes, but still shuns homosexuality. *LA Times*, 30 January.

Stoddart, D. M. (1990) *The Scented Ape: The Biology and Culture of Human Odour*. New York: Cambridge University Press.

Swaab, D. F. and Hofman, M. A. (1990) An enlarged suprachiasmatic nucleus in homosexual men. *Brain Res.*, **24**, 141–8.

Swanson, G. (1960) *Birth of the Gods: The Origin of Primitive Beliefs*. Ann Arbor: University of Michigan Press.

Symons, D. (1979) *The Evolution of Human Sexuality*. New York: Oxford University Press.

Vasey, P. L. (1995) Homosexual behavior in primates: a review of evidence and theory. *Int. J. Primatol.*, **16**, 173–204.

Waal, F. B. M. de. (1989) *Peacemaking Among Primates*. New York: Penguin Books.

Werner, D. (1979) A cross-cultural perspective on theory and research on male homosexuality. *J. Homo.*, **1**, 345–62.

Werner, D. (1981) Are some people more equal than others? Status inequality among the Mekranoti of Central Brazil. *J. Anthro. Res.*, **37**, 360–73.

Werner, D. (1984) *Amazon Journey: An Anthropologist's Year among Brazil's Mekranoti Indians*. New York: Simon and Schuster.

Werner, D. (1996) Leadership. In *Encyclopedia of Cultural Anthropology*, vol 2, eds. D. Levinson and M. Ember, pp. 693–7. New York: Henry Holt & Company.

Werner, D. (1999a) *Sexo, símbolo e solidariedade: ensaios de psicologia evolucionista*. Florianópolis, SC BRASIL: Coleção Ilha: Programa de Pós-Graduação em Antropologia Social, Universidade Federal de Santa Catarina.

Werner, D. (1999b) Evolution: implications for epistemology and cross-cultural variation. In *The Darwinian heritage and sociobiology*, eds. J. M. G. van der Dennen, D. Smillie and D. R. Wilson, pp. 83–100. London: Praeger.

Werner, D. (2003) Princípios morais e a evolução de um senso moral. *Revista de Ciências Humanas*, **34**, 253–81.

Whitam, F. L. (1983) Culturally invariable properties of male homosexuality: tentative conclusions from cross-cultural research. *Arch. Sex. Behav.*, **12**, 207–26.

Whitam, F. L. and Mathy, R. M. (1986) *Male homosexuality in Four Societies: Brazil, Guatemala, Philippines, and the United States.* New York: Praeger.

Whitam, F. L. and Zent, M. (1984) A cross-cultural assessment of early cross-gender behavior and familial factors in male homosexuality. *Arch. Sex. Behav.*, **13**, 427–39.

Wilbert, J. (1972) The fishermen: the Warao of the Orinoco delta, In *Survivors of Eldorado: four Indian cultures of South America*, ed. J. Wilbert, pp. 65–115. New York: Praeger.

Williams, W. L. (1985) Persistence and change in the berdache tradition among contemporary Lakota Indians. *J. Homo.*, **11**, 191–200.

Williams, W. L. (1992) *The Spirit and the Flesh: Sexual Diversity in American Indian Culture.* Boston: Beacon Press.

Wilson, D. S. (1994) Adaptive genetic variation and human evolutionary psychology. *Ethol. Sociobiol.*, **15**, 219–235.

Wrangham, R. and Peterson, D. (1996) *Demonic Males: Apes and the Origins of Human Violence.* Boston: Houghlin Mifflin.

Yamagiwa, J. (1992) Functional analysis of social staring behavior in an all-male group of mountain gorillas. *Primates*, **33**, 523–44.

Young, O. P. (1983) An example of 'apparent' dominance-submission behavior between adult male howler monkeys (*Alouatta palliata*). *Primates*, **24**, 283–7.

Zucker, K. J., Wild, J., Bradley, S. J. and Lowry, C. B. (1993) Physical attractiveness of boys with gender identity disorder. *Arch. Sex. Behav.*, **22**, 23–36.

PART IV   OUTLOOKS: SCIENCE AND BEYOND

14

Where do we go from here? Research on the evolution of homosexual behaviour in animals

PAUL L. VASEY

Judging from the recent spate of popular science articles (Adler, 1997; Silverstone, 2000; Vines, 1999; Wahl, 1999) and wildlife documentaries (McKeown, 2000; Menéndez et al., 2001) on homosexual behaviour in animals, there is a significant amount of interest in this topic. Indeed, Terry (2000) has commented on our collective fascination with 'queer' animals. Curiously, this popular interest has not translated into much research on the part of scientists.

As outlined in Chapter 1, apprehension about homophobic reaction may account for why some researchers steer clear of this subject matter (Wolfe, 1991; Bagemihl, 1999). The repressive effects of homophobia on scholarly careers and research agendas have been well documented across academic disciplines (McNaron, 1997). At the same time, however, it is important to note that research on homosexual behaviour in animals challenges some of our most basic theoretical assumptions about how animals should behave. If researchers do not know how to interpret homosexual behaviour in animals within the context of the larger theoretical frameworks they employ, then they may avoid reporting that such interactions even exist. My personal experience has been characterized by an overwhelmingly positive interest among colleagues and the general public in my research on female homosexual behaviour in Japanese macaques and overtly homophobic encounters have been exceptionally rare.

Choosing the 'Right' model species

While there is little doubt that homophobia plays a role in dissuading some researchers from studying or publishing on this subject, other factors may be equally or even more important in contributing to the lack of research. It is possible, for example, that the lack of research on this topic is due to the fact that relatively few species engage in homosexual behaviour on a routine basis. If animals exhibit homosexual behaviour, but do so infrequently, then the behaviour will not necessarily be amenable to detailed study. As such, it will be important that scholars wishing to investigate spontaneously occurring homosexual behaviour choose their animal models with care so as to identify species in which these sorts of interactions are regularly manifested. In this regard, Bagemihl (1999) has done a great service in providing encyclopedic coverage of homosexual behaviour in the animal kingdom with rough estimates of frequency. His volume makes it clear that there are a number of excellent candidates for study, namely species in which homosexual behaviour appears to occur on a routine basis, but for which we have little more than intriguing anecdotal information. If we focus on mammals alone, promising models for research include certain species of primates, cetaceans, pinnipeds, sirenias and ungulates.

Similarly, researchers would do well to consider focusing more attention on all-male groups. All-male groups are a phylogenetically widespread and species-typical pattern of social organization for numerous animals, yet we know very little about the social dynamics that characterize these groups. Overwhelmingly bisexual groups are the focus of study, while the all-male counterparts are ignored. Research on humans has demonstrated that homosexual behaviour is a salient feature of some all-male groups (Bell and Weinberg, 1978; Williams, 1986). It seems reasonable to assume that homosexual behaviour may also play an important role in structuring relationships within all-male animal groups. Once again, anecdotal evidence seems to support this claim (Vasey, 1995; Bagemihl, 1999), but more empirical research is necessary.

Many evolutionists may consider animal homosexual behaviour an uninteresting subject for study because it does not appear to contribute to reproduction. If evolution is contingent on reproduction, then it follows that some investigators working on issues pertaining to the evolution of behaviour, particularly sexual behaviour, might reason that non-reproductive modes of sexuality, such as homosexual activity, are, from an ultimate perspective, simply irrelevant. Convincing scientists that the study of homosexual behaviour in animals is a worthwhile area of investigation will inevitably involve demonstrating that animal homosexual behaviour is an evolutionarily relevant subject of study.

The evolution of socio-sexual signals

The dominant evolutionary approach to the study of animal homosexual behaviour has been to interpret these interactions as socio-sexual adaptations; that is, interactions that are sexual in terms of their outward form, but enacted to faciliate some sort of functional social goal or reproductive strategy (Wickler, 1967). The concept of socio-sexuality represents an attractive theoretical framework for researchers interested in understanding how homosexual behaviour might be selected for and maintained in a population over evolutionary time. It is entirely reasonable to begin any investigation of animal homosexual activity with the assumption that the behaviour we seek to understand is adaptive. Too often, however, investigators attribute adaptive 'explanations' to homosexual interactions in an offhanded manner and in the absence of any empirical support. Future research on animal homosexual behaviour needs to provide strong tests of adaptive hypotheses. In other words, researchers need to work harder to generate multiple socio-sexual hypotheses for animal homosexual behaviour, which can then be used to produce non-overlapping predictions about how the animals should behave, which can, in turn, be used to eliminate some, or perhaps all, of the hypotheses.

In this volume, a number of contributors provide empirical evidence that homosexual behaviour, as expressed by their study species, is a socio-sexual adaptation. For example, this evidence suggests that flamingos, geese, bottlenose dolphins, rhesus macaques and bonobos employ homosexual activity to facilitate alliance formation. This raises an important question for anyone interested in the functional basis of animal homosexual behaviour; namely, why do some social signals become sexualized in certain species? Why do some species use same-sex mounting to express dominance when physical aggression seems to achieve the same goal? Why do some species use same-sex mounting or courtship to reduce social tension, while others use social affiliation such as grooming? In short, why employ homosexual behaviour to achieve social ends, when non-sexual affiliation would appear to do the job?

One intriguing answer to these questions is that some homosexual interactions may be superior forums for communicating social messages of conformity, coordination and cooperation (Smuts and Watanabe, 1990; Wrangham, 1993; Watanabe and Smuts, 1999). In such instances, the actual form these interactions take can quite literally embody the messages that the participants are attempting to communicate to each other. For example, in some contexts, asymmetrical mounting may actually represent the social message of both the physical domination of the mounter over the mountee and the mountee's acceptance of a subordinate status. Similarly, bi-directional mounting and courtship may exemplify the social message of coordination and cooperation during affiliative interactions such as reconciliation, alliance formation and co-feeding.

Some researchers have argued persuasively that such interactions might be best conceptualized as ritual exchanges (Smuts and Watanabe, 1990; Watanabe and Smuts, 1999). Smith (1977) noted that the formal structure of ritual exchanges necessitates that the participants accommodate each other to begin, sustain and complete the interactions. Each participant conforms to a predetermined set of symmetrical or asymmetrical parts or roles, which they perform according to a predetermined set of movements and responses that proceed in a predictable manner. These reciprocal behavioural sequences fit together in an orderly way within the interactive formal structure of the ritual interaction. Thus, ritualized interactions circumscribe possible behavioural responses and subordinate the participants to the requirements of the ritual's formal structure. By holding individuals to a mutually induced conformity, Rappaport (1979) argued that ritual interactions produce mutual invariance that, in turn, becomes an iconic representation of conformity, coordination and cooperation.

Ritualized interactions that incorporate intimate body contact may be particularly affective mechanism, for communicating one's willingness to engage in cooperative, trusting relationships. By exposing their genitalia and engaging in other forms of intimate body contact, partners demonstrate vulnerability to each other. This willingness to demonstrate vulnerability in each other's presence may signal trust and a readiness to pursue a mutually beneficial relationship despite the short-term costs or long-term asymmetries that such a relationship might entail (Zahavi, 1977; Smuts and Watanabe, 1990; Wrangham, 1993; Watanabe and Smuts, 1999).

The study by Smuts and Watanabe (1990; also see Watanabe and Smuts, 1999) on male savanna baboons is, to date, the most detailed exploration of these ideas. Males that mounted and manipulated each other's genitals more frequently formed the most cohesive and successful alliances against other males. Often, as if to reaffirm their alliance bond, two males would engage in homosexual behaviour just before challenging a rival (for similar observations, see Owen, 1976). The most intensely bonded males encouraged behavioural symmetry in their relationships by actively soliciting each other for mounts and genital fondling. Smuts and Watanabe (1990) suggest that males permit potential rivals intimate contact with their genitalia, putting their future reproductive success literally in the palm of another male's hands, in order to demonstrate a willingness to accept risk, and, thus, genuine interest in forming a reciprocal alliance in spite of whatever short-term costs might be involved.

Wrangham (1993) has suggested that the formal structure of genito–genital rubbing in bonobos embodies social messages of trust and cooperation, because the participants expose intimate body regions during such interactions that can be easily lacerated by potential rivals. Not surprisingly, then, homosexual behaviour between female bonobos has been observed prior to co-feeding in small food patches, when forming alliances during group transfer and resource defense and during reconciliation attempts following conflicts (Chapter 13).

Future research on the evolution of animal homosexual behaviour may benefit from considering whether the formal structure of these interactions renders them better mediums for communicating particular social messages when compared with non-sexualized alternatives. Investigations of this sort may help contribute to a better understanding of the unique role that homosexual behaviour plays in structuring social relationships in some animal societies.

Sexual reward and evolutionary history

Some researchers with an adaptationist perspective, intent on finding a socio-sexual explanation for animal homosexual behaviour, have been quick to dismiss the idea that such behaviour is sexually motivated. As with so much 'research' on this topic, these sorts of dismissals have been typically advanced in the absence of any empirical evidence. Although animal homosexual behaviour may serve some socio-sexual function, this need not negate a sexual component to such interactions (Wickler, 1967). Moreover, the possibility that animal homosexual activity is purely sexual and serves no socio-sexual function(s) needs to be considered as well. Various contributors to this volume make it clear that sexual motivation plays at least some causal role in the expression of homosexual behaviour in their study species (Chapter 4, 7, 8, 10, 11, 12).

As such, future research on animal homosexual behaviour needs to provide strong tests of the 'sexual' hypothesis before rejecting the idea that these interactions are devoid of any sexual motivation. As outlined in Chapter 1, there are a number of ways in which the sexual nature of same-sex courtship, mounting and genital contact might be established or refuted using non-invasive observational techniques. These include drawing parallels between homosexual and heterosexual activity, demonstrating how homosexual behaviour is distinct from same-sex *social* interactions and documenting the occurrences of genital contact and stimulation during homosexual interactions.

Strong tests of the 'socio-sexual' and 'sexual' hypotheses may reveal examples of animal homosexual behaviour that are primarily sexual and not enacted to mediate any type of socio-sexual goal. Available evidence suggests that female homosexual behaviour in Japanese macaques is one such example (Chapter 8) and male homosexual behaviour in feral cats may be another (Chapter 7). In such instances, researchers will need to look beyond a narrow adaptationist perspective to account for the behavioural phenomenon under investigation. This will require two principle changes to our thinking. To begin, there will need to be a greater recognition of the fact that adaptations are not the only products produced by the evolutionary process; functionless by-products of adaptations are also produced (Buss *et al.*, 1998). These are traits that do not affect an individual's reproductive success. They do not evolve to solve adaptive problems, and, thus, do not have a function and are not products of natural selection. Instead, functionless by-products evolve in association with particular adaptations because they happened to be coupled with those adaptations. Functionless by-products of adaptations cannot be explained in functional terms. Instead, Tinbergen's *other* type of evolutionary analysis – namely, *evolutionary* (or *phylogenetic*) *history* – must be invoked to account for their existence. Explanations that invoke evolutionary history focus on reconstructing the evolutionary steps that lead to a behaviour. A greater sensitivity to the dual concepts of functionless by-products of adaptations and evolutionary history may help furnish researchers with a more nuanced understanding of how some forms of homosexual behaviour evolve. Moreover, this expanded perspective should provide researchers with a greater appreciation of the fact that adaptations do not evolve in a vacuum independent of any real-world constraints.

My research on female homosexual behaviour in Japanese macaques helps illustrate these important points. I have speculated previously that female–male mounting is an adaptation that females use to manipulate male mate choice (Vasey, 2002). Female mounting of males seems to focus the male's sexual attention on the mounting female and increase the probablity that the male will, in turn, mount the female (pers. obs., unanalysed data 2001–5). One of the constraints imposed on the adaptive design of female–male mounting was that sexual selection could not favour this adaptation in isolation from the capacity for females to mount in a homosexual context. The adaptation (female–male mounting) could simply not evolve without the by-product (the capacity for female–female mounting). The Japanese macaque example raises another interesting issue. Namely, the evolution of homosexual behaviour may be closely tied to that of other forms of non-conceptive sexuality, especially non-conceptive heterosexual behaviour. As such, researchers interested in understanding the evolution of homosexual behaviour might benefit by articulating the evolutionary relationship between homosexual behaviour and other forms of non-conceptive sex.

Of course, phylogenetic analyses need not be limited to those instances in which homosexual behaviour appears to be functionless. Functional homosexual behaviour can also be investigated from a phylogenetic perspective. Reconstructing the evolutionary history of animal homosexual behaviour that is unifunctional would appear to be a relatively straightforward task. However, how do we account for the evolution of homosexual behaviour in a particular species if it appears to have multiple functions? Bonobo homosexual behaviour provides a good example of this conundrum, because there is evidence that it functions to reduce social tension, reconcile conflicts and facilitate alliance formation (Chapter 13). In attempting to understand why bonobo homosexual behaviour originated and what changes it underwent over time, how do we decide which of these functions was selected for first, second and third? Or, is it erroneous on our part to assume that any one selective pressure was responsible for the origin of bonobo homosexual behaviour? Investigators interested in the evolutionary history of multifunctional animal homosexual interactions will need to develop theoretical tools for understanding the role various selective agents have on the evolution of such behaviour.

Homosexual behaviour and sexual selection

Sexual selection is the key theoretical framework for understanding the evolution of sexual behaviour. Sexual selection depends on the advantages which certain individuals have over other individuals of the same sex and species, in exclusive relation to reproduction (Darwin, 1871). How the sexual selection process occurs has been overwhelmingly conceptualized in terms of two interactive mechanisms: intra-sexual competition among males for female mates (that is, reproductive partners), and choice of male mates by females (Darwin, 1871). Although other mechanisms such as male choice of female mates have long been recognized (Darwin, 1871), they have attracted relatively limited research attention.

The manner in which homosexual interactions impact on the reproductive activities of other group members is a relatively unexplored area of investigation, but one which might prove important in terms of our understanding of sexual selection. For example, male Japanese macaques frequently intrude on pair bonds between anestrous females and compete with one female partner for sexual access to the other (Vasey, 1998). Anestrous females, which are the objects of such competition, frequently ignored the intruding male and continue to engage in sexual activity with their female partners. These findings indicate that Japanese macaques compete inter-sexually for sexual partners, and raise the possibility that competition for *reproductive partners* may occur inter-sexually as well. They also suggest that reproductive value alone may not account for why individuals in some species choose and compete for others.

Theoretically speaking, inter-sexual mate competition occurs when a sexually motivated individual (the reproductive competitor) attempts to acquire or maintain exclusive access to a reproductive partner (the focus of competition), while decreasing or preventing an opposite-sex conspecific (the opposite-sex competitor) from interacting sexually and/or socially with the same individual. Sexual motivation on the part of the opposite-sex competitor is not a prerequisite for the occurrence of inter-sexual mate competition. So long as they are able to constrain reproductive competitors from acquiring or maintaining mates, the potential exists for inter-sexual mate competition to ensue, regardless of the opposite-sex competitors' motivation.

The two facets of these interactions (that is, competition for mates by reproductive competitors and competition for sexual/social partners by opposite-sex competitors) are simultaneously linked in occurrence. Consequently, how one chooses to characterize these competitive exchanges simply depends on which perspective is taken: that of the reproductive, or the opposite-sex, competitor.

No study has been conducted, to date, with the express purpose of investigating inter-sexual competition for mates. Consequently, information on this topic is relatively rare and scattered widely throughout the literature. Potential examples of inter-sexual mate competition can be difficult to verify because researchers often refer to these interactions using cryptic terminology (for example, Orange-fronted parakeets, Hardy, 1963; Roseate cockatoos, Rowley, 1990). Many examples of inter-sexual mate competition may have gone completely unreported to date, simply because researchers lack any sort of theoretical framework for identifying such interactions, let alone interpreting them. Those data that do exist are anecdotal in nature and should be interpreted with caution (Table 14.1).

It must be stressed, for example, that aggressive interactions between males and females in the presence of potential mates do not provide sufficient evidence, in and of themselves, for the existence of inter-sexual mate competition. Before such interactions can be labeled as such, it has to first be demonstrated that aggression between the suspected male and female competitors is not actually some form of sexual coercion that one of the participants performs to obtain reproductive access to the other. Second, it must be shown that aggression is not merely a form of interference or harassment aimed solely at disrupting sexual/social interactions between same-sex partners. Third, it has to be established that such aggression is not simply part of a larger pattern of social competition with the goal of acquiring access to a social partner (for example, a grooming partner) in the form of the focus of competition. Finally, it must be demonstrated that suspected reproductive competitors are competing for mates (that is, reproductive opportunities) and not simply opposite-sex partners with whom they engage in non-reproductive sex.

None of the studies listed in Table 14.1 fulfill all four of the above requirements. Consequently, the functional significance of these competitive interactions has yet to be firmly established. Clearly, the value of the data presented lies not in its ability to furnish us with detailed and definitive

Table 14.1 *Possible examples of intersexual mate competition in animals*

Species	Conditions	Comments	Reference
Mallard duck (*Anas platyrhynchos*)	Wild	Females court female partners while threatening approaching drakes.	Lebret, 1961
Canada goose (*Branta canadensis*)	Captive	Ganders disrupt female–female pairs by driving away one female partner and copulating with the other.	Allen, 1934
Pukeko (*Porphyrio porphyrio melanotus*)	Wild	Males disrupt mounting between females by approaching the pair and courting one of the females.	Jamieson and Craig, 1987
Sage grouse (*Centrocercus urophasianus*)	Wild	Males disrupt mounting between females by attempting to mount one of the female partners.	Patterson, 1952
Common greenshank (*Tringa nebularia*)	Wild	Female vocally challenges male who courts and mounts her male mate.	Nethersole-Thompson, 1951
Rufous kangaroo rat (*Aepyprymnus rufescens*)	Captive	Males intervene in mounting between females in an attempt to deter sexual rivals.	Ganslosser and Fuchs, 1988
Livingstone's fruit bat (*Pteropus livingstonii*)	Captive	Female pursues and mounts other female while fighting off approaching males.	Courts, 1996
Cattle (*Bos taurus*)	Free-ranging	Bulls 'break' into sexually active all-female groups and vigorously court the cows in order that they cease mounting each other and orient their attention towards the bull.	Kilgour *et al.*, 1977
Wapiti (*Cervus elaphus*)	Wild	Bulls separate 'bonded' female pairs by herding one of the partners away from the other.	Franklin and Lieb, 1979; Lieb, 1973
Uganadan kob (*Kobus kob tomasi*)	Wild	Males engage in head-butting contests with female competitor to gain access to her same-sex sexual partners.	Buechner and Schloeth, 1965
Lechwe (*Kobus leche*)	Wild	Males attempt to separate female sexual partners by herding one of the females away from the other.	de Vos and Dowsett, 1966

(cont.)

Table 14.1 (*cont.*)

Species	Conditions	Comments	Reference
Puku (*Kobus vardoni*)	Wild	Males attempt to separate female sexual partners by herding one of the females away from the other.	de Vos and Dowsett, 1966
Brown capuchin (*Cebus appella*)	Captive	Males 'break up' females engaged in homosexual mounting.	Linn *et al.*, 1995
Rhesus macaque (*Macaca mulatta*)	Free-ranging	Males charge female pairs to disrupt their mounting and gain sexual access to one of the females.	Akers and Conaway, 1979
Japanese macaque (*Macaca fuscata*)	Captive, free-ranging, wild	Males intrude on female homosexual consortships displacing or aggressing one female, while approaching and/or soliciting the other.	Vasey, 1998
Mountain gorilla (*Gorilla gorilla beringei*)	Wild	Males aggressively separate females engaged in homosexual mounting and then mount one of the females. Males also threaten females that solicit their female mates for sex.	Harcourt *et al.*, 1981
Chimpanzee (*Pan troglodytes*)	Captive	Males engage in aggressive interactions with one female to gain sexual access to other receptive females.	de Waal, 1982

conclusions about the nature of inter-sexual mate competition. What it can do, however, is alert us to the possibility that inter-sexual mate competition is a real behavioural phenomenon. This can, in turn, prompt us to examine more closely these competitive exchanges and consider the potential role they play in the sexual selection process.

With these caveats in mind, the available evidence suggests some form of inter-sexual mate competition is manifested in 19 species in captivity, under free-ranging conditions and in the wild. These interactions are phylogenetically widespread, occurring in seven avian, and nine non-human mammalian, genera (for examples in humans and for discussion see Sorenson, 1984; Williams, 1992).

By far, the most ubiquitous pattern of inter-sexual mate competition occurs when a sexually motivated male disrupts a female–female dyad engaged in homosexual behaviour (i.e., mounting and/or courtship) and attempts to engage one of the females in sexual activity. This type of inter-sexual mate competition appears to be not uncommon in certain species (for example, Japanese macaques: Vasey, 1998; antelope in the genus Kobus, Buechner and Schloeth 1965; de Vos and Dowsett, 1966).

Like intra-sexual competition for mates, it appears that inter-sexual mate competition involves behaviours that influence the acquisition and retention of reproductive partners through interactions with conspecifics. It is not immediately apparent, however, why reproductive competitors would compete for mates with members of the opposite-sex. After all, opposite-sex competitors and foci of competition cannot reproduce together, so they pose no direct threat to a competitor's reproductive fitness. Moreover, these interactions involve potential costs above and beyond time and energy expenditure, which include the risk of injury, rank decreases and stress-related effects, stemming from the need for increased vigilance against opposite-sex competitors. In light of such considerations, should not reproductive competitors avoid inter-sexual mate competition altogether, and instead simply approach potential mates immediately after their associations with opposite-sex competitors terminate?

This might be the case were it not for the fact that associations between opposite-sex competitors and potential mates occur in the context of intra-sexual mate competition. In such a context, the potential always exists for reproductive competitors to lose copulations to their same-sex rivals. Reproductive competitors able to access mates engaged in sexual/social interactions with opposite-sex individuals, prior to same-sex rivals, would gain a reproductive advantage. To this end, the competitive abilities that evolved for intra-sexual mate competition could easily be co-opted for inter-sexual mate competition and sexual selection would favour reproductive competitors that were adept in this new context. In line with this reasoning, inter-sexual mate competition could be characterized as an exaptation; that is, a characteristic currently favoured by selection, but which was not built by selection for the adaptive role that it currently serves (Gould and Vrba, 1982). In an ultimate sense then, inter-sexual mate competition can be seen as occurring in response to a social environment in which the potential for intra-sexual mate competition is omnipresent.

Comparing the behavioural tactics reproductive competitors employ during intra- and inter-sexual mate competition could help distinguish the type of selective regime that characterizes the latter. The existence of identical behavioural tactics during intra- and inter-sexual mate competition would indicate that those seen in the latter context were simply co-opted from the former. If, on the other hand, reproductive competitors employ behavioural tactics that are unique to an inter-sexual context, this would suggest that inter-sexual mate competition is under direct selection. This would, in turn, raise the possibility that inter-sexual mate competition acts as a selective mechanism (independent of its intra-sexual counterpart) favouring the evolution of adaptive traits that facilitate access to mates, and, ultimately, reproduction.

Sexual selection has favoured individuals that demonstrate flexibility in the means by which they compete for mates. To date, competition for mates has been conceived of as an exclusively intra-sexual phenomenon. In certain species, the ability to compete for mates may be a more generalized capacity that extends into the inter-sexual arena. Successful inter-sexual mate competition could potentially increase the fitness of reproductive competitors at the expense of same-sex rivals, just as successful intra-sexual mate competition does. Our understanding of sexual selection and the evolution of mating systems may be improved by investigating the role that inter-sexual mate competition plays in the acquisition and maintenance of reproductive partners in certain species.

Homosexual behaviour and sexual strategies

The sex differences in patterns of mate choice and competition initially outlined by Darwin (1871) were later explained in terms of Triver's (1972) parental investment theory. In 'typical' species, females provide more parental investment than males and, as such, there will be fewer receptive females than reproductively active males in a population. Consequently, male reproductive success is limited by access to fertile females. Under these conditions it follows that males should copulate with females in a relatively indiscriminate manner and attempt to acquire female reproductive partners through intra-sexual competition and sexual coercion. Females, on the other hand, should carefully discriminate among potential male mates in favour of those that contribute the most to offspring quality and survival.

Because males and females pursue different sexual strategies, it is believed that their primary strategies are compromised within a heterosexual context. For example, female choice is sometimes negated by male sexual coercion (Smuts and Smuts 1991). As such, theoretical predictions concerning the sexes' primary (i.e., unconstrained) sexual strategies cannot be definitively tested with evidence derived from heterosexual interactions. In light of this body of theory, Symons (1979) argued that homosexual interactions represent the 'acid test' for theoretical predictions about sex differences in primary sexual strategies. This is because sexual activity is unconstrained by the opposite-sex within a homosexual context, and, as such, primary sexual strategies should be expressed relatively more often. It follows from this logic, that the sexes' primary sexual strategies will be most constrained during heterosexual interactions requiring a high degree of coordination between opposite-sex partners (e.g., copulation, mounting, etc.), moderately constrained during heterosexual interactions requiring a lesser degree of coordination between opposite-sex partners (e.g., sexual solicitations) and the least constrained during homosexual interactions.

To date, these ideas have been empirically tested using between-subject comparisons in humans. On average, homosexual and heterosexual men do not differ in terms of their interest in uncommitted sex, but homosexual men report having significantly more sexual partners then heterosexual men (Bailey *et al.*, 1994). These empirical results validate predictions derived from Symons' (1979) hypothesis, namely, that heterosexual men are no less interested in cuasal sex then homosexual men, but they differ in their ability to actualize this interest. Homosexual men are, on average, able to indulge their interest in cuasal sex because they seek sexual partners with the same sexual inclinations. However, a lack of interest in casual sex on the part of heterosexual women limits the ability of heterosexual men to indulge this interest in an unconstrained manner.

Research on female homosexual behaviour in Japanese macaques (Vasey, 2004) also lends support to Symons' (1979) theoretical predictions. Females rarely engage in sexual competition or sexual coercion to acquire same-sex sexual partners. They do, however, employ these behaviours to retain their same-sex consort partners when faced with a male competitor's attempts to usurp that partner. In other words, left to their own devices, females did not initiate competition or sexual coercion. However, in the face of a heterosexual constraint (i.e., an intruding male competitor), a consorting female would compete for, or even sexually coerce, her same-sex sexual partner in order to maintain exclusive access to her. Consorting female competitors performed counter-challenges against males that intruded on their homosexual consortships significantly more often than they intruded on the homosexual consortships of other females. This difference is striking because females had far more opportunity to intrude on homosexual consortships then to perform counter-challenges. Intrusions could be performed at any time during the course of a homosexual consortship, but counter-challenges could only be performed following a consort intrusion. Thus, in the absence of any heterosexual constraint, patterns of partner retention by female Japanese macaques appeared to reflect a more idealized female sexual strategy characterized by mate competition avoidance relative to males. Like sex-role-reversed species (Gywnne 1991), species that routinely engage in exclusive or facultative homosexual behaviour can provide an important means by which to assess the theoretical prediction of sexual selection and parental investment theories as they relate to sex differences in primary sexual strategies.

Concluding remarks

Part of our goal in producing this volume has been to demonstrate how homosexual behaviour is an important component of some species' behavioural repertoires and one that has important evolutionary implications for how their social, sexual and reproductive lives are structured. Hopefully, this volume has stimulated the readers' interest in this subject and prompts further research on homosexual behaviour in many of the species that remain understudied and little known.

References

Adler, T. (1997) Animals' fancies. *Science News*, **151**, 8–9.

Akers, J. S. and Conaway, C. H. (1979) Female homosexual behavior in *Macaca mulatta. Arch. Sex. Behav.*, **8**, 63–80.

Allen, A. A. (1934) Sex rhythm in the ruffed grouse (*Bonasa umbellus* Linn.) and other birds. *Auk*, **51**, 180–99.

Bagemihl, B. (1999) *Biological Exuberance: Animal Homosexuality and Natural Diversity.* New York: St Martin's.

Bailey, J. M., Gaulin, S., Agyei, Y. and Gladue, B. A. (1994) Effects of gender and sexual orientation on evolutionarily relevant aspects of human mating psychology. *J. of Pers. and Soc. Psych.*, **66**, 1081–93.

Bell, A. P. and Weinberg, M. (1978) *Homosexualities: A Study of Diversity in Men and Women.* New York: Simon and Shuster.

Buechner, H. K. and Schloeth, R. (1965) Ceremonial mating behavior in Uganda kob (*Adenota kob thomasi* Neumann). *Z. Tierpsychol.*, **22**, 209–25.

Buss, D., Haselton, M. G., Shackelford, T. K., Bleske, A. L. and Wakefield, J. C. (1998) Adaptations, exaptations, and spandrels. *Am. Psychol.*, **53**, 533–48.

Courts, S. E. (1996) An ethogram of captive Livingstone's fruit bats *Pteropus livingstonii* in a new enclosure at Jersey Wildlife Preservation Trust. *Dodo*, **32**, 15–37.

Darwin, C. (1871) *The Descent of Man and Selection in Relation to Sex.* London: J. Murray.

de Vos, A. and Dowsett, R. J. (1966) The behavior and population structure of three species of the Genus Kobus. *Mammalia*, **30**, 30–55.

de Waal, F. M. B. (1982) *Chimpanzee Politics.* New York: Harper & Row.

Franklin, W. L. and Lieb, J. W. (1979) The social organization of a sedentary population of North American elk: a model for understanding other populations. In *North American Elk*, eds. M. S. Boyce and L. D. Hayden-Wing, pp. 185–98. Laramie: University of Wyoming.

Ganslosser, U. and Fuchs, C. (1988) Some quantitative data on social behavior of rufous rat-kangaroos (*Aepyprymnus rufescens* Gray, 1837 [Mammalia: Potoroidae]) in captivity. *Zool. Anz.*, **220**, 300–12.

Gould, S. J. and Vrba, E. S. (1982) Exaptation – a missing term in the science of form. *Paleobiology*, **8**, 4–15.

Gywnne, D. T. (1991) Sexual competition among females: what causes courtship-role reversal? *Trends Ecol. Evol.*, **6**, 118–21.

Harcourt, A. H., Stewart, K. J. and Fossey, D. (1981) Gorilla reproduction in the wild. In *Reproductive Biology of the Great Apes*, ed. C. E. Graham, pp. 265–79. New York: Academic Press.

Hardy, J. W. (1963) Epigamic and reproductive behavior of the orange-fronted parakeet. *Condor*, **65**: 169–99.

Jamieson, I. G. and Craig, J. L. (1987) Male–male and female–female courtship and copulation behavior in a communally breeding bird. *Anim. Behav.*, **35**, 1251–3.

Kilgour, R., Skarsholt, B. H., Smith, J. F., Bremner, K. J. and Morrison, M. C. L. (1977) Observations on the behaviour and factors influencing the sexually-active group in cattle. *Proc. New Zealand Soc. Anim. Product.*, **37**, 128–35.

Lebret, T. (1961) The pair formation in the annual cycle of the mallard, *Anas platyrhynchos* L.). *Ardea*, **49**, 97–157.

Lieb, J. W. (1973) Social behavior in Roosevelt Elk Cow groups. Unpublished Master's thesis, Humboldt State University.

Linn, G. S., Mase, D., LaFrançois, D., O'Keeffe, R. T. and Lifshitz, K. (1995). Social and menstrual cycle phase influences on the behavior of group-housed *Cebus apella*. *Am. J. Primatol.*, **35**, 41–57.

McKeown, K. (2000) *Gay Creatures* [documentary film]. Toronto: Discovery Channel Canada.

McNaron, T. A. H. (1997) *Poisoned Ivy: Gay and Lesbian Academics Confronting Homophobia*. Philadelphia: Temple University Press.

Menéndez, J., Alexandresco, S. and Loyer, B. (2001) *Out in Nature: Homosexual Behaviour in the Animal Kingdom* [documentary film]. Marseilles, France: St. Thomas Productions.

Nethersole-Thompson, D. (1951) *The Greenshank*. London: Collins.

Owen, N. W. (1976) The development of sociosexual behavior in free-living baboons, *Papio anubis*. *Behaviour*, **57**, 241–59.

Patterson, R. L. (1952) *The Sage Grouse in Wyoming*. Denver: Sage Books.

Rappaport, R. A. (1979) *Ecology, Meaning, and Religion*. Berkeley: North Atlantic Books.

Rowley, I. (1990) *Behavioral Ecology of the Galah*, Eolophus roseicapillus, *in the Wheatbelt of Western Australia*. Chipping Norton, NSW: Surrey Beatty.

Silverstone, M. (2000). Animal outings. *Equinox*, **110**, 24–35.

Smith, W. J. (1977). *The Behavior of Communicating: A Ethological Approach*. Cambridge, MA: Harvard University Press.

Smuts, B. B. and Smuts, R. W. (1991) Male aggression and sexual coercion of females in nonhuman primates and other mammals: evidence and theoretical implications. *Adv. Study Behav.*, **22**, 1–63.

Smuts, B. B. and Watanabe, J. M. (1990) Social relationships and ritualized greetings in adult male baboons (*Papio cynocephalus anubis*). *Int. J. Primatol.*, **11**, 147–72.

Sorenson, A. P. (1984) Linguistic exogamy and personal choice in the northwest Amazon. In *Marriage Practices in Lowland South America*, ed. K. M. Kensinger, pp. 180–193. Urbana: University of Illinois Press.

Symons, D. (1979) *The Evolution of Human Sexuality*. New York: Oxford University Press.

Terry, J. (2000) Unnatural acts in nature: a look at the scientific facination with queer animals. *GLQ: A Journal of Lesbian and Gay Studies*, **6**, 151–93.

Thompson-Handler, N., Malenky, R. K. and Badrain, N. (1984) Sexual behavior of *Pan paniscus* under natural conditions in the Lomako Forest, Equateur, Zaire. In *The Pygmy Chimpanzee*, ed. R. L. Susman. pp. 347–68. New York: Plenum.

Trivers, R. L. (1972) Parental investment and sexual selection. In *Sexual Selection and the Descent of Man, 1871–1971*, ed. B. Campbell, pp. 136–79. Chicago: Aldine.

Vasey, P. L. (1995) Homosexual behavior in primates: a review of evidence and theory. *Int. J. Primatol.*, **16**, 173–204.

Vasey, P. L. (1998) Female choice and inter-sexual competition for female sexual partners in Japanese macaques. *Behaviour*, **135**, 579–97.

Vasey, P. L. (2002) Same-sex sexual partner preference in hormonally and neurologically unmanipulated animals. *Ann. Rev. Sex. Res.*, **13**, 141–79.

Vasey, P. L. (2004) Sex differences in sexual partner acquisition, retention and harassment during female homosexual consortship in Japanese macaques. *Am. J. Primatol.*, **64**, 397–409.

Vines, G. (1999) Queer creatures. *New Scientist*, **2198**, 32–5.

Wahl, N. (1999) Birds and bees. *Outsmart*, **7**, 46–51.

Watanabe, J. M. and Smuts, B. B. (1999) Explaining religion without explaining it away: trust, truth, and the evolution of cooperation in Roy A. Rappaport's 'The Obvious Aspects of Ritual'. *Am. Anthropol.*, **101**, 98–112.

Wickler, W. (1967) Socio-sexual signals and their intra-specific imitation among primates. In *Primate Ethology*, ed. D. Morris, pp. 69–147. London: Weidenfeld & Nicolson.

Williams, W. L. (1986) *The Spirit and the Flesh: Sexual Diversity in American Indian Culture*. Beacon Press: Boston.

Wolfe, L. D. (1991) Human evolution and the sexual behavior of female primates. In *Understanding Behavior: What Primate Studies Tell us about Human Behavior*, eds. J. D. Loy and C. B. Peters, pp. 121–151. New York: Oxford University Press.

Wrangham, R. W. (1993) The evolution of sexuality in chimpanzees and bonobos. *Hum. Nat.*, **4**, 47–79.

Zahavi, A. (1977) The testing of a bond. *Anim. Behav.*, **25**, 246–7.

15

Against nature?! An epilogue about animal sex and the moral dimension

VOLKER SOMMER

The rights and wrongs of sex

People tend to be opinionated about issues related to sex. There is almost no aspect of sex that is not implicitly or explicitly judged as 'right' or 'wrong', be it the age at first sexual encounters, masturbation, the age gap between sex partners, incest, the frequency of encounters, techniques and positions, contraception or pornography.

This is perhaps not surprising for evolutionary biologists, since reproductive sex is the central event that drives evolutionary change. All organisms are descended from ancestors that replicated their genetic material through reproduction. In this way, organisms inherit an evolved predisposition to reproduce and rules concerning how and when reproduction should occur. Neither plants nor animals are thus expected to be 'laid back' about the question of biological reproduction, and the moral ruminations of humans about sexual behaviour might just be an extension of this primeval design.

Moralistic attitudes about sex often become particularly fierce with regard to the question of whether or not same-sex inclinations should be permissible. Hardly a day goes by without the Western media carrying a news item pertaining to whether same-sex couples should be allowed to marry or have the right to adopt children. For example, *The San Diego Tribune* carried news about legal battles related to gay issues almost every day during December, 2004. Gay rights issues are viewed with bewilderment or open hostility by many religious communities. Many believers, be they Christian, Muslim, Jewish, Hindu or Buddhist, adhere to a conservative view that sex between two men or two women is a sin, and that it should be punished (for variations of this theme, see e.g. Trillhaas, 1966; Solimeo, 2004).

Recently, the presidents of Israel, Nigeria and Zimbabwe joined a millennia-old tradition when they publicly spoke out and proclaimed that 'homosexuality is against nature'. 'Nature' has long been a joker in the moralistic card game that people engage in when they argue about the rights and wrongs of patterns of sexual behaviour.

An editor of a volume about homosexual behaviour in animals would not normally be expected to take issue with such opinions. However, while scientific enquiry does not lend itself to judgements concerning whether a particular sexual behaviour is 'right' or 'wrong', it does allow one to evaluate the 'right' and 'wrong' of certain moral criteria if they are based on assumptions about what is going on in nature.

The 'against nature' argument is often founded on the assumption that animals do not engage in homosexual behaviour. This position is based on the 'romantic' notion that cultural decadence has corrupted the original innocence and purity of humankind's sexuality. It therefore reflects the moral motto of 'back to nature', a Rousseauic view which holds up the 'natural' behaviour of animals as a 'good example' (see Sommer, 2000).

While less prominent, there exists, ironically, another line of traditional argument that emphasizes just the opposite, namely those who engage in same-sex sexual behaviour do, indeed, behave against nature but their behaviour is therefore not only acceptable but morally laudable! This is because humans are supposed to rise above their 'natural' heritage and shed 'animalistic' tendencies to achieve that predestined special status that places them above other creatures. Homosexual love is, therefore, considered as a 'special achievement' of specially blessed humans who are no longer fettered by the forces of nature.

History shows that the conflicting imperatives – 'Back to nature!' and 'Away from nature!' – produced wildly different moral perspectives on the links between 'nature' and homosexual behaviour (cf. Boswell, 1980; Weinrich, 1982; Ruse, 1988; Sommer, 1990).

How morals (ab)use nature

Argument 1: Animals don't engage in homosexual behaviour, and, therefore, human homosexuality is to be condemned.
'A perverse custom it is to prefer boys to girls / Since this type of love rebels against nature. / The wildness of beasts despises and flees this passion. / No male animal submits to another. / Animals curse and avoid evil caresses, / While man, more bestial than they, approves and pursues such things' (Boswell, 1980, pp. 389f).

This judgement, part and parcel of a twelfth century Christian sermon, reflects the perhaps most common inference made between animal behaviour and homosexuality: animals don't do it and, therefore, humans who do it are perverted. The argument has many successors and precursors. Greek philosopher Plato, for example, voiced a similar opinion in his later works: 'Our citizens should not be worse than birds or other animals who mate . . . male with female, and female with male' (Platon, *Nomoi*, 840 d-e, cit. in Dover, 1983 [1978], p. 147).

Argument 2: Animals engage in homosexual behaviour, and human homosexuality is therefore to be condemned.
Another line of reasoning assumes the opposite pattern, yet arrives at the identical moral conclusion: Animals do it – but we have to condemn homosexual behaviour *because* it is natural. A prominent example of this type of argument has its roots in the Jewish book *Leviticus*, which declares certain animals off limits for human consumption. Believers are, for example, warned against eating the meat of hyenas, since this might turn them into adulterers or sexually seductive characters (*Leviticus* 11, 5; 29; see Boswell, 1980, pp. 137f). Why would this be so? The *Old Testament* asserts that hyenas change their sex: they can be male one year and a female the next. This misconception about hyena sexual plasticity may stem from the fact that hyena genital morphology is unusal in that female hyenas have extremely large clitorises that look like penises. A heterosexual copulation between a male and a female with such a 'pseudo-penis,' might easily be mistaken for a homosexual interaction between two males. The *Physiologus* – a compilation of moralizing fables and allegories from the second century AD, popular far into the medieval ages – reflects a view expressed by Mosaic laws: 'Don't be like the hyena. . . . These people have already been condemned by the divine apostle who said: "Men have done evil with men"' (*Physiologus*, ch. 21, ch. 24, cit in Treu, 1981, pp. 41, 47).

Argument 3: Animals engage in homosexual behaviour and human homosexuality should, therefore, be condoned.
One of the works of André Gide, nobel-prize winning French author, carries the name of *Corydon*. *Corydon*, a fictitious scholar, possesses a detailed knowledge of natural history, which comes relatively close to what is nowadays perceived as conventional wisdom. In short, he recognizes that animal sexual behaviour is varied and encompasses same-sex inclinations. Because homosexual behaviour is found in nature, Corydon concludes, humans should strive to cultivate and refine this natural gift. This opinion is perhaps not surprising, given that Gide himself was gay – albeit during a time when such orientation was still very much

stigmatized. *Corydon* was first published, anonymously, in 1911, and reflects the obvious attempts of the author to 'justify' his own sexual orientation.

Argument 4: Animals do not engage in homosexual behaviour and human homosexuality is therefore to be condoned.

This argument, however bizarre it might seem at first, is a reflection of the idea that humans should strive to 'do better' than what is granted to them by nature. Pseudo-Lucian, a writer of the second century AD, invented a dialogue that plays in ancient Greece and centres around the question of whether heterosexual or homosexual love is more valuable. Gay love clearly wins out: 'Is it any wonder that, since animals have been condemned by nature not to receive from the bounty of providence any of the gifts afforded by intellect, they have with all else also been deprived of gay desires? Lions do not have such a love, because they are not philosopshers either. Bears have no such love, because they are ignorant of the beauty that comes from friendship. But for humans wisdom coupled with knowledge has after frequent experiments chosen what is best, and has formed the opinion that gay love is the most stable of loves' (*Affairs of the Heart*, cit. in Boswell, 1980, p. 153). A contemporary of Pseudo-Lucian was even more adamant: 'All irrational animals merely copulate, but we rational ones / Are superior in this regard to all other animals: / We discovered homosexual intercourse. / Men under the sway of women / Are no better than dumb animals' (cit. in Boswell, 1980, p. 153).

'Nature does not bestow virtue'

The American sex researcher James Weinrich, while referring to this tangled web of moral arguments about the nature of homosexuality, rightly pointed out that, 'If animals do something what we like, we call it natural. If they do what we don't like, we call it animalistic' (Weinrich, 1982, p. 203). Plutarch, two millennia earlier, was even more succinct: 'It is ridiculous to cite the behaviour of irrational animals in one place as an example and to reject it as irrelevant in another place' (cit. in Boswell, 1980, p. 154).

The juxtaposition of these differing perspectives make it clear that 'nature' does not provide a blueprint for moral judgements. Is the study of nature there-fore completely irrelevant for moral questions? As every so often, the answer is 'No' – and 'Yes'.

The 'No' answer relates to the ideal that scientific inquiry consciously abstains from moral inquiry. In this respect, the mission of science is only to unravel and describe patterns observed in the natural world. While a sentence such as 'sea levels should rise if the burning of fossil fuels continues' carries a 'should', this is not a 'normative' 'should' of the category 'ought to', but a 'predictive' 'should' of the category 'expected to'. Clearly, while scientific inquiry might be able to predict what is likely to happen in nature, this inquiry does not in itself 'justify' the phenomena it describes. This is obvious for a profession such as seismologists or meteorologists, because it would seem non-sensical to accuse scientists who study earthquakes or tornadoes of justifying the loss of human life associated with such forces of nature.

However, the slope becomes slippery, if scientists study phenomena such as the conditions under which human males are likely to kill infants unrelated to them, or the conditions under which soldiers will be inclined to torture prisoners. Do such studies not justify or at least excuse the behaviour? Scientists will adamantly reject such notions, even if they recognize the emotionally charged nature of such research.

Examples such as these demonstrate that, in reality, it is hard for science and scientists to escape entanglement with ethical and moral questions. This was true for those who discovered nuclear fission ('Should an atomic bomb be built to defeat an evil dictator?'), for those who describe the complex mental capabilities of apes ('Should people be allowed to eat chimpanzees and use them in biomedical research?') as well as for those who describe the biodiversity of tropical ecosystems ('Would it not be better for us to protect rainforests from logging?').

Science therefore has at least implicit link to ethical agendas, since the study of nature can provide information on which moral judgements may be based. For example, as has been shown above, the argument 'homosexual behaviour is against nature' is based on misconceptions and misinformation about animals. One of the pioneers of modern sex research, Alfred Kinsey, made this point in 1949. Based on the information available to him at the time, Kinsey described same-sex sexual activities amongst animals. He emphasized that he was not interested whether the existence of such behaviour in animals rendered similar human activities right or wrong. However, he stressed his belief that sex acts that form part of the phylogenetic heritage of humans should not be considered 'against nature', 'abnormal' or 'perverted' (Kinsey et al., 1949, p. 24).

Thus, scientists can at least point out that gays and lesbians do not act 'against nature' – and they probably 'should' speak out against such uninformed nonsense. But should scientists not go a step further? Would it be correct to state that human homosexual behaviour is more 'good' than 'bad', given the widespread occurance of homosexual behaviour among animals? Probably not. It is certainly true that non-human animals behave in ways that many of us would describe as 'good' – such as when mothers take care of their babies, or when youngsters play with each other. However, 'if the gay cause is somehow boosted by parallels from nature, then so are the causes of child-killers, thieves and adulterers' (Coyne, 2004). Clearly, animals display many facets of 'natural' behaviour not easily condoned as morally acceptable when displayed by humans. We could easily compile a long list, and add cannibalism, sex with immature individuals, genocide and failure to care for the elderly (cf. Sommer, 2000). Sweeping moral judgements become even more problematic because although such behaviours may appear abhorrent to the average European or North American citizen, they have been deemed perfectly acceptable or even morally desirable at other times and in other places. This underscores the fact that what is morally right or wrong is highly susceptible to historical and cultural relativism (e.g., Greenberg, 1988).

Therefore, the most valuable service that the scientific study of behaviour can provide to ongoing ethical discourses in societies would seem to be the provisioning of information. André Gide, in his 1922 prologue to *Corydon*, clearly envisioned this link: 'I do not consider it to be the ultimate wisdom to give in to nature and allow one's drives to reign freely. However, I consider it important, to understand them properly, before one attempts to suppress or tame them – because many disharmonies from which we suffer, do only appear to exist and are only the result of wrong perceptions' (Gide, 1966 [1925], pp. 9f).

Or, as Seneca, the foremost representative of Stoic philosophy remarked 2000 years ago: 'Nature does not bestow virtue; it is an art to become good' (cit. in Boswell, 1980, p. 150). Seneca's statement reflects a classic stance of how to derive ethical criteria: we should take aesthetic dimensions into account – because good is what is beautiful.

Given this, we may dare to ask: is the diversity of sexual behaviour that we can observe in nature anything other than mindbogglingly beautiful?

References

Boswell, J. (1980) *Christianity, Social Tolerance, and Homosexuality: Gay People in Western Europe from the Beginning of the Christian Era to the Fourteenth Century.* Chicago: University of Chicago Press.

Coyne, J. (2004) Charm schools. (Review of J. Roughgarden 2004. *Evolution's Rainbow: Diversity, Gender and Sexuality in Nature and People.* University of California Press.) *Times Literary Supplement*, 30 July, 5.

Dover, K. J. (1983) *Homosexualität in der griechischen Antike.* Munich: C. H. Beck. – Original: *Greek Homosexuality.* Cambridge, MA: Harvard University Press, 1978.

Gide, A. (1966) *Corydon. Vier sokratische Dialoge.* Frankfurt a.M.: Suhrkamp. (Original: Paris: Librairie Gallimard, 1925)

Greenberg, D. F. (1988) *The Construction of Homosexuality.* Chicago, London: University of Chicago Press.

Kinsey, A. C., Pomeroy, W. B., Martin C. E. and Gebhard, P. H. (1949) Concepts of normality and abnormality in sexual behavior. In *Psychosexual Development in Health and Disease*, eds. P. H. Hoch and J. Zubin, pp. 11–32. New York: Grune & Stratton.

Ruse, M. (1988) *Homosexuality: A Philosophical Inquiry*. New York: Basil Blackwell.

Solimeo, L. S. (2004) *A Higher Law: Why We Must Resist Same-Sex 'Marriage' and the Homosexual Movement*. Spring Grove, PE: The American TFP.

Sommer, V. (1990) *Wider die Natur? Homosexualität und Evolution*. Munich: C. H. Beck.

Sommer, V. (2000) The holy wars about infanticide: which side are you on? And why? In *Infanticide by Males and its Implications*, eds. C. van Schaik and C. Janson, pp. 9–26. Cambridge: Cambridge University Press.

Treu, U. (ed.) (1981) *Physiologus. Naturkunde in frühchristlicher Deutung*. Hanau: Werner Dausien.

Trillhaas, W. (1966) *Sexualethik*. Göttingen: Vandenhoeck & Ruprecht.

Weinrich, J. D. (1982) Is homosexuality natural? In *Homosexuality: Social, Psychological, and Biological Issues*, eds. W. Paul, J. D. Weinrich, J. C. Gonsiorek and M. E. Hotved, pp. 197–208. Beverley Hills, CA: Sage.

Name and person index

Location index

Subject index

Made in the USA
Lexington, KY
16 June 2011